CHEMISTRY OF COMPLEX EQUILIBRIA

THE VAN NOSTRAND REINHOLD
SERIES IN ANALYTICAL CHEMISTRY

Edited by

DR. R. A. CHALMERS

Department of Chemistry
University of Aberdeen

This series is designed as a coverage of reliable analytical information of value to chemists in research, industry and teaching. Each volume is carefully selected and planned as a modern treatment of a topic of importance to analytical chemists today. New volumes will be added from time to time

---

Chemistry of Complex Equilibria	M. T. BECK
Solvent Extraction of Metals A.K.DE, S.M.KHOPKAR, R.A.CHALMERS	
Advanced Analytical Chemistry	S. HARGREAVES, N. HALSTEAD
Laboratory Handbook of Chromatographic Methods	O. MIKEŠ
Flame Photometry Theory	E. PUNGOR
Modern Methods in Organic Microanalysis	JEAN P. DIXON
Errors, Measurement and Results in Chemical Analysis	K. ECKSCHLAGER

Additional titles will be listed and announced as published

CHEMISTRY
OF COMPLEX EQUILIBRIA

M. T. BECK D. Sc.

Professor of Physical Chemistry
Kossuth Lajos University
Debrecen (Hungary)

Translation Editor:

R. A. CHALMERS

Department of Chemistry
University of Aberdeen

VAN NOSTRAND REINHOLD COMPANY
LONDON
NEW YORK TORONTO MELBOURNE

VAN NOSTRAND REINHOLD COMPANY LTD
Windsor House, 46 Victoria Street, London S. W. 1

INTERNATIONAL OFFICES
New York Toronto Melbourne

A/541.2242

Library of Congress Catalog Card No. 67-27964

PRINTED IN HUNGARY

To Jannik Bjerrum
and
to the memory of
Niels Bjerrum

PREFACE

During the last two decades an enormous amount of information concerning complex equilibria has been published. Many experimental methods have been devised to obtain quantitative data on complex formation, and calculation methods have been developed for the evaluation of stability constants from such data. In addition, thousands of important papers, and some excellent reviews and books have appeared. The most comprehensive of these books is *The Determination of Stability Constants* by *F. J. C. Rossotti* and *H. Rossotti*, a fundamental mathematical treatment of most aspects of complex equilibria in solution, with which I in no way intend to compete here. I feel, however, that perhaps none of these publications gives a sufficiently broad and realistic account of complex equilibria, the chemistry often remaining hidden behind the algebraic equations. One fact in particular has not been adequately pointed out, viz. that the formation of mononuclear complexes with one kind of ligand only, a problem considered in most of the papers, occurs only under fairly artificial conditions. In fact, the only complex equilibrium which does not involve species other than mononuclear complexes with a completely homogeneous co-ordination sphere is the formation of a co-ordinatively saturated complex in a single step. In the present treatment, therefore, a much stronger emphasis is laid on mixed ligand, protonated, polynuclear and outer-sphere type complexes, which are treated in some detail. These species are still frequently regarded as somewhat exotic, but their existence must be taken into consideration in general.

This book is based on a Hungarian version originally published in 1965, but has been completely rewritten and the literature intended to be covered up to 1968.

I should like to express my gratitude to *Professor P. Huhn* for many discussions, to *Professor J. Bjerrum* for his critical comments on a considerable part of the manuscript, to *Professor J. Rydberg* for an unpublished figure, to *Dr. R. A. Chalmers* and *Dr. D. A. Durham* for improving the English text and to *Mrs. E. K. Kállay-Tóth* for her careful editorial work.

My thanks are due to many Publishing Houses and many authors for their permission to reproduce previously published figures. Acknowledgement of the source is given under the appropriate figures.

I am much obliged to all those authors who have kindly provided me with reprints and preprints of their papers.

Debrecen, March, 1969   *Mihály T. Beck*

CONTENTS

9

Chapter 1

INTRODUCTION

The manifold research work done in the field of co-ordination chemistry has played a decisive role in the recent extraordinarily fast development of inorganic chemistry.*

As is well known, the foundation of co-ordination chemistry is connected with the name and work of Alfred Werner. He dealt mostly with so-called substitution inert† complexes, which made possible the preparation of well defined isomeric complexes. The anticipation of the octahedral configuration of hexaco-ordinated complexes led to the discovery of the optical isomerism of certain compounds, which provided an unquestionable proof of the correctness of his co-ordination theory. As kinetically stable complexes were studied, equilibrium problems either did not arise or caused only minor difficulty.

Between the two World Wars interest in inorganic chemistry was rather limited. The flourishing of inorganic chemistry in general and of co-ordination chemistry in particular began about a quarter of a century ago. This coincidence itself appears to indicate the great importance of co-ordination chemistry in modern inorganic chemistry. There is no doubt that the explosion-like development in this field was initiated by Jannik Bjerrum's dissertation [3], published in 1941. In this book Bjerrum elaborated a general method for the determination and calculation of stability constants of metal ammine complexes. It is puzzling why the solution of the problem of successive equilibria took so long to appear, when all the necessary

---

* Including co-ordination chemistry in the domain of inorganic chemistry is not quite correct. In most of the co-ordination compounds the organic constituents dominate and there is no sharp boundary between organometallic and co-ordination compounds. The mechanism of reaction of co-ordination compounds is related to that of characteristic organic compounds. Nevertheless, the behaviour of complexes is mainly determined by the central metal ion (cf. the last paragraph of this chapter). In the history of chemistry two tendencies can always be observed: differentiation and integration [1]. As manifested by the development of quantum chemistry on one hand and co-ordination chemistry on the other, at present the tendency for integration is more pronounced.

† The term inert (or robust) refers to the kinetic behaviour of complexes and must not be confused with the thermodynamic stability [2]. As a matter of course there is a very broad range of inertness. In common usage the term may be applied to complexes which are formed and decompose slowly enough for the reactions to be studied by classical kinetic methods.

conditions had already been given for the quantitative treatment of these complex systems. Albeit the recognition of the principle of electrolytic dissociation and the knowledge of the law of mass action are evidently prerequisites of the treatment of complex equilibria, one may find some results among the earlier studies indicating that in the solutions of certain metal salts the formation of some sort of 'complex'* must be considered. The most important among these observations was Hittorf's discovery [5] that the values of the transport numbers defined by him depend on the concentration and may be negative, which means — according to present terminology — that under certain conditions negatively charged complex ions, moving towards the anode during electrolysis, are formed from positively charged metal ions by overcompensation of the charges by the anions bound to these cations.

Arrhenius's theory of electrolytic dissociation was connected with the law of mass action by Ostwald [6]. He calculated the acid dissociation constants of a number of weak acids on the basis of conductometric data. The existence of complex ions was evidenced by distribution experiments [7], solubility behaviour [8] and kinetic properties. Quantitative or semi-quantitative data were sporadically published on the stability of some complex ions.

The formation of a complex MA_n can be expressed by the equation

$$M + nA = MA_n.$$ (1.1)

Applying the law of mass action:

$$K_n = \frac{[MA_n]}{[M][A]^n}.$$ (1.2)

In the solution containing metal ion M and ligand A, not only one complex species exists, but a whole series of complexes of the general composition M_iA_j. Although the stepwise formation of the complexes was suspected, it seemed difficult, and to many chemists impossible, to determine exactly the composition and particularly the stability constants of these species by using the chemical methods known at the time. However, under extreme conditions it may happen that only one of these species is present in considerable quantity, and Bodländer [9] pointed out the possibility of determination of the composition and stability of these species. For this purpose he used both solubility and potentiometric methods. The essence of the latter method is as follows [10]. By use of a suitable concentration cell the concentration of 'free' metal ion can be determined. At an extremely high concentration of the ligand A only one complex is present in considerable amount, namely the one which is the richest in A; we will call this M_qA_r. Applying the law of mass action to the dissociation of this complex, we have

$$K_{qr} = \frac{[M]^q[A]^r}{[M_q A_r]}.$$ (1.3)

* For the first use of the term complex see ref. [4].

For two solutions of different metal ion concentration

$$\frac{[M]_1^q}{[M]_2^q} = \frac{[M_q A_r]_1}{[M_q A_r]_2} \cdot \frac{[A]_2^r}{[A]_1^r} \qquad (1.4)$$

and if A is present in great excess, then

$$[A]_1 = T_{A_1} \text{ and } [A]_2 = T_{A_2} \qquad (1.5)$$

where T represents the analytical concentration of the species indicated by the subscript.

If the concentration of A is kept constant:

$$\frac{[M]_1^q}{[M]_2^q} = \frac{[M_q A_r]_1}{[M_q A_r]_2} \qquad (1.6)$$

and e.m.f. of the concentration cell, E, is given by

$$E = \frac{RT}{nF} \ln \frac{[M]_1}{[M]_2} = \frac{RT}{nF} \ln \left(\frac{[M_q A_r]_1}{[M_q A_r]_2} \right)^{1/q}. \qquad (1.7)$$

Due to the great relative and absolute concentration of the ligand, practically the whole of M is present in the form of $M_q A_r$ and

$$[M_q A_r] = T_M. \qquad (1.8)$$

Thus the value of q can be calculated. By varying the ligand concentration at constant total metal ion concentration, the value of r/q can be evaluated. The value of the overall dissociation constant can also be estimated, but this method does not provide any information on the composition and stability of the complexes present under different, but not extreme, conditions.

The stepwise formation of complexes was first proved by Niels Bjerrum. Bjerrum studied the chemistry of chromium(III) complexes very carefully and thoroughly. He published in 1915 a long paper [11] on the kinetic and equilibrium study of chromium(III)-thiocyanate complexes. These species are formed and dissociate fairly slowly, i.e., they are inert complexes. This allows the determination of each of the complexes by the usual chemical methods of separation of the constituents of the system, provided that the formation and dissociation reactions are slow enough, compared to the manipulations in the analysis, for the equilibrium not to be disturbed. Bjerrum applied the following ingenious analytical procedure. After the equilibrium had been established at 50°C, it was frozen by cooling the solution to room temperature. Two complex anions, $Cr(SCN)_5^{2-}$ and $Cr(SCN)_6^{3-}$, were precipitated with quinolinium sulphate. The ratio of the amounts of these complexes was obtained by the determination of thio cyanate and chromium after decomposition of the mixture with alkali.

$([Cr(SCN)_6^{3-}] = T_{SCN} - 5 T_{Cr}; [Cr(SCN)_5^{2-}] = 6 T_{Cr} - T_{SCN}.)$ When the acidic solution was extracted with ether the neutral $Cr(SCN)_3$ and $HCr(SCN)_4$ went into the organic phase. These two complexes could be separated by stripping the tetrathiocyanato complex into a neutral (buffered) aqueous phase. From the aqueous phase of the first extraction the free chromium(III) was precipitated as chrome alum by adding potassium sulphate and ethanol. From the remaining solution the concentration of $Cr(SCN)_2^+$ and $CrSCN^{2+}$ is given by the determination of total chromium and thiocyanate $([Cr(SCN)_2^+] = T_{SCN} - T_{Cr}; [CrSCN^{2+}] = 2 T_{Cr} - T_{SCN}).$ Though the study of successive complex formation by suitable analytical procedures is possible in the case of any other inert system, nevertheless this method is of very limited applicability. The fundamental importance of Bjerrum's investigations is the furnishing of clearcut evidence for stepwise complex formation. He dealt very briefly with the mathematical description of equilibrium relations because 'to him it was quite obvious, that once one knows what the law of mass action is and knows the formation constants for the various complexes, it is a matter of simple algebra to calculate the concentration of each complex in any solution, and it would be a waste of paper, and an insult to the reader to give any detail of the calculations' [12].

Jaques was the first to point out in the appendix of his book [13] published in 1914, that from a series of electrode potential measurements the compositions, dissociation constants and concentrations of all the complexes present in the solution can be determined. The total concentration of the metal ion is evidently the sum of the concentrations of each species:

$$T_M = [M] + [MA] + \ldots + [MA_N], \tag{1.9}$$

where N denotes the maximum number of A ligands coordinated to the central ion. Considering the overall dissociation constants (K_i) we get the following set of equations:

$$T_M = [M]_1 \left(1 + \frac{[A]_1}{K_1} + \frac{[A]_1^2}{K_1 K_2} + \ldots + \frac{[A]_1^N}{K_1 K_2 \ldots K_N}\right) \tag{1.10}$$

$$T_M = [M]_2 \left(1 + \frac{[A]_2}{K_1} + \frac{[A]_2^2}{K_1 K_2} + \ldots + \frac{[A]_2^N}{K_1 K_2 \ldots K_N}\right) \tag{1.11}$$

$$\cdot$$
$$\cdot$$
$$\cdot$$

$$T_M = [M]_N \left(1 + \frac{[A]_N}{K_1} + \frac{[A]_N^2}{K_1 K_2} + \ldots + \frac{[A]_N^N}{K_1 K_2 \ldots K_N}\right). \tag{1.12}$$

Hence, the measurement of the free metal ion concentration at N different ligand concentrations enables one to calculate the unknown dissociation constants. Jaques was aware of some difficulties of his 'theoretical method', namely that small errors in the potential measurements become greatly magnified in the calculation, and that the ligand has to be added in great excess, otherwise the basic condition $[A] \simeq T_A$ would not be fulfilled. Naturally the problem of activity, which nowadays results in many conflicts in the interpretation of complex equilibria, could not bother Jaques in 1914. Jaques's method was not realized until the early 1940's. His name did not appear after 1914, and his book published just before the First World War did not receive as much attention as it deserved. After the war the investigations of electrolyte solutions were directed mostly towards the problem of activity. These investigations had been initiated among others also by Niels Bjerrum [14]. These studies soon led to superb results and the work on complex equilibria was overshadowed [15] by the success of the Debye–Hückel theory [16]. At that time it was generally thought that the formation of complexes has to be considered rather as an exception than a rule and the deviations from the laws of ideal solutions can be interpreted in terms of long-range interactions between the ions instead of their association. For a while most chemists forgot what McBain and Van Rysselberghe [17] correctly stressed in 1928, that 'it must remain a primary aim [of the study of electrolyte solutions] to determine the actual molecular species and their real concentrations'.

Between the World Wars only a few articles were published on the quantitative study of complex equilibria. The most important among these are Jannik Bjerrum's [18] papers on the stability of copper(II) ammine complexes and Møller's [19] dissertation on the iron(III) thiocyanate complexes, which, however, created only very little stir at that time.

The discovery of the glass electrode made the measurement of pH very easy and this gave Jannik Bjerrum the idea of potentiometric investigation of complex equilibria. Simultaneously with the publication of Bjerrum's dissertation, Ido Leden published a paper [20] on the determination of the stability constants of cadmium(II) halide complexes, by means of e.m.f. measurements. Leden's work has essentially solved the problem originally raised by Jaques. The fact that Bjerrum's dissertation was republished in 1957, which is fairly rare among doctoral theses, illustrates well its importance.

After the articles of Bjerrum and Leden had appeared there began a development unmatched in the history of inorganic chemistry. This can be judged by comparing the very limited number of quantitative equilibrium data known before 1941 with the immense number of data tabulated in the two volumes of 'Stability Constants' published in 1957–58 [21], and by considering that the second edition [22] of this wonderful collection of equilibrium data, published in 1964, contains about three times as many data as the first. In this revival of co-ordination chemistry the discovery of new elements, the production of different compounds in very pure form and, perhaps to the greatest extent, the new and very high demands on analytical chemistry, have played the most important role. The development of the chemistry of complex equilibria initiated research work in the other fields of co-ordination chemistry. This led to important discoveries on the nature of the chemical bond and on the structure and reactivity of compounds. The present status of co-ordination chemistry perhaps may be characterized by the twofold meaning of the term 'co-ordination' in chemistry. First, co-ordination refers to the structural arrangement of the atoms of compounds, secondly it reflects the connection of the different branches of chemistry. In this connection the share of co-ordination chemistry is considerable indeed.

REFERENCES

1. NYHOLM, R. S., *J. Chem. Educ. 34*, 166 (1957).
2. TAUBE, H., *Chem. Rev. 50*, 69 (1950).
3. BJERRUM, J., *Metal Ammine Formation in Aqueous Solution*, Haase, Copenhagen, 1941.
4. SZABADVÁRY, F. and BECK, M. T., *Chelates in Analytical Chemistry*, (Edited by A. J. Barnard and H. Flaschka), Vol. 1, p. 1, Dekker, New York, (1967).
5. HITTORF, J. H., *Pogg. Ann. 89*, 177 (1853).
6. OSTWALD, W., *Z. phys. Chem. 2*, 36 (1888).
7. DAWSON, H. M. and McCRAE, J., *J. Chem. Soc. 77*, 1239 (1900).
8. BODLÄNDER, G., *Z. phys. Chem. 9*, 770 (1882).
9. BODLÄNDER, G. and EBERLEIN, W., *Ber. 36*, 3945 (1903).
10. BODLÄNDER, G., *Festschrift zum 70. Geburtstag von R. Dedekind*, Brunswick, (1901).
11. BJERRUM, N., *Kgl. Danske Videnskab. Selskab. Naturvidenskab. math. Afdel.* (7) *12*, 147 (1915) (in Danish). *Z. anorg. Chem. 118*, 131 (1921); *119*, 39, 54, 179 (1921) (in German).
    A long abstract is given in English in the volume of Niels Bjerrum's selected papers, pp. 279–285, Munksgaard, Copenhagen, (1949).
12. SILLÉN, L. G., *Essays in Co-ordination Chemistry* (Edited by W. Schneider, G. Anderegg and R. Gut) p. 24, Birkhäuser Verlag, Basle, (1964).
13. JAQUES, A., *Complex Ions in Aqueous Solutions*. Longmans, London, (1914).
14. BJERRUM, N., *Fys. Tidsskr. 15*, 59 (1916) (in Danish).
    English translation is given in the volume of Niels Bjerrum's selected papers, p. 58, Munksgaard, Copenhagen, (1949).
    German translation: *Z. Elektrochem. 24*, 321 (1918).
15. SILLÉN, L. G., *J. Inorg. Nucl. Chem. 8*, 176 (1958).
16. DEBYE, P. and HÜCKEL, E., *Phys. Z. 24*, 185, 305 (1923).
17. McBAIN, J. W. and VAN RYSSELBERGHE, P. J., *J. Am. Chem. Soc. 50*, 3009 (1928).
18. BJERRUM, J., *Kgl. Danske Videnskab. Selskab. Mat.-fys. Medd. 11*, No. 5 (1931); *11*, No. 10 (1932); *12*, No. 15 (1934).
19. MØLLER, M., *Studies on Aqueous Solutions of the Iron Thiocyanates*, Dana Bogtrykkeri, Copenhagen, (1937).
20. LEDEN, I., *Z. phys. Chem. A 188*, 160 (1941); *Potentiometrisk Undersökning av några kadmiumsalters komplexitet*, Gleerupska Univ.-Bokhandeln, Lund (1943).
21. BJERRUM, J., SCHWARZENBACH, G. and SILLÉN, L. G., *Stability Constants Part I, Organic Ligands; Part II, Inorganic Ligands*, Chemical Society, London, (1957 and 1958).
22. SILLÉN, L. G. and MARTELL, A. E., *Stability Constants of Metal-Ion Complexes*, Special Publication No. 17, Chemical Society, London, (1964).

Chapter 2

COMPLEX EQUILIBRIA AND EQUILIBRIUM CONSTANTS

In the following chapters the different aspects of the chemistry of complex equilibria will be dealt with in detail. It seems appropriate to give first of all a survey of complex equilibria and to treat some fundamental concepts in connection with equilibrium constants.

2.1 Types of complex equilibria in solution

Considering the charge and the size of the metal ions it is obvious that in solution they cannot exist 'freely', but that they are associated with the counter ion(s) or other components of the solution having non-bonding electron pair(s). The term *ligand* refers to any anion or neutral molecules which can be co-ordinated to a metal ion. In nearly all cases the donor property of the solvent itself is considerable. If so, the solvated metal ions are well defined chemical species: *solvo complexes*. If the solvent is water, *aquo complexes* are formed. The number of directly co-ordinated water molecules is usually equal to the *maximum co-ordination number N*.

If the solution also contains another ligand, the stepwise substitution of the ligand molecules for the co-ordinated water molecules occurs. If this second ligand is also a neutral molecule, the charge of the successive complexes is the same as that of the central ion. For example, increasing the concentration of ammonia in the solution of cobalt(II) perchlorate causes the following reactions to take place:

$$Co(H_2O)_6^{2+} + NH_3 \rightleftharpoons Co(H_2O)_5NH_3^{2+} + H_2O \qquad (2.1)$$

$$Co(H_2O)_5NH_3^{2+} + NH_3 \rightleftharpoons Co(H_2O)_4(NH_3)_2^{2+} + H_2O \qquad (2.2)$$

$$\vdots$$

$$CoH_2O(NH_3)_5^{2+} + NH_3 \rightleftharpoons Co(NH_3)_6^{2+} + H_2O \qquad (2.3)$$

The intermediate species $Co(H_2O)_n(NH_3)_{(6-n)}^{2+}$ where $n = 1, 2, 3, 4, 5$, are in fact *mixed ligand complexes*. However, as will be pointed out later, in dilute solutions the co-ordination of the solvent molecules need not be considered, so this type of complex is regarded as a *mononuclear binary complex* and the water molecules are omitted from the formulae. Nevertheless, we have to bear in mind that it means a simplification which cannot always be applied: the co-ordination of the solvent molecule has always to be considered in the case of inert complexes, but only under certain conditions in the case of labile complexes. It may be mentioned that it is not necessarily true that the binding of *one unidentate* (see later for definition) ligand results in the displacement of *one* solvent molecule. For brevity the mononuclear binary complexes may be termed *parent complexes:* all types of complexes can be derived from these species.

If the ligand is an anion, then during stepwise complex formation the positive charge of the metal ion is gradually neutralized, and it often happens that an overcompensation of the charge of the central ion occurs and complex anions are formed:

$$Al^{3+} + F^- \rightleftharpoons AlF^{2+} \tag{2.4}$$

$$AlF^{2+} + F^- \rightleftharpoons AlF_2^+ \tag{2.5}$$

$$\cdot$$
$$\cdot$$
$$\cdot$$

$$AlF_5^{2-} + F^- \rightleftharpoons AlF_6^{3-} \tag{2.6}$$

If the ligand has more than one atom or group with donor properties (*multidentate ligands*), complexes of cyclic structure may be formed. This kind of complex was termed by Morgan and Drew [1] a *chelate* ($\chi\eta\lambda\dot{\eta}$ = crab's claw). (Originally the term was used for the ring itself [*chelate ring*], but soon also to describe the entire structure [*metal chelate* or simply *chelate*]; nowadays it is even used as a verb.) The chelate-forming ligands may be either anions, as for example oxalate, or neutral molecules, e.g. ethylene diamine or may contain both ionic and neutral donor groups as in the case of glycine. Naturally the number of different chelating agents is extremely large, and day by day dozens of new compounds are prepared.

In the case of multidentate ligands the maximum number of ligands which may be co-ordinated is evidently smaller than the maximum co-ordination number. If the number of donor groups is big enough and their steric arrangement is suitable, very stable complexes with 1 : 1 metal : ligand ratio may be formed.

The successive complex formation is quite analogous to the stepwise protonation of the ligands:

$$^-OOC-COO^- + H^+ \rightleftharpoons HOOC-COO^- \tag{2 7}$$

$$HOOC-COO^- + H^+ \rightleftharpoons HOOC-COOH \tag{2.8}$$

From this fact two important conclusions must be drawn.

(1) The mathematical treatment of the acid–base equilibria is the same as that of the successive complex equilibria.

(2) There is a competition between protons and metal ions for the ligand. In the case of a few ligands the conjugate acids areso strong (e.g., HCl, HBr, HI) that this competition can be disregarded. In general, however, the dissociation of the conjugate acids of the ligands is of fundamental importance in the treatment of complex equilibria.

The protonation of a co-ordinated ligand does not necessarily mean the breaking of the bond between the metal ion and the ligand. Thus there is a possibility for the formation of *protonated complexes*. For example, the protonation of the complexes of multidentate ligands may take place according to the following reaction:

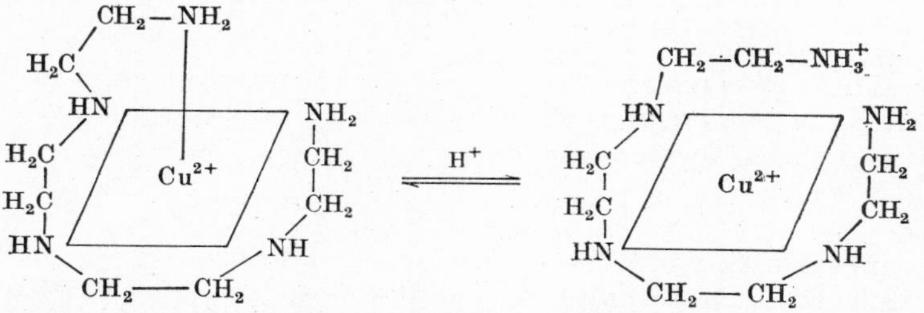

$$\text{(2.9)}$$

Protonation of complexes of unidentate ligands may also occur. For example, the protonation of hydroxo complexes results in the formation of aquo complexes, because of the donor property of the water — the conjugate acid of the hydroxide ion:

$$Fe(H_2O)_5OH^{2+} + H^+ = Fe(H_2O)_6^{3+} \qquad (2.10)$$

It follows from this reaction that the aquo complexes (aquo cations) are weak acids. The aquo cations, however, are much stronger acids than the water itself. This increasing acidity is the result of the repulsion between the central ion and the protons of the co-ordinated water.

The donor capacity of the ligand is frequently not exhausted by being co-ordinated to a certain metal ion. Thus, there is a possibility of it forming a co-ordinative bond with another metal ion. The ligand acts as a *bridge* between the two central ions. The complexes having more than one central ion are termed *polynuclear complexes*. (In the most recent literature the term *multicentre* is sometimes used, to avoid the confusion between the atomic nucleus and the central ion of the complex.) Both unidentate and multidentate ligands may serve as bridges in polynuclear complexes:

$$ZnOH^+ + Zn^{2+} \rightleftharpoons [Zn\overset{\text{H}}{O}Zn]^{3+} \qquad (2.11)$$

$$NH_2-CH_2-CH_2-NH_2Ag^+ + Ag^+ \rightleftharpoons [AgNH_2-CH_2-CH_2-NH_2Ag]^{2+}$$
$$\qquad (2.12)$$

In naming these polynuclear complexes the Greek letter μ is written before the bridging ligand.

The polynuclear complexes may be *homo-* or *heteropolynuclear* depending on whether the metal centres are the same or different.

Association between a co-ordinatively saturated complex and a ligand may also occur:

$$Co(NH_3)_6^{3+} + Cl^- = Co(NH_3)_6 \cdot Cl^{2+} \qquad (2.13)$$

$$Co(en)_3^{3+} + S_2O_3^{2-} = Co(en)_3 \cdot S_2O_3^+ \qquad (2.14)$$

These species are the *outer-sphere complexes*. Their formation also takes place in a stepwise manner.

Besides all these formation reactions the so-called substitution reactions are very frequent and important. In the case of a *ligand substitution reaction* the displacement of one co-ordinated ligand by another occurs:

$$HgCl_4^{2-} + 4\ I^- = HgI_4^{4-} + 4\ Cl^- \qquad (2.15)$$

(As was pointed out, even the simple formation reactions mean the displacement of the coordinated solvent molecules.) The central ion may also be displaced by another suitable metal ion:

$$CdI^+ + Hg^{2+} = HgI^+ + Cd^{2+} \qquad (2.16)$$

As appears from this brief summary, in a solution of several metal ions and ligands a fairly large number of different species may exist.[*] As the physical, chemical and biological properties of a given solution are determined by the nature and the quantity of the different constituents, the fundamental importance of reliable methods to establish the composition and the concentration of these species is obvious. The quantitative treatment of these equilibria is based on the law of mass action.

2.2 Types of complex equilibrium constants

The law of mass action strictly determines the concentration relations of the reactants and products in every reversible chemical reaction. The numerical value of the equilibrium constants depends on the concentration scale applied. Let us regard the simplest complex formation reaction:[†]

$$Me + L = MeL. \qquad (2.17)$$

According to the law of mass action:

$$K = \frac{[MeL]}{[Me][L]}. \qquad (2.18)$$

The dimensions of the equilibrium constant (K) are therefore the reciprocal of those of the concentration. In complex equilibrium studies the most frequently used concentration scale is the molarity (moles of solute per

---

[*] Even more equilibria have to be considered if two (or more) phases are involved. This is the case when the ligand is volatile, or one of the complexes forms a solid phase, or ion-exchange resins are applied, or in the case of liquid–liquid distribution. These more complicated systems will be treated in detail in the subsequent chapters.

[†] Me is the metal ion, L is the ligand, charges are omitted, and [X] is the concentration of X.

1000 ml of solution); sometimes, however, the molality scale (moles of solute per 1000 g of solvent) and the dimensionless mole-fraction scale is also used. In the case of dilute solutions ($< 10^{-1}$ M) the difference between the molarity and molality is negligible; in more concentrated solutions molarity (M) can be simply obtained from molality (m) by using the relationship:

$$M = \frac{1000\,dm}{1000 + m \cdot w}$$
(2.19)

where d is the density of the solution and w is the molecular weight of the particular solute. For fairly dilute solutions the mole-fraction of a certain solute can be simply converted into molarity by multiplying the mole-fraction (x) by the molarity of the pure solvent (M_s):

$$M = M_s x.$$
(2.20)

In dilute aqueous solutions $M = 55.51x$.

The interconversion of the equilibrium constants based on different concentration scales can be made accordingly, taking into account the resultant power to which the concentration is raised.

The stepwise formation of parent complexes can be described by the following set of equilibrium constants:

$$K_1 = \frac{[\text{MeL}]}{[\text{Me}][\text{L}]}$$
(2.21)

$$K_2 = \frac{[\text{MeL}_2]}{[\text{MeL}][\text{L}]}$$
(2.22)

$$\vdots$$

$$K_N = \frac{[\text{MeL}_N]}{[\text{MeL}_{N-1}][\text{L}]}.$$
(2.23)

These equilibrium constants characterize the stability of the complexes and are usually called *stability constants*. Sometimes, especially in the older literature, the reciprocal values of these constants are used and these are called instability constants. The products of the individual stability constants also give characteristic constants, called overall or cumulative formation constants or stability products, usually denoted by β:

$$\beta_1 = K_1 = \frac{[\text{MeL}]}{[\text{Me}][\text{L}]}$$
(2.24)

$$\beta_2 = K_1 K_2 = \frac{[\text{MeL}_2]}{[\text{Me}][\text{L}]^2}$$
(2.25)

$$\vdots$$

$$\beta_N = K_1 K_2 \ldots K_N = \frac{[\text{MeL}_N]}{[\text{Me}][\text{L}]^N}.$$
(2.26)

In general $\quad \beta_n = \prod_1^n K_i.$

Unfortunately, there is no unified symbolism in connection with even these simplest complex equilibrium constants. Instead of K, k or b is also frequently used and the stability product is sometimes denoted by K or $\varkappa$. Although to the present author it seems to be more logical to use K and $\varkappa$ for the stability constant and product respectively, it is proposed to use the symbols recommended by the IUPAC Commission on Equilibrium Data [2]. The meaning of the subscripts and superscripts may differ from author to author and it is always very important to define the symbols used.

For reactions in which the metal ion reacts with the protonated ligand

$$MeL_{n-1} + HL \rightleftharpoons ML_n + H^+ \tag{2.27}$$

or

$$Me + nHL \rightleftharpoons ML_n + nH^+ \tag{2.28}$$

it is sometimes practical to use the following constants:

$$*K_n = \frac{[MeL_n][H^+]}{[MeL_{n-1}][HL]} \tag{2.29}$$

and

$$*\beta_n = \frac{[MeL_n][H^+]^n}{[Me][HL]^n}. \tag{2.30}$$

These constants are composite ones, $*K_n$ being the product of the stability constant and the acidity constant (see later):

$$*K_n = K_n \cdot K_a \tag{2.31}$$

$$*\beta_n = \beta_n \cdot K_a^n. \tag{2.32}$$

The equilibrium constant of the ligand substitution reaction

$$MeL + X \rightleftharpoons MeX + L \tag{2.33}$$

is evidently the ratio of the corresponding stability constants:

$$K = \frac{[MeX][L]}{[MeL][X]} = \frac{K_{MeX}}{K_{MeL}}. \tag{2.34}$$

Analogously, the equilibrium constant of the central ion exchange reaction

$$MeL + M \rightleftharpoons ML + Me \tag{2.35}$$

is the ratio of the stability constants of the complexes in question:

$$K = \frac{[ML][Me]}{[MeL][M]} = \frac{K_{ML}}{K_{MeL}}. \tag{2.36}$$

Symbols referring to polynuclear and mixed ligand complexes will be treated in the corresponding chapters.

Just as complex equilibria can be characterized by stability and instability constants, the acid-base equilibria of the ligands can be treated by *protonation* and *acidity constants*, respectively. The *acidity constant* $(K_{a,n})$ is the equilibrium constant for the splitting off the nth proton from a charged or uncharged acid, to be defined. The *protonation constant* $(K_{H,n})$ is the equilibrium constant for the addition of the nth proton to a charged or uncharged ligand, to be defined. The following equations define these constants and show their interrelation:

$$K_{a,1} = \frac{[H_{Z-1}L][H]}{[H_ZL]} = \frac{1}{K_{H,Z}} \tag{2.37}$$

$$\vdots$$

$$K_{a,2} = \frac{[H_{Z-2}L][H]}{[H_{Z-1}L]} = \frac{1}{K_{H,Z-1}} \tag{2.38}$$

$$\vdots$$

$$K_{a,Z+1-n} = \frac{[H_{n-1}L][H]}{[H_n L]} = \frac{1}{K_{H,n}} \tag{2.39}$$

$$\vdots$$

$$K_{a,Z} = \frac{[L][H]}{[HL]} = \frac{1}{K_{H,1}}. \tag{2.40}$$

The value of Z, the number of dissociable protons, is not free from some ambiguity. For example, tartaric acid has two carboxyl and two alcoholic hydroxyl groups. These latter groups are such weak acids that their acidic character is usually not considered. In complex formation reactions, however, they play an important role, so it is not permitted to disregard these equilibria. The following abbreviations are variously used for the tartrate ion as ligand and for its conjugate acid:

$$T^{2-}, \; T^{4-}, \; H_2T, \; H_2T^{2-}, \; H_4T.$$

The interrelation between the acidity and protonation constant depends on the value of Z. If tartaric acid is defined as H_4T, the relations are as follows:

$$K_{a,1} = \frac{1}{K_{H,4}}; \; K_{a,2} = \frac{1}{K_{H,3}}; \; K_{a,3} = \frac{1}{K_{H,2}}; \; K_{a,4} = \frac{1}{K_{H,1}}. \tag{2.41}$$

However, if tartaric acid is defined as H_2T, the following relations are valid:

$$K_{a,1} = \frac{1}{K_{H,2}}; \; K_{a,2} = \frac{1}{K_{H,1}}. \tag{2.42}$$

To avoid confusion it is imperative to define the formula of the ligand and its conjugate acid unambiguously.

2.3 Equilibrium constants involving concentrations and activities

The law of mass action is strictly valid only when activities are used instead of concentrations. Because the activity of a species is equal to the product of its concentration and its activity coefficient, there is a simple relationship between the stoichiometric equilibrium constants (involving concentrations) and the thermodynamic equilibrium constants (involving activities), $°K_n$:

$$°K_n = \frac{\{MeL_n\}}{\{MeL_{n-1}\}\{L\}} = K_n \frac{f_{MeL_{n-1}} \cdot f_L}{f_{MeL_n}} \tag{2.43}$$

where the activities are written in { } brackets, and f denotes an activity coefficient. The activity of a particular solute approaches the concentration as the system approaches a certain limiting state chosen arbitrarily. In the case of the standard activity scale this limiting state is the pure solvent. Thermodynamically, equally well defined and useful activity scales can be obtained by choosing a mixed solvent or a solution of a salt as the limiting state. Of course, only equilibrium constants referring to the same activity scale can be directly compared.

The different experimental methods provide information on either concentrations or activities. In general it can be said that electrochemical methods give activities, while optical methods give concentrations. In the evaluation of the results the conversion between activities and concentrations has to be carried out by means of the activity coefficients. Sometimes it is useful to apply mixed equilibrium constants, which were first recommended by Brønsted. In this case the concentration of one of the species is replaced by its activity. This species is most frequently the proton because its activity can be directly obtained from potentiometric pH determination.

It is evident that the calculation of thermodynamic equilibrium constants requires either the knowledge of the activity coefficients or at least that they should be kept constant. If the constancy of the activity coefficients, or more correctly the constancy of their combination written in Eq. 2.43, is secured, $°K_n$ is proportional to K_n.

Thermodynamic equilibrium constants referring to the standard activity scale can be obtained by (1) studying very dilute solutions (total electrolyte concentrations less than 10^{-3} M) in which the combinations of the activity coefficients are practically equal to unity; (2) studying fairly dilute solutions where the activity coefficients can be obtained theoretically or semiempirically; (3) determining the equilibrium constants at different ionic strengths and then extrapolating to infinite dilution.

The applicability of the first method is evidently limited to the investigation of very stable complexes. The basis of the second and third methods is the Lewis-Randall principle [3] according to which in dilute solutions the activity coefficient of a given solute is the same in all solutions of the same ionic strength. The ionic strength is defined by the equation

$$I = \frac{1}{2} \Sigma \, c_i z_i^2 \qquad (2.44)$$

(where c_i is the concentration and z_i the charge of the ith species). This principle was later theoretically corroborated by the Debye–Hückel theory. If the concentration of the complexes in the system is not negligible in comparison with the concentration of the inert electrolyte, the change of the ionic strength due to complex formation has to be considered. This change may be quite large. For example, in a 0.1 M solution of cadmium chloride the ionic strength calculated by assuming total dissociation is 0.3, whereas if complex formation is taken into account the value is only 0.11.

When the value of $°K_n$ is obtained by extrapolation of a series of K_n values determined at different ionic strengths, the problem arises as to which function of the ionic strength should be used in the extrapolation, that is, are the log K_n values to be plotted against I, $I^{1/2}$ or $I^{1/3}$ etc.* The deviation of the values of $°K_n$ obtained by different extrapolation methods may amount to several hundred per cent [4]!

The validity of the original and of the different extended forms of the Debye–Hückel equation is fairly limited. The most frequently applied formulae are summarized in Table 2.1. High ligand concentrations must be used when working with weak or moderately stable complexes and so relia-activity coefficients cannot be calculated. It is frequently necessary to work with a high concentration of an inert electrolyte present, i.e., to use another activity scale. Moreover, this is the only effective method for the majority of complex systems, and in principle, if the necessary precautions are taken, it is as sound theoretically as the study of dilute solutions where the standard activity scale is applied. The basis of the constant ionic medium method (first used by Bodländer) is Brønsted's principle of the constant ionic environment [14], which states that the activity coefficient of all solutes present as small fractions of the total electrolyte concentration is constant at constant total electrolyte concentration. For the development and detailed analysis of the constant ionic medium method the fundamental paper by Biedermann and Sillén [15] should be consulted. The inert electrolyte has to meet the following requirements:

(1) it must be a strong electrolyte;
(2) its cation must not associate with the ligand and with the complex species;
(3) its anion must not associate with the central metal ion and with the complex species;
(4) redox reaction must not occur between the constituents of the inert electrolyte and the central ion or ligand;
(5) its solubility has to be large enough;
(6) its contribution to the measured physical or chemical property must be negligible.

Among the very few salts which satisfy these demands sodium perchlorate is most frequently used. According to experiments, up to 20 per cent of

---

* Evidently, in the case of uncharged ligands log K_n must be plotted against I.

TABLE 2.1

*Equations for the calculation of mean activity coefficients**

Equation	Range of validity	Author	Ref.
$-\log f_{\pm} = A \mid Z_+ Z_- \mid \sqrt{I}$	$< 10^{-3}$ M	Debye and Hückel	[5]†
$-\log f_{\pm} = A \mid Z_+ Z_- \mid \dfrac{\sqrt{I}}{1 + B \mathring{a} \sqrt{I}}$	$< 10^{-2}$ M	Debye and Hückel	[5]§
$-\log f_{\pm} = A \mid Z_+ Z_- \mid \dfrac{\sqrt{I}}{1 + B a \sqrt{I}} + B' I$	up to a few molar	Hückel	[7]‡
$-\log f_{\pm} = A \mid Z_+ Z_- \mid \dfrac{\sqrt{I}}{1 + \sqrt{I}}$	< 0.1	Güntelberg	[8]
$-\log f_{\pm} = A \mid Z_+ Z_- \mid \dfrac{\sqrt{I}}{1 + \sqrt{I}} + B' I$	< 0.5	Guggenheim	[9]‡
$-\log f_{\pm} = A \mid Z_+ Z_- \mid \left(-\dfrac{\sqrt{I}}{1 + \sqrt{I}} - 0.2\, I \right)$	< 0.2	Davies	[10, 11]
$-\log f_{\pm} = A \mid Z_+ Z_- \mid \dfrac{\sqrt{I}}{1 + 1.5\sqrt{I}}$	< 0.2	Scatchard	[12]
$-\log f_{\pm} = A \mid Z_+ Z_- \mid \dfrac{\sqrt{I}}{1 + \sqrt{I}} + B' I + C' I^{3/2}$	< 0.2	Datta and Grzybowski	[13]‡

* $f_{\pm}$ is the geometric mean of the thermodynamically indeterminate individual ionic activity coefficients.

† A is a constant. $A = (2N \pi e^6)^{1/2}/2.3026 (10k \varepsilon T)^{3/2}$ where N is Avogadro's number. For aqueous solutions at 25°C, $A = 0.509$.

§ B is a constant. $B = 50.3 (\varepsilon T)^{-1/2}$. For aqueous solutions at 25°C $B = 0.328$. $\mathring{a}$ is an adjustable parameter, and represents the mean effective diameter of the hydrated ions in Ångström units. The value of $\mathring{a}$ was estimated by Kielland [6] for 130 ions.

‡ B' and C' are adjustable parameters. The values of B' and C' are different in the different equations.

the sodium perchlorate may be replaced by the salt (or the conjugate acid) of the ligand anion without considerable change in the activity coefficients, if the total electrolyte concentration is kept at 3 M. The potential oxidizing property of the perchlorate ion has always to be borne in mind. For a while it was thought that perchlorate is not co-ordinated by any metal ion at all. However, it was found that Fe(III) [16], Ce(III) [17], Hg(I) and Hg(II) [18], Nd(III) [19], La(III), Tl(I), Cd(II), Mg(II) and Mn(II) [20] form weak complexes with perchlorate ions. The interaction is, however, so small that even in the case of most of the metal ions mentioned above there is no appreciable complex formation in a 3 M sodium perchlorate solution. In the evaluation of results obtained with such concentrated solutions it must be reckoned that not only the sodium perchlorate present in high concentration is responsible for the observed effect, but also some of its impurities. In sodium perchlorate 'low in chloride', produced by BDH, the maximum amount of chloride is less than 0.002% [21]. This means that in a 3 M sodium perchlorate solution the chloride ion concentration may amount to 2.4×10^{-4} M, and the concentration of sulphate may be much higher. These concentrations are not always negligible, especially considering the increase of the stability of complexes in highly concentrated solutions, due to the much lower water activity [22].

Just as the study of dilute solutions cannot furnish data on the water content of the species, the constant ionic medium method cannot provide any information on the number of units of either constituent of the medium associated with the complex species [23, 24]. In equilibrium analysis the concentration of MeL_n is the sum of the concentrations of all the mononuclear species in which the ratio L : Me equals n. For example, the concentration of the copper(II) diammine complex in a sodium perchlorate medium is as follows:

$$[Cu(NH_3)_2^{2+}] = \Sigma\Sigma\Sigma\ [Cu(NH_3)_3(H_2O)_x(Na^+)_y(ClO_4^-)_z^{2+y-z}]. \qquad (2.45)$$

There are two immediate consequences of this situation: (1) the equilibrium constants (K_n) determined in different ionic media do not refer to the same chemical species MeL_n; (2) the relation of the different ligands and the corresponding conjugate acids, i.e., the distribution of the protons among the different constituents associated with the metal ion, is not exactly defined. For example, the quantity $[Hg(NH_3)_2^{2+}]$ in the equilibrium analysis made in a constant concentration ammonium nitrate medium in fact includes the whole group of species $Hg(NH_3)_2(H_2O)_x(NH_4^+)_y(NO_3^-)_z$. By choosing $z = 0$, and $(x, y) = (2, -2), (1, -1)$ or $(0, -1)$ one finds that such different species as $Hg(OH)_2$, $HgNH_3OH^+$ and $HgNH_2^+$ cannot be distinguished on the basis of a study performed in a single ionic medium [24, 25].

The complexes to be studied are sometimes so weak that it is not possible to work with a constant ionic medium. In such cases the introduction of the *effective activity coefficient quotient* and the *effective stability constant* may be expedient [26]. The thermodynamic stability constant is the product of a concentration quotient (K_n) and of an activity coefficient quotient (F_n):

$$°K_n = K_n F_n. \qquad (2.46)$$

The effective activity coefficient (F_n') is the product of the activity quotient and the mean ionic activity coefficient of the salt of the ligand:

$$F_n' = F_n \cdot f_{\pm CL}. \qquad (2.47)$$

Marcus [26]. who developed the concept of the effective stability constant, suggested calculating F_n' by using Hückel's equation, and derived the following approximate equation:

$$\log F_n' = -[2(n - m) - 1]\ \frac{A\sqrt{C_{CL}}}{1 + B_a^o\sqrt{C_{CL}}} + C_{CL}\Sigma\ b_i z_i \qquad (2.48)$$

where m is the charge on the metal ion, C_{CL} is the concentration of the ligand and b is the adjustable parameter of Hückel's equation. This method was successfully applied in a number of cases. Nevertheless, the chemical meaning of the constants obtained by this method is not clear and these constants may be regarded as apparent ones which allow description of the behaviour of a given system. However, if the counter ion of the ligand is varied, the same function cannot be applied to describe the system.

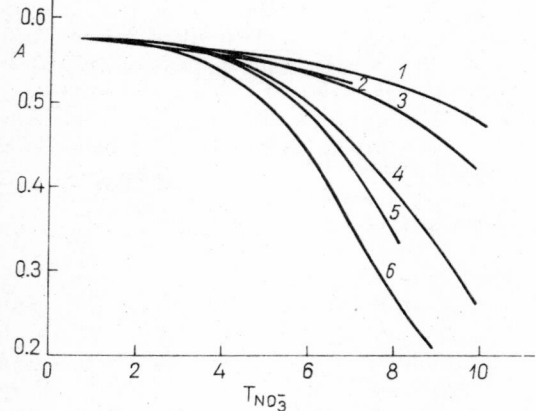

Fig. 2.1 The absorbance of Pu(VI) at 948 mμ as a function of nitrate concentration for several aqueous nitrate salt solutions. Curve $1 - \mathrm{Ca(NO_3)_2}$; Curve $2 - \mathrm{NaNO_3}$; Curve $3 - \mathrm{NH_4NO_3}$; Curve $4 - \mathrm{LiNO_3}$; Curve $5 - \mathrm{Al(NO_3)_3}$; Curve $6 - \mathrm{HNO_3}$. (Reproduced with permission from *J. Phys. Chem. 65*, 1099 (1961))

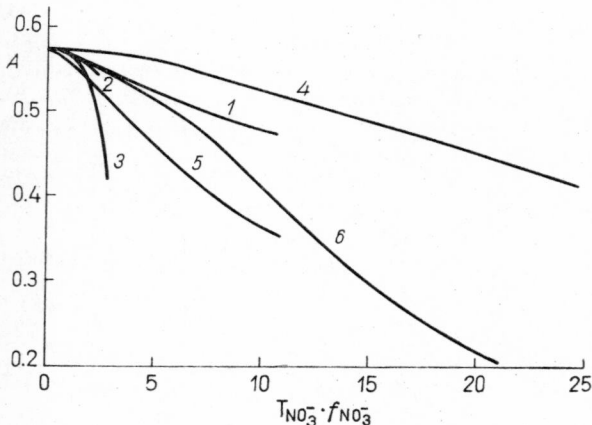

Fig. 2.2 The absorbances of Fig. 2.1 plotted as a function of $T_{\mathrm{NO_3^-}} \cdot f_{\mathrm{NO_3^-}}$. Curve $1 - \mathrm{Ca(NO_3)_2}$; Curve $2 - \mathrm{NaNO_3}$; Curve $3 - \mathrm{NH_4NO_3}$; Curve $4 - \mathrm{LiNO_3}$; Curve $5 - \mathrm{Al(NO_3)_3}$; Curve $6 - \mathrm{HNO_3}$. (Reproduced with permission from *J. Phys. Chem. 65*, 1099 (1961))

This is evident from Ryan's experiments [27] on the formation of nitrate complexes of plutonium(VI). In Fig. 2.1 the absorbance at 948 mμ is plotted against nitrate molarity. In Fig. 2.2 the absorbance at the same wavelength is plotted against the product of the nitrate molarity and the mean ionic activity coefficient. If the concept of effective stability constant were correct all the experimental points would be on the same curve, which is far from being the case.

REFERENCES

1. MORGAN, T. and DREW, H. D. K., *J. Chem. Soc. 117*, 1456 (1920).
2. *IUPAC Information Bulletin* No. *26.*, p. 25, (August 1966).
3. LEWIS, G. N. and RANDALL, M., *J. Am. Chem. Soc. 43*, 1140 (1921).
4. ADITYA, S., NANDA, R. K. and DAS, R. C., *Z. physik. Chem. Frankfurt 48*, 126 (1966).
5. DEBYE, P. and HÜCKEL, E., *Phys. Z. 24*, 185, 305 (1923).
6. KIELLAND, J., *J. Am. Chem. Soc. 59*, 1675 (1937).
7. HÜCKEL, E., *Phys. Z. 26*, 119 (1925).
8. GÜNTELBERG, E., *Z. phys. Chem. 123*, 199 (1926).
9. GUGGENHEIM, E. A., *Phil. Mag. 19*, 588 (1935).
10. DAVIES, C. W., *J. Chem. Soc.* 2093 (1938).
11. DAVIES, C. W., *Ion Association*, p. 41, Butterworths, London (1962).
12. SCATCHARD, G., *Chem. Rev. 19*, 309 (1936).
13. DATTA, S. P. and GRZYBOWSKI, A. K., *Trans. Faraday Soc. 54*, 1179 (1958).
14. BRØNSTED, J. N. and PEDERSEN, K., *Z. phys. Chem. 103*, 307 (1922).
15. BIEDERMANN, G. and SILLÉN, L. G., *Arkiv Kemi 5*, 425 (1952).
16. SYKES, K. W., *J. Chem. Soc.* 2473 (1959).
17. HEIDT, L. J. and BERESTECKI, J., *J. Am. Chem. Soc. 77*, 2049 (1954).
18. HIETANEN, S. and SILLÉN, L. G., *Arkiv Kemi 10*, 103 (1956).
19. KRUMHOLZ, P., *J. Phys. Chem. 63*, 1313 (1959).
20. JONES, M. M., JONES, E. A., HARMON, D. F. and SEMMES, R. T., *J. Am. Chem. Soc. 83*, 2083 (1961).
21. *BDH Laboratory Chemicals Catalogue* p. 233, (1964).
22. COLL, H., NAUMANN, R. V. and WEST, P. W., *J. Am. Chem. Soc. 81*, 1284 (1959).
23. LEDEN, I., *Potentiometrisk Undersökning av några kadmiumsalters komplexitet*, pp. 40–41, Gleerupska Univ. Bokhandeln, Lund (1943).
24. SILLÉN, L. G., *J. Inorg. Nucl. Chem. 8*, 176 (1958).
25. BJERRUM, J., *Metal Ammine Formation in Aqueous Solution*, p. 15, Haase, Copenhagen, (1941).
26. MARCUS, Y., *Record Chem. Progr. 27*, 105 (1966).
27. RYAN, J. L., *J. Phys. Chem. 65*, 1099 (1961).

Chapter 3

THE COMPLEX FORMATION FUNCTION

3.1 Basic principles

Most of the numerical and graphical methods developed for the calculation of equilibrium constants are based on or related to the quantity called the complex formation function. Therefore, before treatment of the different methods for the evaluation of equilibrium constants, the complex formation function and some related general problems will be discussed.

As we have seen, the stepwise formation of the mononuclear binary complexes can be described by a set of equilibrium constants (Eqs. 2.21–2.23). The individual stability constants cannot be directly calculated, and to obtain them it is necessary to find a suitable relationship between the individual constants (or their products) and the experimentally measurable parameters. The total concentrations of the central metal ion and of the ligand are evidently known, and the concentrations of the free (uncomplexed) metal ion and of the free ligand appear to be more or less readily accessible, for example by potentiometric measurements.

The total concentration T_{Me} of the central metal ion Me is the sum of the concentrations of the different species containing it:

$$T_{Me} = [Me] + [MeL] + \ldots + [MeL_N] = \sum_{i=0}^{N} MeL_i \qquad (3.1)$$

where L is the ligand. Similarly, the total concentration of the ligand is the weighted sum of the concentrations of the species containing it:

$$T_L = [L] + [MeL] + 2[MeL_2] + \ldots + N[MeL_N] = [L] + \sum_{i=1}^{\lceil N \rceil} i[MeL_i]. \qquad (3.2)$$

Considering these mass-balances and Eqs. 2.21–2.26 we obtain for the total concentrations the following expressions:

$$T_{Me} = [Me] \sum_{i=0}^{N} \beta_i [L]^i \qquad (3.3)$$

$$T_L = [L] + [Me] \sum_{i=1}^{N} i \beta_i [L]^i. \qquad (3.4)$$

The extent of complex formation can be characterized by the complex formation function or the average co-ordination number introduced by Niels Bjerrum [1]:

$$\bar{n} = \frac{[MeL] + 2\,[MeL_2] + \ldots + N\,[MeL_N]}{[Me] + [MeL] + [MeL_2] + \ldots + [MeL_N]} = \frac{T_L - [L]}{T_{Me}} \ . \quad (3.5)$$

Therefore, taking into account Eqs. 3.3 and 3.4:

$$\bar{n} = \frac{\displaystyle\sum_1^N i\,\beta_i\,[L]^i}{\displaystyle\sum_0^N \beta_i\,[L]^i} = \frac{\displaystyle\sum_1^N i\,\beta_i\,[L]^i}{1 + \displaystyle\sum_1^N \beta_i\,[L]^i} \ . \quad (3.6)$$

Two fundamental conclusions can be drawn from Eq. 3.3. First, that measuring the free ligand concentration for at least N different total ligand concentrations permits the calculation of the complete set of stability constants; secondly, that the average co-ordination number is independent of the total concentrations of both central ion and ligand. Solutions in which the total concentrations (T_{Me} and T_L) are different but $\bar{n}$ and $[L]$ are the same, are termed *corresponding solutions* [2].

The degree of formation of the system is defined by the ratio

$$\alpha = \frac{\bar{n}}{N} \quad (3.7)$$

while the degree of formation of the nth complex is given by

$$\alpha_n = \frac{[MeL_n]}{T_{Me}} = \frac{\beta_n\,[L]^n}{\displaystyle\sum_0^N \beta_i\,[L]^i} \ . \quad (3.8)$$

Sometimes α_n is called the mole fraction of the nth complex, but this term is not quite correct. Throughout this book α_n and analogous quantities will be referred to as partial mole fractions. It follows from Eq. 3.8 that the measurement of the partial mole fraction of any of the successive complexes, including uncomplexed metal ion, makes possible the calculation of the stability constant and that the values of the partial mole fractions are independent of the total metal ion and ligand concentrations.

3.2 Statistical considerations

The detailed analysis of the complex formation function was given by Jannik Bjerrum in his thesis [3] (see also the more easily available paper by Sen [4]). In the discussion of stepwise complex formation the earlier results of Adams [5], Wegscheider [6] and N. Bjerrum [7] on the gradual dissociation of polybasic acids were considered. The absolute values and the ratios of the stability constants of the successive complexes are deter-mined by a number of factors, some of which are still unknown. If we assume that the co-ordination sites are strictly equivalent and they remain so during the gradual formation of complexes, the ratio of the successive stability constants is determined statistically alone. This means that the tendency of the MeL_n species to split off a (unidentate) ligand is pro-portional to the number of occupied sites (n), while its tendency to take up a ligand is proportional to the number of available co-ordination sites $(N - n)$; that is, the following proportionalities hold:

$$K_1 : K_2 = \frac{N}{1} : \frac{N-1}{2}; \quad K_n : K_{n+1} = \frac{(N-n+1)}{n} : \frac{(N-n)}{(n+1)};$$

$$K_{N-1} : K_N = \frac{2}{(N-1)} : \frac{1}{N}. \tag{3.9}$$

The ratio of the stability constants of two successive complexes is

$$\frac{K_n}{K_{n+1}} = \frac{(n+1)(N-n+1)}{n(N-n)} \equiv f_n. \tag{3.10}$$

In the case of multidentate ligands these ratios are evidently different. For bidentate ligands, assuming octahedral configuration, the ratio of the three successive constants is as follows:

$$K_1 : K_2 : K_3 = \frac{12}{5} : \frac{5}{2} : \frac{4}{15}. \tag{3.11}$$

In the case of $N = 4$ two configurations (square planar and tetrahedral) are possible, but the ratio of the two successive constants is independent of the configuration:

$$K_1 : K_2 = \frac{2}{1} : \frac{1}{4}. \tag{3.12}$$

In Table 3.1 are shown the statistical ratios of successive stability constants calculated from Eq. 3.10.

TABLE 3.1

*The values of the statistical ratios
of the successive stability constants*

N	K_1/K_2	K_2/K_3	K_3/K_4	K_4/K_5	K_5/K_6
2	4	—	—	—	—
3	3	3	—	—	—
4	2.67	2.25	2.67	—	—
5	2.50	2	2	2.50	—
6	2.40	1.88	1.78	1.88	2.40

It is clear that the statistical values of K_n/K_{n+1} change regularly, and are all of the same order of magnitude. In Table 3.2 the ratios of the successive stability constants for a few complex systems are summarized:

TABLE 3.2

The logarithms of the ratios of successive stability constants for a few complex systems

System	N	$\log \frac{K_1}{K_2}$	$\log \frac{K_2}{K_3}$	$\log \frac{K_3}{K_4}$	$\log \frac{K_4}{K_5}$	$\log \frac{K_5}{K_6}$
$Ag^+ - NH_3$	2	−0.63	—	—	—	—
$Ag^+ - SO_4^{2-}$	2	0.18	—	—	—	—
$Hg^{2+} - SO_4^{2-}$	2	0.20	—	—	—	—
$TiO^{2+} - SO_4^{2-}$	2	1.20	—	—	—	—
$Cd^{2+} - I^-$	4	1.31	−1.38	0.67	—	—
$Hg^{2+} - I^-$	4	1.92	7.28	1.30	—	—
$VO^{2+} - F^-$	4	1.02	0.48	1.47	—	—
$Pd^{2+} - Cl^-$	4	0.80	0.86	1.00	—	—
$Al^{3+} - F^-$	6	1.11	1.17	1.11	1.11	1.16
$Zr^{4+} - F^-$	6	1.48	1.50	0.97	0.20	0.67
$Hg^{4+} - F^-$	6	1.22	0.70	0.20	1.22	1.01
$Bi^{3+} - Cl^-$	6	1.22	−0.71	1.10	0.13	0.78

Comparing the values of Table 3.2 with the statistical ones, it appears that besides the statistical effects other factors have to be taken into consideration. These factors will be dealt with in Chapter 11 in detail. Formally these effects can be taken into account by the spreading factor (x) introduced by Jannik Bjerrum [3]:

$$\frac{K_n}{K_{n+1}} = f_n\, x^2 . \tag{3.13}$$

The value of the spreading factor may assume all values between 0 and ∞, but for the statistical case $x = 1$. The spreading factor may, but not necessarily, be constant for the whole system of successive complexes. Before treating the general case, some characteristics of the simplest successive complex system ($N = 2$) will be discussed. For $N = 2$, Eq. 3.13 gives

$$\frac{K_1}{K_2} = 4x^2 . \tag{3.14}$$

K_1 and K_2 can be expressed by the spreading factor and the average stability constant $(\bar{K})$:

$$K_1 = 2x\bar{K} \tag{3.15}$$

$$K_2 = \frac{1}{2x}\bar{K}.$$

3*

Combining Eqs. 3.6 and 3.15:

$$\bar{n} = \frac{2x\bar{K}\,[\mathrm{L}] + 2\bar{K}^2\,[\mathrm{L}]^2}{1 + 2x\bar{K}[\mathrm{L}] + \bar{\bar{K}}^2[\mathrm{L}]^2}. \tag{3.16}$$

Thus for $\bar{n} = 1$ the product of the average constant and the free ligand concentration is unity; that is, the average constant is equal to the reciprocal free ligand concentration when the degree of formation is 0.5. The shape of the complex formation function (plot of $\bar{n}$ vs. log [L]) depends on the value of the spreading factor. This appears strikingly from Fig. 3.1 where

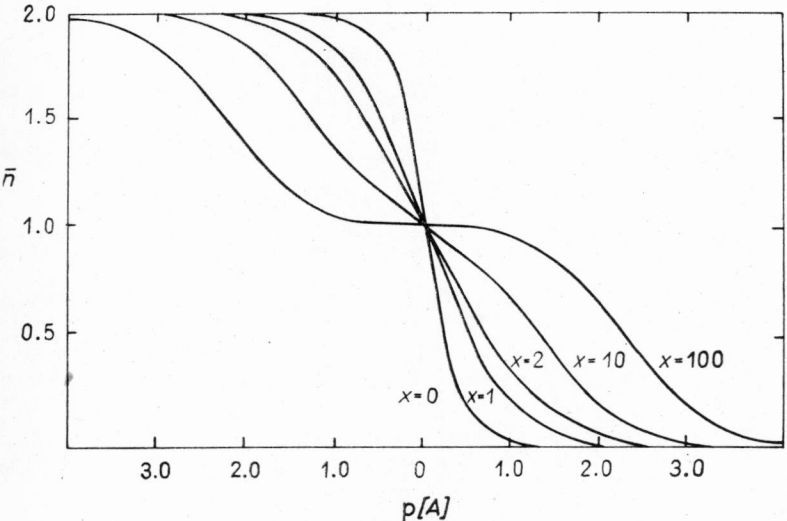

Fig. 3.1 $\bar{n}$ as a function of the negative logarithm of the free ligand concentration at $\bar{N} = 2$, $K = 1$ and varying values of the spreading factor

n is plotted as a function of x, for $N = 2$, $\bar{K} = 1$. If $x \geqslant 100$, the two complexes are formed in fully separated stages; as the spreading factor gradually decreases the two steps come nearer, and at $x \leqslant 2$ it is difficult to distinguish the two successive steps, while if $x = 0$ the uptake of both ligands occurs in a single step.

The relationship between the spreading factor and the slope of the formation curve is defined by the equation:

$$\delta = -\frac{d\bar{n}}{d\ln[\mathrm{L}]} = -\,0.4343\,\frac{d\bar{n}}{d\mathrm{p}[\mathrm{L}]}. \tag{3.17}$$

The quantity δ can be obtained by differentiating the formation function with respect to the free ligand concentration:

$$\delta = \frac{2x\bar{K}[L] + 4\bar{\bar{K}}^2[L]^2 + 2x\bar{K}^3[L]^3}{(1 + 2x\bar{K}[L] + \bar{\bar{K}}^2[L]^2)^2} .$$

(3.18)

A characteristic quantity — the slope at the mid-point of the formation curve (Δ) — can be obtained by inserting $K[L] = 1$ into Eq. 3.18:

$$\Delta = \frac{1}{1 + x}$$

(3.19)

$$x = \frac{1 - \Delta}{\Delta} .$$

Equation 3.19 refers to the curve obtained by plotting $\bar{n}$ against $d\ln[L]$; analogously, the relationship between the spreading factor and the mid-point slope (Δ') of the curve obtained by plotting $\bar{n}$ as a function of $dp[L]$ is:

$$\Delta' = -2.303 \frac{1}{1 + x}$$

(3.20)

$$x = \frac{-2.303\,\Delta' - 1}{\Delta'} .$$

By combining Eqs. 3.14 and 3.20 a simple relationship can be obtained [8] between the mid-point slope and K_1/K_2:

$$\Delta' = \frac{-4.606}{2 + K_1/K_2} .$$

(3.21)

Interdependent values of Δ' and $\log K_1/K_2$ are given in Table 3.3.

TABLE 3.3

Interdependent values of the mid-point slopes and $\log K_1/K_2$ ratios at $N = 2$

$\log K_1/K_2$	5	4	3	2	1	0
Δ'	−0.0145	−0.0451	−0.137	−0.384	−0.892	−1.535
$\log K_1/K_2$	−1	−2	−3	−4	−5	−6
Δ'	−1.989	−2.193	−2.267	−2.292	−2.300	−2.303

In the general case the relationship between the slope of the formation curve and the spreading factor can be obtained by the rearrangement and differentiation of Eq. 3.6:

$$\frac{d\bar{n}}{d[L]}\left(1 + \sum_1^N \beta_i[L]^i\right) + \bar{n}\sum_1^N i\,\beta_i[L]^{(i-1)} = \sum_1^N i^2\beta_i[L]^{(i-1)}.$$

(3.22)

From Eq. 3.17 the slope is

$$\delta = \frac{\sum_1^N i(i - \bar{n}) \beta_i [L]^i}{1 + \sum_1^N \beta_i [L]^i} .$$

(3.23)

Assuming that the whole system can be characterized by a single spreading factor, i.e. that Eq. 3.13 is valid, the values of the successive stability constants and products are given by the relationships:

$$K_n = \frac{(N - n + 1)}{n} \bar{\bar{K}} x^{(N+1-2n)}$$

(3.24)

$$\beta_n = \frac{N!}{n!(N-n)!} \bar{\bar{K}}^n x^{n(N-n)} \equiv \binom{N}{n} \bar{\bar{K}}^n x^{n(N-n)}.$$

(3.25)

Inserting Eq. 3.25 into Eq. 3.6, we obtain for the formation function

$$\bar{n} = \frac{\sum_1^N i \dfrac{N!}{(N - i)! i!} \bar{K}^i [L]^i x^{i(N-i)}}{1 + \sum_1^N \dfrac{N!}{(N - i)! i!} \bar{K}^i [L]^i x^{i(N-i)}} .$$

(3.26)

The analysis of Eq. 3.26 shows that at the mid-point of the formation curve ($\bar{n} = N/2$) the product of the average constant and the reciprocal of the free ligand concentration is unity, which, as was pointed out for $N = 2$, is generally true. Therefore, for the mid-point slope we have

$$\Delta = \frac{\sum_1^N i \left(i - \dfrac{N}{2} \right) \binom{N}{i} x^{i(N-i)}}{1 + \sum_1^N \binom{N}{i} x^{i(N-i)}} .$$

(3.27)

3.3 Particular values of the complex formation function

3.3.1 HALF-VALUES OF THE FORMATION FUNCTION

If the successive steps of complex formation are distinct, that is if $x \geqslant 100$, only two complex species exist simultaneously in considerable concentration:

$$T_{Me} = [MeL_{n-1}] + [MeL_n] .$$

(3.28)

If

$$\bar{n} = n - \frac{1}{2} \equiv \bar{n}^* $$

(3.29)

i.e. at the half-value of the average co-ordination number

then

$$[MeL_{n-1}] = [MeL_n] \tag{3.30}$$

$$K_n = \frac{1}{[L]_{\bar{n}*}} . \tag{3.31}$$

It is easy to realize [9] that Eq. 3.30 may be true even if the value of the spreading factor is not big enough, provided at the half-value points the average co-ordination number for the other complexes is also equal to $(n - 1/2)$:

$$\frac{\displaystyle\sum_{(n-1) \neq i \neq n}^{N} i \, [MeL_i]}{\displaystyle\sum_{(n-1) \neq i \neq n}^{N} [MeL_i]} = n - \frac{1}{2} . \tag{3.32}$$

It requires that both of the following equalities be valid:

$$\beta_n \, [L]_{\bar{n}*}^{n} = \beta_{n-1} \, [L]_{\bar{n}*}^{(n-1)} \tag{3.33}$$

$$\frac{\displaystyle\sum_{(n-1) \neq i \neq n}^{N} i\beta_i \, [L]_{\bar{n}*}^{i}}{\displaystyle\sum_{(n-1) \neq i \neq n}^{N} \beta_i \, [L]_{\bar{n}*}^{i}} . \tag{3.34}$$

Therefore, $1/K_n$ for any value of n is the root of the equation

$$\sum_{i=0}^{N} (2n - 1 - 2i) \, \beta_i \, [L]_{\bar{n}*}^{i} = 0 . \tag{3.35}$$

Equation 3.31 can be written in the form

$$\sum_{i=0}^{N} (2n - 1 - 2i) \frac{\beta_i}{K_n^i} = 0 . \tag{3.36}$$

That is, Eq. 3.31 may be valid either if the spreading factor is large or if there is a definite relationship among the successive constants. This relationship is particularly simple if $\bar{n} = \bar{n}*$ at the half-value points for the pairs of complexes MeL_{n-1-j} and MeL_{n+j}. This means that the successive complexes form a geometric series, expressed by the equations

$$\beta_n \, [L]_{\bar{n}*}^{n} = \beta_{n-1} \, [L]_{\bar{n}*}^{(n-1)}$$

$$\beta_{n-1} \, [L]_{\bar{n}*}^{(n+1)} = \beta_{n-2} \, [L]_{\bar{n}*}^{(n-2)} \tag{3.37}$$

$$\vdots$$

$$\beta_{n+j} \, [L]_{\bar{n}*}^{(n+j)} = \beta_{n-1-j} \, [L]_{\bar{n}*}^{(n-1-j)},$$

that is,

$$\left(\frac{\beta_n}{\beta_{n-1}}\right)^3 = \frac{\beta_{n+1}}{\beta_{n-2}} \tag{3.38}$$

and

$$\left(\frac{\beta_n}{\beta_{n-1}}\right)^{(2j+1)} = \frac{\beta_{n+j}}{\beta_{n-1-j}} \tag{3.39}$$

which are equivalent to

$$K_n^2 = K_{n-1} \cdot K_{n+1} \tag{3.40}$$

and

$$K_n^{(2j+1)} = \Pi \, K_{n-j} \cdot K_{n-j+1} \cdot \; \cdots \; \cdot K_{n+j}. \tag{3.41}$$

(It is obvious that from Eq. 3.32 on, $2 \leqslant n \leqslant N$.) The meaning of these equations is that the ratio of the successive constants is constant for the whole system. Such systems can be characterized by only two constants (any individual constant and the ratio of two successive constants, i.e. the quotient of the geometric series) instead of N constants. This condition

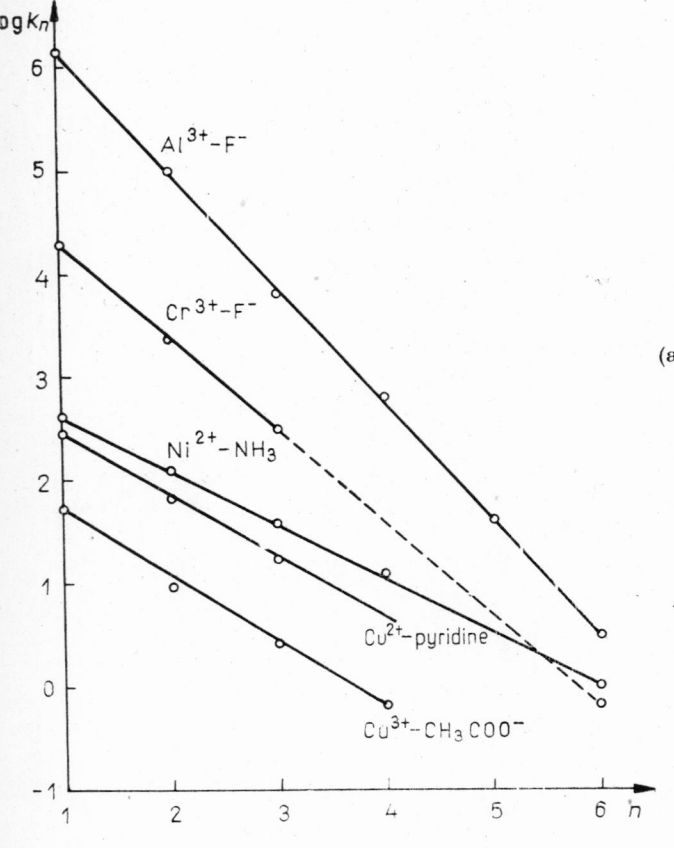

(a)

Fig. 3.2 (a) Log K_n as a function of n for several complex systems (b) Log K_n as a function of n for several complex systems predicted by Eq. 3.44. Dotted lines represent extrapolated values

is the basis of the Dyrssen–Sillén [10] two-parameter equation and some other methods for the calculation of successive constants. Van Panthaleon van Eck [11] found that the following relationship holds for many complex systems:

$$\log K_n = \log K_1 - 2\,\lambda\,(n-1) \tag{3.42}$$

which is equivalent to the constancy of the ratio of successive constants:

$$\frac{K_n}{K_{n+1}} = 10^{2\lambda}. \tag{3.43}$$

If Eqs. 3.10 or 3.13 were valid, the value of $\log K_n$ would be given by the expression

$$\log K_n = \log K_1 - \log Nn + \log(N - n + 1) \tag{3.44}$$

instead of by Eq. 3.42. As is evident from Fig. 3.2, the van Panthaleon van Eck equation is valid instead of Eq. 3.44 for some complex systems. This means that for these and similar systems the effect of the statistical factor is just compensated by other effects, and that these systems cannot be characterized by a single spreading factor, but the products of the corresponding statistical and spreading factors are constant:

$$\frac{K_n}{K_{n+1}} = f_n\,x_n^2 = 10^{2\lambda}. \tag{3.45}$$

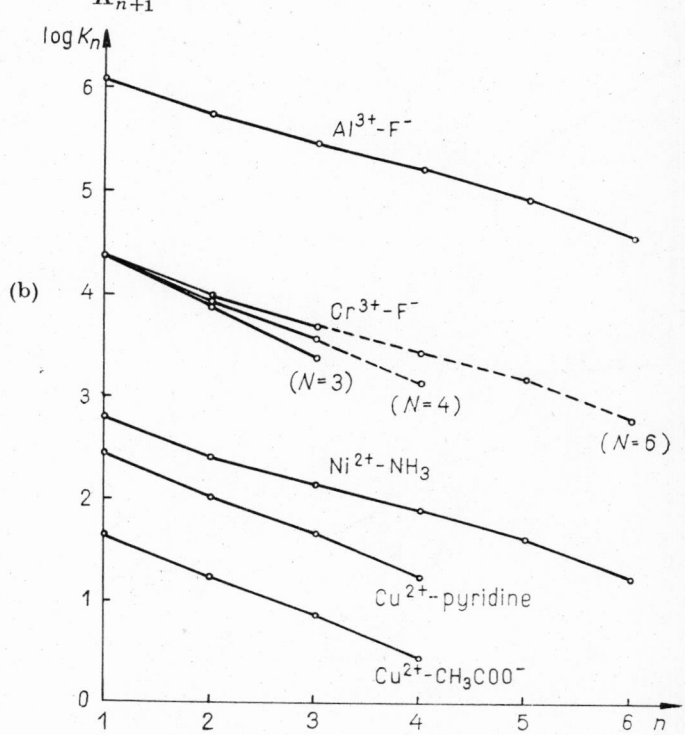

In general, the validity of Eq. 3.42 is limited to complexes in which polarization effects do not dominate. However, even in these cases the relationship may be partially valid, as is seen from Fig. 3.3 where data for mercury(II)-halide complexes are plotted according to both Eqs. 3.42 and 3.46, the latter of which can be obtained from the former by taking the stability products into consideration:

$$\frac{\log \beta_n}{n} = \log \beta_1 - \lambda(n-1). \tag{3.46}$$

It is worth mentioning that where Eqs. 3.42 and 3.44 are not valid, that is for complexes with preponderantly covalent bonding, the K_n/K_{n+1} ratio varies with temperature to a much greater extent than in the case of complexes in which electrostatic forces dominate. As illustrated by Fig. 3.4 the relationship is valid for the cadmium(II)-chloride [12] and tin(II)-chloride [13] systems at a certain temperature only.

It is illuminating to analyse the relationship between the ratio of the successive stability constants and the ratio of the free ligand concentrations at the half-values of $\bar{n}$. This can be simply obtained for $N = 2$ by the

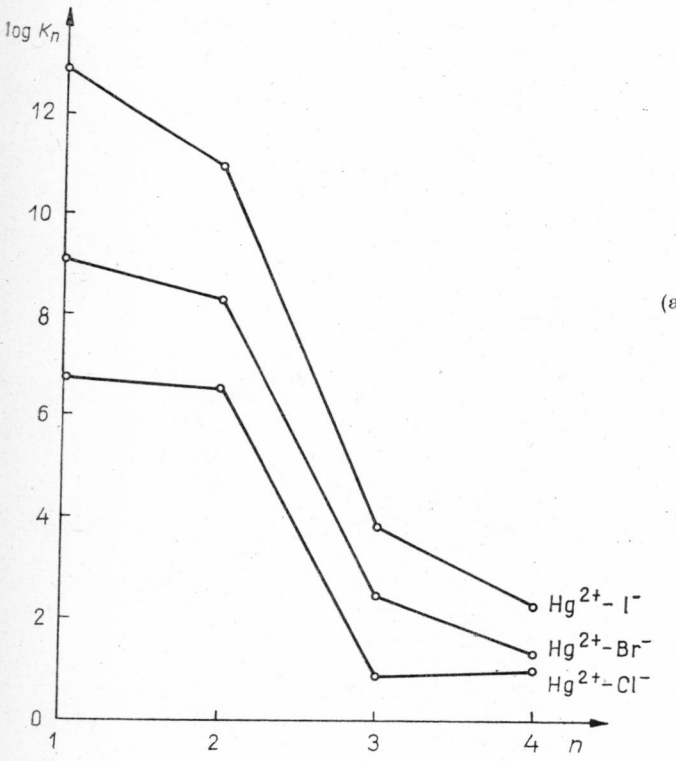

(a)

Fig. 3.3 (a) Log K_n as a function of n for halide complexes of mercury (II). (b) Log β_n/n as a function of n for halide complexes of mercury (II). (Reproduced with permission from Rec. Trav. Chim. 72, 529 (1953))

solution of two quadratic equations written for $\bar{n} = 0.5$ and 1.5:

$$1.5\ K_1 K_2 [L]_{0.5}^2 + 0.5\ K_1 [L]_{0.5} - 0.5 = 0$$

$$0.5\ K_1 K_2 [L]_{1.5}^2 - 0.5\ K_1 [L]_{1.5} - 1.5 = 0.$$

This leads to [14]

$$\frac{K_1}{K_2} = \frac{([L]_{1.5} - 3[L]_{0.5})^2}{[L]_{1.5} \cdot [L]_{0.5}}. \tag{3.47}$$

If $K_1/K_2 = 10^3$, the ratio

$$\frac{[L]_{1.5}}{[L]_{0.5}} \simeq \frac{K_1}{K_2}. \tag{3.48}$$

As the ratio of the constants is gradually decreased, the ratio of the ligand concentrations at the half-value points also decreases (Fig. 3.5), but by definition $K_n \geqslant 0$, so

$$\frac{[L]_{1.5}}{[L]_{0.5}} \geqslant 3. \tag{3.49}$$

(b)

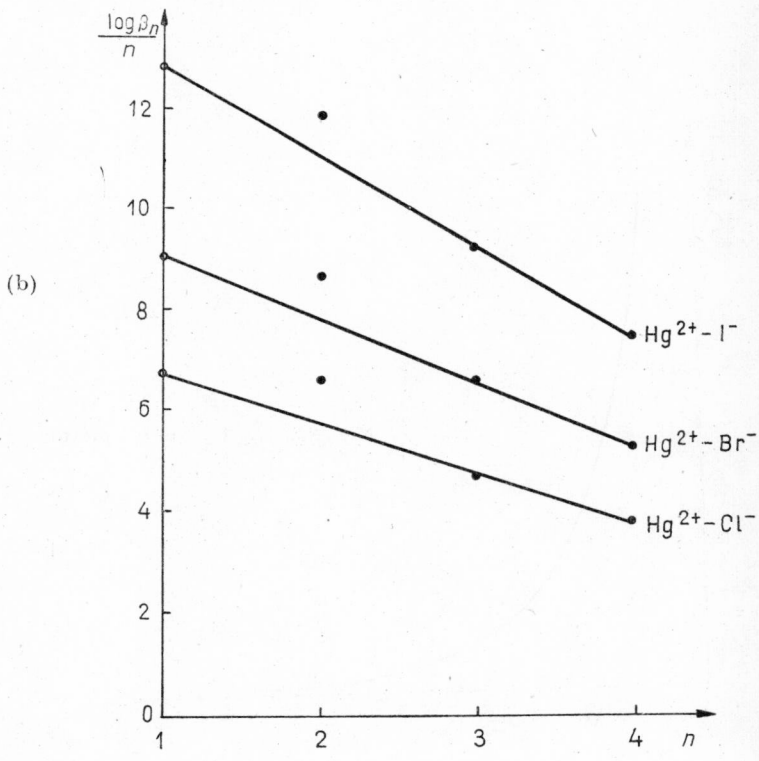

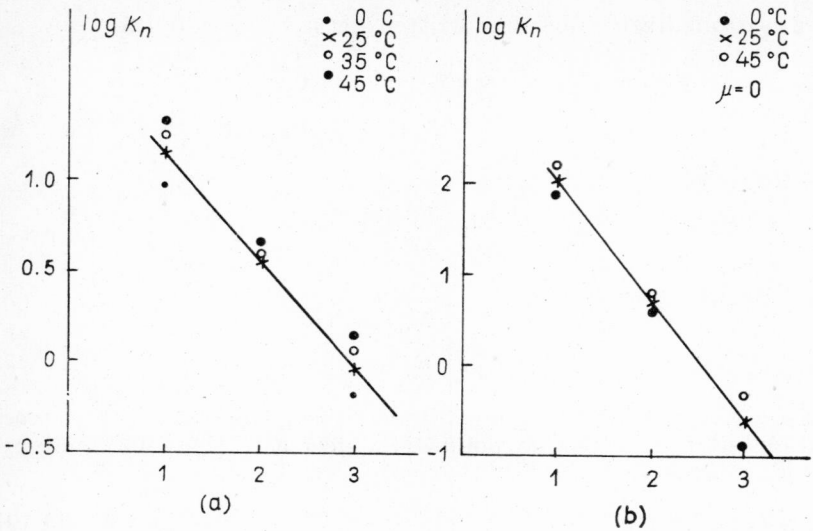

Fig. 3.4 Log K_n as a function of n at different temperatures. (a) Sn(II)–Cl$^-$ system (b) Cd(II)–Cl$^-$ system

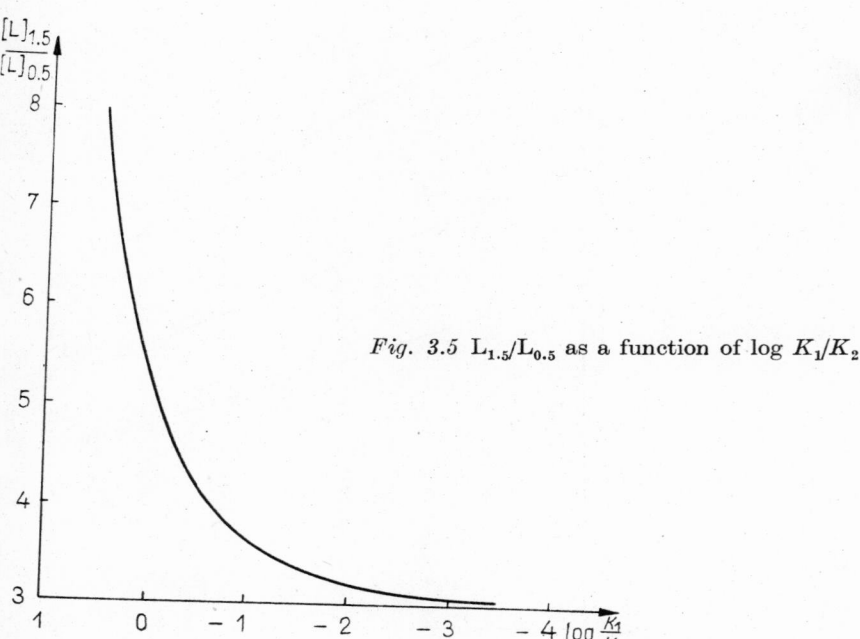

Fig. 3.5 $L_{1.5}/L_{0.5}$ as a function of $\log K_1/K_2$

If the ratio of the ligand concentrations was equal to 3, it would mean that $K_1 = 0$, $K_2 = \infty$, and β_2 is of finite value.*

However, when $N > 2$, and $K_N \gg K_{N-2}, \ldots, K_1$, the ratio $[L]_{1.5}/[L]_{0.5}$ may be less than 3. The minimum value of this ratio can be obtained by assuming that $K_1 = K_2 = \ldots K_{N-1} = 0$, i.e. that the complex formation function can be written as

$$n = \frac{N \beta_N [L]^N}{1 + \beta_N [L]^N}. \tag{3.50}$$

Then

$$\left(\frac{[L]_{1.5}}{[L]_{0.5}}\right)_{min} = \sqrt[N]{3 \; \frac{N - 0.5}{N - 1.5}}. \tag{3.51}$$

Equation 3.51 gives the following minimum values for the ratios of free ligand concentrations at $\bar{n} = 1.5$ and 0.5 half-values:

	$N = 2$	$N = 3$	$N = 4$	$N = 5$	$N = 6$
$\left(\dfrac{[L]_{1.5}}{[L]_{0.5}}\right)_{min} =$	3	1.71	1.43	1.31	1.25

Consequently, if an experimentally found ligand concentration ratio is less than 3, this can be considered as an indication of an extreme relation of the successive constants.

The number of systems in which $K_n < K_{n+1}$ is rather limited. This order is typical for silver(I)–ammine complexes [15–18], and it was also found for the systems Hg(II)–OH⁻ [19], Zn(II)–NH₃ [20], Co(II)–thioglycollic acid [21], Cu(II)–diethylbiguanide [22], Cd(II)[Mg(II), Ni(II)]–o-ethyl-thiobenzoic acid [23], Co(II)–1-nitroso-2-naphthol [24], bivalent transition metals with some pyrazolone derivatives [25–27], Fe(II)–1,10-phenan-throline [28, 29] and derivatives [30, 31], Cu(II)–dimethylglyoxime [32, 33] and derivatives [34], etc.

3.3.2 INTEGER VALUES OF THE FORMATION FUNCTION

Whether the half-values have physical meaning or not depends on the ratios of the stability constants of the successive complexes. However, the integer values of the formation function possess a definite chemical meaning independently of the values and ratios of the successive constants.

---

* There is no reason to assume that any member is missing in the series of successive complexes. If the equilibrium analysis shows that the value of K_n is negative or near zero, it means only that the term $\beta_n [L]_n$ is negligible compared to the other terms in the complex formation function. It does not mean, however, that MeL_n does not exist at all. When a property of the system is studied for which the corresponding intensive factor is great enough to compensate or even overcompensate the minuteness of $[MeL_n]$, the contribution of this 'missing' complex to the measured quantity may be considerable.

Namely, at the integer point $\bar{n} = n$, the partial mole fraction of the complex MeL_n is of maximum value. This can be simply proved [3] by rearranging and differentiating Eq. 3.8:

$$\frac{d\alpha_n}{d[\text{L}]}\left(1 + \sum_1^N \beta_i[\text{L}]^i\right) + \alpha_n \sum_1^N i\beta_i[\text{L}]^{(i-1)} - n\beta_n[\text{L}]^{(n-1)} = 0. \qquad (3.52)$$

Taking into account Eq. 3.6 and the relationships

$$\frac{d\log\alpha_n}{d\text{p}[\text{L}]} = \frac{-d\ln\alpha_n}{d\ln[\text{L}]} = \frac{[\text{L}]d\alpha_n}{\alpha_n d[\text{L}]}$$

we can write Eq. 3.52 as follows:

$$\bar{n} = n + \frac{d\log\alpha_n}{d\text{p}[\text{L}]}. \qquad (3.53)$$

Evidently the value of the second term on the right-hand side is zero if the partial mole fraction is of maximum value. Equation 3.53 is valid for the intermediary complexes, that is those for which $1 \leqslant n \leqslant N - 1$.

There are some experimental possibilities for the determination of the integer points. These will be discussed together with some potential calculation methods in Chapter 4.

REFERENCES

1. BJERRUM, N., *Z. anorg. Chem. 119*, 179 (1921).
2. BJERRUM, J., *Kgl. Danske Videnskab. Selskab. Mat.-Fys. Medd. 21*, No. 4. (1944).
3. BJERRUM, J., *Metal Ammine Formation in Aqueous Solution*, Haase, Copenhagen (1941).
4. SEN, B., *Anal. Chim. Acta 27*, 515 (1962).
5. ADAMS, E. Q., *J. Am. Chem. Soc. 38*, 1503 (1916).
6. WEGSCHEIDER, R., *Monatsh. Chem. 16*, 153 (1895).
7. BJERRUM, N., *Z. phys. Chem. 106*, 219 (1923).
8. IRVING, H. and ROSSOTTI, H. S., *J. Chem. Soc.* 3397 (1953).
9. BECK, M. T. and HUHN, P., *Acta Chim. Acad. Sci. Hung. 20*, 285 (1959).
10. DYRSSEN, D. and SILLÉN, L. G., *Acta Chem. Scand. 7*, 663 (1953).
11. VAN PANTHALEON VAN ECK, C. L., *Rec. Trav. Chim. 72*, 529 (1953).
12. VANDERZEE, C. E. and DAWSON, J. A., *J. Am. Chem. Soc. 75*, 5659 (1953).
13. VANDERZEE, C. E. and RHODES, D. E., *J. Am. Chem. Soc. 74*, 3552 (1952).
14. SCHRØDER, K. H., *Acta Chem. Scand. 20*, 1401 (1966).
15. Ref. [3], p. 130.
16. DATTA, S. P. and GRZYBOWSKI, A. K., *J. Chem. Soc.* 1091 (1959).
17. BRUEHLMAN, R. J. and VERHOEK, F. H., *J. Am. Chem. Soc. 70*, 1401 (1948).
18. MURMAN, R. K. and BASOLO, F., *J. Am. Chem. Soc. 77*, 3484 (1955).
19. HIETANEN, S. and SILLÉN, L. G., *Acta Chem. Scand. 6*, 747 (1952).
20. Ref. [3], p. 152.
21. LEUSSING, D. L., *J. Am. Chem. Soc. 80*, 4180 (1958).

22. RAY, A. K., *Z. anorg. Chem. 305*, 207 (1960).
23. IRVING, R. J., *J. Phys. Chem. 60*, 1427 (1956).
24. CALLAHAN, C. M., FERNELIUS, W. F. and BLOCK, B. P., *Anal. Chim. Acta 16*, 101 (1957).
25. SNAVELY, F. A., FERNELIUS, W. C. and BLOCK, B. P., *J. Am. Chem. Soc. 79*, 1028 (1957).
26. SNAVELY, F. A. and KRECKER, B. D., *J. Am. Chem. Soc. 81*, 4199 (1959).
27. SNAVELY, F. A., KRECKER, B. D. and CLARK, C. G., *J. Am. Chem. Soc. 81*, 2337 (1959).
28. LEE, T. S., KOLTHOFF, I. M. and LEUSSING, D. L., *J. Am. Chem. Soc. 70*, 2348 (1948)
29. SCHILT, A. A., *J. Phys. Chem. 60*, 1546 (1956).
30. BRANDT, W. W. and GULLSTROM, D. K., *J. Am. Chem. Soc. 74*, 3582 (1952).
31. BANKS, C. V. and BYSTROFF, R. C., *J. Am. Chem. Soc. 81*, 6153 (1959).
32. HAINES, R. A., RYAN, D. E. and CHENEY, G. E., *Can. J. Chem. 40*, 1149 (1962).
33. DYRSSEN, D., *Kgl. Tek. Hogsk. Hand.* No. 220 (1964).
34. BURGER, K. and RUFF, I., *Talanta 10*, 329 (1963).

Chapter 4

NUMERICAL AND GRAPHICAL METHODS FOR CALCULATION OF COMPLEX STABILITY CONSTANTS

Besides the complex formation function, other relationships have been found between the stability constants and the extent of complex formation. These correlations can be used for the calculation of the stability constants. Leden [1] introduced the function

$$F(L) = \frac{T_{Me} - [Me]}{[Me][L]} = \sum_{1}^{N} \beta_i [L]^{i-1} \tag{4.1}$$

while Fronaeus [2] defined another function

$$X(L) = 1 + \beta_1 [L] + \ldots + \beta_N [L]^N \equiv 1 + \sum_{1}^{N} \beta_i [L]^i . \tag{4.2}$$

There is a simple relationship between these two functions:

$$X(L) = [L]F(L) + 1. \tag{4.3}$$

However, there is no simple transformation between either of these functions and the average co-ordination number. The combination of Eqs. 3.6 and 4.1 gives [3]

$$F(L) = \frac{\left[\sum_{1}^{N} i [MeL_i] \right] - \bar{n} [Me]}{\bar{n} [Me][L]} . \tag{4.4}$$

Most of the methods recommended for the calculation of the stability constants of the parent complexes are based on Bjerrum's, Leden's and Fronaeus's functions. In choosing the calculation method, it has first to be considered which data ($\bar{n}$, α_n, [L], [Me]) are provided by the experimental studies.

4.1 Methods based on the half-values

As was shown in Chapter 3, if the ratio of the successive constants is great enough ($> 10^4$), the reciprocal value of the free ligand concentration at $\bar{n} = n - \dfrac{1}{2}$ is just equal to K_n. It is always disadvantageous to calculate

48

a stability constant from only one result, when the experimental errors may result in considerable error in the stability constant, whereas if the constants are calculated from several experimental points on the formation curve, the experimental errors may compensate each other. If $n > \bar{n} > > (n - 1)$ the ratio of the concentrations of two successive complexes is simply given as

$$\frac{[\text{MeL}_n]}{[\text{MeL}_{n+1}]} = \frac{\bar{n} + 1 - n}{n - \bar{n}}. \tag{4.5}$$

Thus

$$K_n = \frac{(\bar{n} + 1 - n)}{(n - \bar{n})} \frac{1}{[\text{L}]_{\bar{n}}}. \tag{4.6}$$

The corresponding formula for $1 > \bar{n} > 0$ is

$$K_1 = \frac{\bar{n}}{(1 - \bar{n})} \frac{1}{[\text{L}]_{\bar{n}}}$$

and for $2 > \bar{n} > 1$

$$K_2 = \frac{(\bar{n} - 1)}{(2 - \bar{n})} \frac{1}{[\text{L}]_{\bar{n}}}. \tag{4.8}$$

If the ratio of successive constants is less than 10^4, the calculated values of the stability constants are not equal to the true values, and they must be considered as preliminary ones from which the correct constants can be obtained by successive approximation. In the general case the approximation formulae are [4]:

$$K_n = \frac{1}{[\text{L}]} \frac{\displaystyle\sum_{t=0}^{t=n-1} \frac{\bar{n} - n + 1 + t}{[\text{L}]^t K_{n-1} K_{n-2} \ldots K_{n-t}}}{\displaystyle\sum_{t=0}^{t=N-n} (n - \bar{n} + t) [\text{L}]^t K_{n+1} K_{n+2} \ldots K_{n+t}} \tag{4.9}$$

and

$$K_n = \frac{1}{[\text{L}]_{\bar{n}*}} \frac{1 + \displaystyle\sum_{t=1}^{t=n-1} \frac{1 + 2t}{[\text{L}]_{\bar{n}*}^t K_{n-1} K_{n-2} \ldots K_{n-t}}}{1 + \displaystyle\sum_{t=1}^{t=N-n} (1 + 2t) [\text{L}]_{\bar{n}*}^t K_{n+1} K_{n+2} \ldots K_{n+t}}. \tag{4.10}$$

In the case of $N = 2$ the approximation formulae are rather simple:

$$K_1 = \frac{1}{[\text{L}]_{\bar{n}}} \frac{\bar{n}}{(1 - \bar{n}) + (2 - \bar{n}) K_2 [\text{L}]_{\bar{n}}} \tag{4.11}$$

and

$$K_2 = \frac{1}{[\text{L}]_{\bar{n}}} \frac{\bar{n} + (\bar{n} - 1) K_1 [\text{L}]_{\bar{n}}}{(2 - \bar{n}) K_1 [\text{L}]_{\bar{n}}}. \tag{4.12}$$

TABLE 4.1

Stability constants calculated from half-values of $\bar{n}$ and refined
by successive approximation until the ratio of the values is constant within ± 0.001

No. of approximations necessary	$[L]_{\bar{n}=1/2} \times 10^6$	$[L]_{\bar{n}=3/2} \times 10^6$	$\dfrac{[L]_{\bar{n}=1/2}}{[L]_{\bar{n}=1/2}}$	Starting values		'refined'		'true'	
				$K_1 \times 10^{-6}$	$K_2 \times 10^{-6}$	K_1	K_2	$K_1 \times 10^{-6}$	$K_2 \times 10^{-6}$
4	0.125	1.000	8.000	8.000	1.000	5.000×10^6	1.600×10^6	5.000	1.600
5	0.143	1.000	7.000	7.000	1.000	4.000×10^6	1.750×10^6	4.000	1.724
5	0.167	1.000	6.000	6.000	1.000	3.001×10^6	2.000×10^6	2.980	2.010
7	0.200	1.000	5.000	5.000	1.000	2.001×10^6	2.500×10^6	2.000	2.500
11	0.250	1.000	4.000	4.000	1.000	1.001×10^6	3.997×10^6	1.000	4.000
17	0.286	1.000	3.500	3.500	1.000	5.013×10^5	6.984×10^6	0.4960	7.040
998	0.333	1.000	3.000	3.000	1.000	1.499×10^3	2.003×10^9	0	∞

For the half-values, Eqs. 4.11 and 4.12 can be written as

$$K_1 = \frac{1}{[L]_{\bar{n}=0.5}} \frac{1}{1 + 3K_2 [L]_{\bar{n}=0.5}} \tag{4.13}$$

and

$$K_2 = \frac{1}{[L]_{\bar{n}=1.5}} \left(1 + \frac{3}{K_1 [L]_{\bar{n}=1.5}}\right). \tag{4.14}$$

Evidently the smaller the ratio of the successive constants, the greater the number of approximation steps necessary to obtain correct values for the equilibrium constants. This is well illustrated by the data in Table 4.1 where the results of some model calculations [5] are summarized.

As pointed out by Schrøder [5], in the case of $N = 2$ the successive constants can be obtained directly, without approximations, by means of the equations:

$$K_1 = \frac{[L]_{\bar{n}=1.5} - 3[L]_{\bar{n}=0.5}}{[L]_{\bar{n}=1.5} [L]_{\bar{n}=0.5}} \tag{4.15}$$

and

$$K_2 = \frac{1}{[L]_{\bar{n}=1.5} - 3[L]_{\bar{n}=0.5}}. \tag{4.16}$$

4.2 Correction methods

In the case of $N = 2$, the theoretical formation curve is symmetrical about its mid-point. Therefore, considering that*

$$\log K_1 K_2 = 2\mathrm{pL}_{\bar{n}=1} \tag{4.17}$$

it follows that

$$\mathrm{pL}_{1-d} + \mathrm{pL}_{1+d} = \log K_1 K_2 \tag{4.18}$$

where $1 > d > 0$ (practically $0.9 > d > 0$).

Substituting $1 - d$ and $1 + d$, respectively for $\bar{n}$ in the logarithmic form of Eqs. 4.11 and 4.12, after simplification we obtain

$$\log K_1 = \mathrm{pL}_{1-d} + \log \frac{2(1 - d)}{d + \sqrt{d^2 + 4(1 - d^2) K_2/K_1}} \tag{4.19}$$

and

$$\log K_2 = \mathrm{pL}_{1+d} - \log \frac{2(1 - d)}{d + \sqrt{d^2 + 4(1 - d^2) K_2/K_1}}. \tag{4.20}$$

The value of the second term of Eqs. 4.19 and 4.20, the so-called correction term

$$y = \log \frac{2(1 - d)}{d + \sqrt{d^2 + 4(1 - d^2) K_2/K_1}} \tag{4.21}$$

* Here, $\mathrm{pL} = -\log [L]$.

depends on the ratio of the successive constants. If $K_1 \gg K_2$, the value of the correction term is equal to $\log (1 - d)/d$ and Eqs. 4.19 and 4.20 are simplified to

$$\log K_1 = pL_{1-d} + \log \frac{1 - d}{d} \tag{4.22}$$

and

$$\log K_2 = pL_{1+d} - \log \frac{1 - d}{d} \tag{4.23}$$

which are analogous to Eqs. 4.7 and 4.8. The correction term can be evaluated from symmetrical points of the formation curve

$$pL_d \equiv pL_{1-d} - pL_{1+d} = \log \frac{K_1}{K_2} + 2y . \tag{4.24}$$

The corresponding values of y, pL_d and $\log (K_1/K_2)$ can be calculated by means of Eqs. 4.21 and 4.24. For nine points of the complex formation function the pairs of y and pL_d are summarized in Table 4.2 [6].

The value of the correction term can also be calculated from Eq. 4.25

$$y = \log \frac{1 - d}{d} + \log \left\{ 1 - \frac{(1 + d)[L]_{1-d}}{(1 - d)[L]_{1+d}} \right\} . \tag{4.25}$$

Essentially the Irving–Rossotti correction term method [6] is based on the fact that the distance along the pL axis between any given two symmetrical points of the formation curve is determined by the ratio of the

TABLE 4.2

The corresponding values of the correction term, the ratio of successive constants and the logarithmic distance between symmetrical points

log (K_1/K_2)	$d = 0.1$		$d = 0.2$		$d = 0.3$		$d = 0.4$		$d = 0.5$	
	y	pL_d	y	pL_d	y	pL_d	y	pL_d	y	pL_d
5.000	0.954	3.092	0.602	3.796	0.368	4.264	0.176	4.648	0.000	5.000
4.000	0.954	2.092	0.602	2.796	0.368	3.264	0.176	3.648	0.000	4.000
3.523	0.942	1.639	0.599	2.325	0.366	2.789	0.175	3.173	0.000	3.523
3.000	0.916	1.167	0.592	1.816	0.364	2.272	0.174	2.652	−0.001	3.003
2.523	0.861	0.802	0.574	1.376	0.355	1.812	0.169	2.185	−0.004	2.531
2.000	0.746	0.507	0.523	0.954	0.330	1.341	0.158	1.684	−0.012	2.025
1.523	0.593	0.336	0.430	0.663	0.273	0.976	0.120	1.283	−0.035	1.592
1.000	0.388	0.225	0.274	0.452	0.159	0.682	0.036	0.928	−0.094	1.188
0.523	0.178	0.167	0.093	0.336	0.004	0.515	−0.085	0.693	−0.196	0.916
0.000	−0.065	0.130	−0.131	0.262	−0.202	0.405	−0.278	0.556	−0.362	0.725
−0.477	−0.296	0.116	−0.354	0.231	−0.413	0.349	−0.477	0.477	−0.549	0.621
−1.000	−0.551	0.101	−0.602	0.204	−0.656	0.312	−0.715	0.430	−0.778	0.556
−1.477	−0.786	0.095	−0.835	0.193	−0.885	0.294	−0.940	0.402	−1.000	0.523
−2.000	−1.046	0.091	−1.092	0.184	−1.141	0.282	−1.193	0.287	−1.251	0.502

TABLE 4.2 *(continued)*

log (K_1/K_2)	$d = 0.6$		$d = 0.7$		$d = 0.8$		$d = 0.9$	
	y	pL_d	y	pL_d	y	pL_d	y	pL_d
5.000	−0.176	5.352	−0.368	5.736	−0.602	6.205	−0.954	6.908
4.000	−0.176	4.352	−0.368	4.736	−0.602	5.205	−0.954	5.908
3.523	−0.176	3.876	−0.368	4.260	−0.602	4.727	−0.954	5.431
3.000	−0.177	3.354	−0.368	3.737	−0.602	4.204	−0.954	4.909
2.523	−0.178	2.880	−0.369	3.261	−0.603	3.728	−0.954	4.432
2.000	−0.184	2.367	−0.372	2.745	−0.604	3.209	−0.955	3.910
1.523	−0.198	1.918	−0.381	2.285	−0.609	2.741	−0.957	3.437
1.000	−0.238	1.477	−0.407	1.815	−0.615	2.230	−0.964	2.928
0.523	−0.317	1.158	−0.465	1.452	−0.662	1.846	−0.980	2.482
0.000	−0.459	0.917	−0.582	1.164	−0.749	1.497	−1.032	2.064
−0.477	−0.633	0.789	−0.736	0.996	−0.879	1.281	−1.123	1.770
−1.000	−0.852	0.705	−0.944	0.887	−1.066	1.133	−1.279	1.557
−1.477	−1.069	0.661	−1.154	0.831	−1.268	1.060	−1.459	1.441
−2.000	−1.317	0.634	−1.398	0.796	−1.506	1.012	−1.680	1.359

successive stability constants. As pointed out by Gergely, Nagypál and Mojzes [7], this ratio can also be evaluated from the integral of the formation curve, namely, the area between the formation curve, the pL axis and ordinates at $pL = pL_{\bar{n}_1}$ and $pL = pL_{\bar{n}_2}$, depends only on the ratio K_1/K_2 and the values of $\bar{n}_1$ and $\bar{n}_2$. The integration of the formation function gives:

$$\int_{pL_{\bar{n}}=0}^{pL_{\bar{n}}} \bar{n}\,dpL = \left[-\log\left(1 + K_1[L] + K_1 K_2 [L]^2\right) \right]_{[L]=0}^{[L]_{\bar{n}}}. \qquad (4.26)$$

[L] and [L]² can be expressed from the inverse function of Eq. 3.6:

$$[L] = \frac{(1 - \bar{n})K_1 - \sqrt{(\bar{n}-1)^2 K_1^2 - 4\bar{n}(\bar{n}-2)K_1 K_2}}{2(\bar{n}-2)K_1 K_2} \qquad (4.27)$$

and

$$[L]^2 = \frac{(1-\bar{n})^2 K_1^2}{4(\bar{n}-2)K_1^2 K_2^2} - $$
$$- \frac{2(1-\bar{n})K_1\sqrt{(\bar{n}-1)^2 K_1^2 - 4\bar{n}(\bar{n}-2)K_1 K_2} + (\bar{n}-1)^2 K_1^2 - 4(\bar{n}-2)K_1 K_2}{4(\bar{n}-2)K_1^2 K_2^2}. $$
$$(4.28)$$

Inserting Eqs. 4.27 and 4.28 into Eq. 4.26 and simplifying, we get

$$\int_{pL_{\bar{n}=0}}^{pL_{\bar{n}}} \bar{n}\,dpL = \log\left(1 + \frac{\bar{n}}{2-\bar{n}} + \frac{\bar{n}-1}{2(\bar{n}-2)^2}\frac{K_1}{K_2} + \right.$$
$$\left. + \frac{1}{2(\bar{n}-2)^2}\sqrt{(\bar{n}-1)^2\left(\frac{K_1}{K_2}\right)^2 - 4(\bar{n}-2)\bar{n}\frac{K_1}{K_2}}\right). \qquad (4.29)$$

TABLE 4.3 *Corresponding values of* $\log (K_1/K_2)$ *and the*

$\bar{n}$	$\log (K_1/K_2)$									
	0.0	0.2	0.4	0.6	0.8	1.0	1,2	1.4	1.6	1.8
	$\displaystyle\int_0^{\bar{n}} \bar{n}\,d\mathrm{pL}$									
0.1	0.0420	0.0431	0.0439	0.0446	0.0450	0.0452	0.0454	0.0455	0.0456	0.0457
0.2	0.0829	0.0863	0.0892	0.0915	0.0932	0.0945	0.0953	0.0959	0.0962	0.0965
0.3	0.1240	0.1304	0.1361	0.1411	0.1451	0.1482	0.1504	0.1520	0.1530	0.1537
0.4	0.1663	0.1759	0.1852	0.1937	0.2011	0.2071	0.2117	0.2150	0.2174	0.2189
0.5	0.2103	0.2234	0.2368	0.2497	0.2616	0.2719	0.2803	0.2867	0.2914	0.2947
0.6	0.2565	0.2735	0.2915	0.3096	0.3272	0.3434	0.3574	0.3689	0.3780	0.3844
0.7	0.3056	0.3268	0.3498	0.3739	0.3984	0.4223	0.4445	0.4640	0.4803	0.4931
0.8	0.3582	0.3839	0.4124	0.4434	0.4761	0.5096	0.5428	0.5743	0.6029	0.6276
0.9	0.4150	0.4456	0.4803	0.5189	0.5610	0.6062	0.6534	0.7016	0.7494	0.7955
1.0	0.4771	0.5132	0.5545	0.6015	0.6544	0.7128	0.7767	0.8458	0.9196	0.9975
1.1	0.5458	0.5877	0.6365	0.6930	0.7575	0.8308	0.9132	1.0052	1.1076	1.2207
1.2	0.6227	0.6712	0.7284	0.7951	0.8725	0.9614	1.0628	1.1773	1.3056	1.4474
1.3	0.7105	0.7665	0.8329	0.9110	1.0020	1.1070	1.2267	1.3612	1.5098	1.6713
1.4	0.8126	0.8772	0.9542	1.0447	1.1502	1.2712	1.4076	1.5587	1.7225	1.8968
1.5	0.9348	1.0097	1.0986	1.2030	1.3234	1.4599	1.6113	1.7756	1.9505	2.1333
1.6	1.0869	1.1741	1.2771	1.3966	1.5326	1.6837	1.8481	2.0230	2.2060	2.3946
1.7	1.2876	1.3904	1.5100	1.6463	1.7981	1.9630	2.1385	2.3219	2.5108	2.7036
1.8	1.5801	1.7033	1.8433	1.9983	2.1661	2.3438	2.5287	2.7188	2.9124	3.1082
1.9	2.1060	2.2584	2.4241	2.6002	2.7841	2.9734	3.1665	3.3620	3.5592	3.7574

Equation 4.29 clearly shows that the value of the integral is independent of the absolute values of the successive constants. Table 4.3 summarizes the corresponding values of $\log (K_1/K_2)$ and the integrals for several values of $\bar{n}$. The area between the formation curve and two given points on the pL axis, that is the integral, can be simply and correctly determined by graphical integration.

For the symmetrical points of the formation curve the values of the integral and those of their distance apart are the same. That is, writing the integral between two symmetrical points ($\bar{n}_1 = 1 - d$ and $\bar{n}_2 = 1 + d$) of the formation curve, one gets

$$\int_{\mathrm{pL}_{1-d}}^{\mathrm{pL}_{1+d}} \bar{n}\,d\mathrm{pL} = \log \frac{1 + K_1[\mathrm{L}]_{1+d} + K_1 K_2[\mathrm{L}]_{1+d}^2}{1 + K_1[\mathrm{L}]_{1-d} + K_1 K_2[\mathrm{L}]_{1-d}^2}. \qquad (4.30)$$

Considering that for the symmetrical points

$$K_1 K_2 = \frac{1}{[\mathrm{L}]_{1-d}\,[\mathrm{L}]_{1+d}},$$

Eq. 4.30 can be written as

integrals (Eq. 4.29) at given values of the formation function

n	\multicolumn{11}{c}{$\log (K_1/K_2)$}
	2.0	2.2	2.4	2.6	2.8	3.0	3.2	3.4	3.6	3.8	4.0
	\multicolumn{11}{c}{$\overset{\bar{n}}{\underset{0}{\int}} \bar{n}d\mathrm{pL}$}
0.1	0.0457	0.0457	0.0457	0.0457	0.0458	0.0458	0.0458	0.0458	0.0458	0.0458	0.0458
0.2	0.0967	0.0968	0.0968	0.0968	0.0969	0.0969	0.0969	0.0969	0.0969	0.0969	0.0969
0.3	0.1541	0.1544	0.1546	0.1547	0.1548	0.1548	0.1549	0.1549	0.1549	0.1549	0.1549
0.4	0.2200	0.2207	0.2211	0.2214	0.2216	0.2217	0.2217	0.2218	0.2218	0.2218	0.2218
0.5	0.2969	0.2984	0.2993	0.2999	0.3003	0.3006	0.3008	0.3009	0.3009	0.3010	0.3010
0.6	0.3890	0.3921	0.3942	0.3956	0.3964	0.3970	0.3973	0.3975	0.3977	0.3978	0.3978
0.7	0.5026	0.5094	0.5141	0.5172	0.5192	0.5206	0.5214	0.5219	0.5223	0.5225	0.5226
0.8	0.6478	0.6634	0.6749	0.6831	0.6886	0.6923	0.6947	0.6964	0.6973	0.6980	0.6983
0.9	0.8381	0.8762	0.9085	0.9346	0.9547	0.9694	0.9797	0.9868	0.9915	0.9946	0.9965
1.0	1.0792	1.1640	1.2516	1.3415	1.4333	1.5266	1.6213	1.7170	1.8136	1.9108	2.0086
1.1	1.3452	1.4815	1.6296	1.7889	1.9584	2.1364	2.3210	2.5107	2.7038	2.8994	3.0965
1.2	1.6021	1.7682	1.9439	2.1271	2.3159	2.5083	2.7034	2.9003	3.0983	3.2970	3.4962
1.3	1.8435	2.0242	2.2111	2.4025	2.5969	2.7933	2.9909	3.1895	3.3885	3.5879	3.7876
1.4	2.0792	2.2674	2.4596	2.6546	2.8514	3.0493	3.2480	3.4472	3.6467	3.8463	4.0461
1.5	2.3219	2.5144	2.7095	2.9064	3.1044	3.3032	3.5024	3.7019	3.9016	4.1014	4.3012
1.6	2.5872	2.7825	2.9794	3.1774	3.3762	3.5754	3.7749	3.9746	4.1743	4.3742	4.5742
1.7	2.8990	3.0960	3.2941	3.4930	3.6922	3.8917	4.0914	4.2912	4.4910	4.6909	4.8909
1.8	3.3056	3.5038	3.7029	3.9022	4.1018	4.3015	4.5013	4.7012	4.9011	5.1011	5.3011
1.9	3.9562	4.1555	4.3550	4.5547	4.7545	4.9544	5.1544	5.3543	5.5543	5.7543	5.9543

$$\int\limits_{\mathrm{pL}_{1-d}}^{\mathrm{pL}_{1+d}} \bar{n}d\mathrm{pL} = \log \frac{[\mathrm{L}]_{1+d}}{[\mathrm{L}]_{1-d}}. \tag{4.31}$$

The correction term methods provide a convenient and reliable possibility for the calculation of successive constants. Their application is limited to systems where either $N = 2$, or (although $N > 2$) the formation curve consists of distinct steps with only two overlapping complex equilibria. Both correction term methods can be applied even if only the first part of the formation curve ($n < 1$) can be determined, provided that the formation curve is symmetrical and only two complex species (MeL and MeL$_2$) exist under the given conditions. In such cases, however, the constants may be used — with some reservation — only to describe the system up to the highest value of $\bar{n}$ determined, and it is not permitted to extrapolate to higher values of $\bar{n}$.

4.3 Graphical methods

Various graphical methods are widely used in equilibrium studies. These methods make possible (and often necessitate) the simultaneous consideration of many experimental data and they clearly indicate the experimental errors. The most serious disadvantage of these methods is that the stability constants of complicated systems can be evaluated only in several steps and an accumulation of errors occurs. Three types of graphical methods will be treated: *(i)* linear extrapolation; *(ii)* the elimination method; *(iii)* curve-fitting methods.

4.3.1 LINEAR EXTRAPOLATION

Writing Leden's function in the form

$$F(L) = \frac{T_{Me} - [Me]}{[Me][L]} = \beta_1 + \sum_2^n \beta_i [L]^{i-1} \tag{4.32}$$

we see that if we plot F[L] against [L], β_1 can be obtained as the graphical limit

$$\lim_{[L] \to 0} F(L) = \beta_1. \tag{4.33}$$

If [L] = 0, both the numerator and denominator of Leden's function are equal to zero, and the limit appears to be indeterminate; however, Sullivan and Hindman [3] proved that the limit is the desired quantity:

$$\lim_{[L] \to 0} \left\{ \frac{T_{Me} - [Me]}{[Me][L]} \right\} = \lim_{[L] \to 0} \left\{ \frac{\sum_1^n [MeL_i]}{[Me][L]} \right\} = \lim_{[L \to 0]} \left\{ \frac{\dfrac{d}{d[L]} \sum_1^N [MeL_i]}{[Me]} \right\} =$$

$$= \lim_{[L] \to 0} \{ \beta_1 + 2\beta_2 [L] \} = \beta_1. \tag{4.34}$$

If $N = 2$, F[L] *vs.* [L] is a straight line, the slope of which gives β_2. If $N > 2$ a curve is obtained, and to evaluate β_2 a new function is introduced:

$$F'(L) = \frac{F(L) - \beta_1}{[L]} = \beta_2 + \sum_3^N \beta_i [L]^{i-1}. \tag{4.35}$$

Evidently β_2 can be obtained as the limit of the curve of F'(L) *vs.* [L]. The further constants can be evaluated by a continuation of the procedure. Leden's method can be applied if [Me] and [L] are the measurable quantities. (The method can be used even if only [Me] is measurable; in such cases a successive approximation is applied to evaluate [L].) There are a number of graphical methods which need knowledge of $\bar{n}$ as a function of [L]. By differentiating Fronaeus's function (Eq. 4.3) we get

$$\frac{dX(L)}{d[L]} = \sum_1^N i\beta_i [L]^{i-1} \tag{4.36}$$

and from Eq. 3.7

$$\frac{\bar{n}}{[L]} = \frac{\dfrac{dX(L)}{d[L]}}{X(L)} = \frac{d}{d[L]} \log \left(1 + \sum_1^N \beta_i [L]^i\right) \qquad (4.37)$$

that is

$$\frac{d \log X(L)}{d \log [L]} = \bar{n}. \qquad (4.38)$$

Thus $X(L)$ can be evaluated by graphical integration of the right-hand side of the equation

$$\ln X(L) = \int_0^L \frac{\bar{n}}{[L]} \, d[L]. \qquad (4.39)$$

$X(L)$ having been found, the stability constants can be evaluated [2] according to the procedure with Leden's function already described because Eq. 4.3 provides a simple relationship between $X(L)$ and $F(L)$.

The method given by Rossotti and Rossotti [8, 9] is based on a rearrangement of the expression of the complex formation function, leading to

$$\frac{\bar{n}}{(1 - \bar{n})[L]} = \beta_1 + \frac{2 - \bar{n}}{1 - \bar{n}} \beta_2 [L] + \sum_3^N \frac{i - \bar{n}}{1 - \bar{n}} \beta_1 [L]^{i-1}. \qquad (4.40)$$

Thus the initial part of the plot of $\bar{n}/(1 - \bar{n})[L]$ against $(2 - \bar{n})[L]/(1 - \bar{n})$ is a straight line of intercept β_1 and of slope β_2. The further constants can be obtained, by using a generalized form of Eq. 4.40:

$$\sum_0^{n-1} \left(\frac{\bar{n} - i}{n - \bar{n}}\right) \beta_i [L]^{i-n} = \beta_n + \sum_{n+1}^N \left(\frac{i - \bar{n}}{n - \bar{n}}\right) \beta_i [L]^{i-n}. \qquad (4.41)$$

If values of $\beta_1, \beta_2, \ldots \beta_{n-1}$ have previously been calculated, the left-hand side of Eq. 4.41 is known. If this is plotted against $(n + 1 - \bar{n})[L]/(n - \bar{n})$, β_n is obtained as the intercept and β_{n+1} as the limiting slope. When $n = = N - 1$, Eq. 4.41 is written as

$$\sum_0^{N-2} \left(\frac{\bar{n} - i}{N - 1 - \bar{n}}\right) \beta_i [L]^{i-N+1} = \beta_{N-1} + \beta_N \frac{N - \bar{n}}{N - 1 - \bar{n}} [L]. \qquad (4.42)$$

Plotting the left-hand side of Eq. 4.42 against $(N - \bar{n})[L]/(N - 1 - \bar{n})$ we get a straight line with intercept β_{N-1} and slope β_N. As in the case of Leden's method, the errors in the values of the successive constants β_1, $\beta_2, \ldots \beta_{n-1}$ will accumulate in the value of β_n. Thus it is helpful to evaluate the constants by means of an alternative arrangement of the complex formation function:

$$\left(\frac{N-\bar{n}}{N-1-\bar{n}}\right)[L] = \frac{\beta_{N-1}}{\beta_N} + \frac{\beta_{N-2}}{\beta_N}\left(\frac{\bar{n}-N+2}{N-1-\bar{n}}\right)[L]^{-1} +$$

$$+ \sum_0^{N-3}\left(\frac{\bar{n}-i}{N-1-\bar{n}}\right)\frac{\beta_i}{\beta_N}[L]^{i-N+1} . \qquad (4.43)$$

Then, on plotting $(N-\bar{n})[L]/(N-1-\bar{n})$ against $(\bar{n}-N+2)/(N-1-\bar{n})[L]$, the intercept is β_{N-1}/β_N and the slope of the initial part is β_{N-2}/β_N. All the ratios β_n/β_N can be analogously obtained.

If $N = 1$, Eq. 4.40 is reduced to

$$\frac{\bar{n}}{1-\bar{n}} = \beta_1[L] , \qquad (4.44)$$

which is equivalent to the equations used for the determination of the dissociation of monobasic acids [10, 11] and the equation obtained by Block and McIntyre [12] (see later).

If $N = 2$, the following equation is obtained

$$\frac{\bar{n}}{(1-\bar{n})[L]} = \beta_1 + \beta_2\frac{(2-\bar{n})[L]}{(1-\bar{n})} \qquad (4.45)$$

which was suggested earlier by Irving and Rossotti [6] for the calculation of stability constants and — in another form — by Speakman [13] for the calculation of dissociation constants of dibasic acids.

Another arrangement of the complex formation function was suggested by Olerup [14]:

$$G(L) = \frac{\bar{n}}{[L]} = (1-\bar{n})\beta_1 + (2-\bar{n})\beta_2[L] + \ldots + (N-\bar{n})\beta_N[L]^{N-1} =$$

$$= \sum_1^N (i-\bar{n})\beta_i[L]^{i-1}. \qquad (4.46)$$

β_1 is obtained as the intercept of the plot of $G(L)$ *vs.* $[L]$, while β_2 can be calculated from the initial slope. These graphs are rather curved and it is better to obtain the constants from plots of $G(L)$ *vs.* $\bar{n}$. Fomin and Majrova [15] used Olerup's method to obtain β_1 by plotting $(G(L) - \beta_1)/[L]$ against $[L]$ (to calculate β_2), the intercept giving the value of $(2\beta_2 - \beta_1^2)$:

$$\frac{G(L) - \beta_1}{[L]} = \frac{\bar{n} - \beta_1[L]}{[L]^2} =$$

$$= \frac{(2\beta_2 - \beta_1^2) + (3\beta_3 - \beta_1\beta_2)[L] + \sum_4^N (i\beta_i - \beta_1\beta_i)[L]^{i-2}}{\sum_0^N \beta_i[L]^i} . \qquad (4.47)$$

As pointed out by Irving [16] Olerup's function is monotonic only if $\beta_1^2 > 2\beta_2$, that is if $K_1 > 2K_2$. If the relation of the successive constants is the opposite of this, a maximum is found on the graph. Therefore a maximum on the graph indicates an unusual ratio of the successive constants. In the system where $N = 2$, it is possible to calculate the ratio K_1/K_2 from the position of the maximum on the curve $\bar{n}/[L]$ vs. [L]. Differentiating Eq. 4.46 we get the following condition for the extremum:

$$(2\beta_2 - \beta_1^2) - 2\beta_1\beta_2[L]_* - 2\beta_2^2[L]_*^2 = 0 \tag{4.48}$$

and thus

$$[L]_* = \frac{-\beta_1 + \sqrt{(4\beta_2 - \beta_1^2)}}{2\beta_2}. \tag{4.49}$$

From Eqs. 4.46, 4.48 and 4.49, the value of the average co-ordination number at the maximum is given by

$$\bar{n}_* = 1 - \beta_1(4\beta_2 - \beta_1^2)^{-1/2} \tag{4.50}$$

whence

$$\frac{K_1}{K_2} = \frac{\beta_1^2}{\beta_2} = \frac{4(1 - \bar{n}_*)^2}{1 + (1 - \bar{n}_*)^2}. \tag{4.51}$$

Another convenient function was introduced by Scatchard [17]

$$Q(L) = \frac{\bar{n}}{(N - \bar{n})[L]} \tag{4.52}$$

and first applied by Edsall and his coworkers [18] to calculate the stability constants of copper(II) and zinc(II) imidazole complexes. $Q(L)$ is the 'equilibrium constant' for the equilibrium

unoccupied site $+ L \rightleftharpoons$ occupied site.

However, because the ease with which a site can be occupied depends on the number of neighbouring sites already occupied, $Q(L)$ is not a constant. It is easy to realize that by extrapolating the curve for $Q[L]$ vs. [L], or more conveniently for $Q[L]$ vs. $\bar{n}$, to $[L] = 0$ or $\bar{n} = 0$, K_1/N is obtained, while extrapolation to $[L] = \infty$ or $\bar{n} = N$ leads to the value of NK_N. In principle, K_2 and K_{N-1} can be evaluated from the slope of the curve for $\log Q(L)$ vs. $\bar{n}$ at $\bar{n} = 0$ and $\bar{n} = N$, respectively, but this method provides only uncertain values. If Eq. 3.43 is valid for the system studied it can be written that

$$\frac{K_1}{K_N} = 10^{2\lambda N}. \tag{4.53}$$

As pointed out by Poe [19], Scatchard's function is very suitable for evaluating all the stepwise constants by a rapid successive approximation procedure.

Figure 4.1 illustrates the application of some linear extrapolation methods to the copper(II)-threonine system [20a].

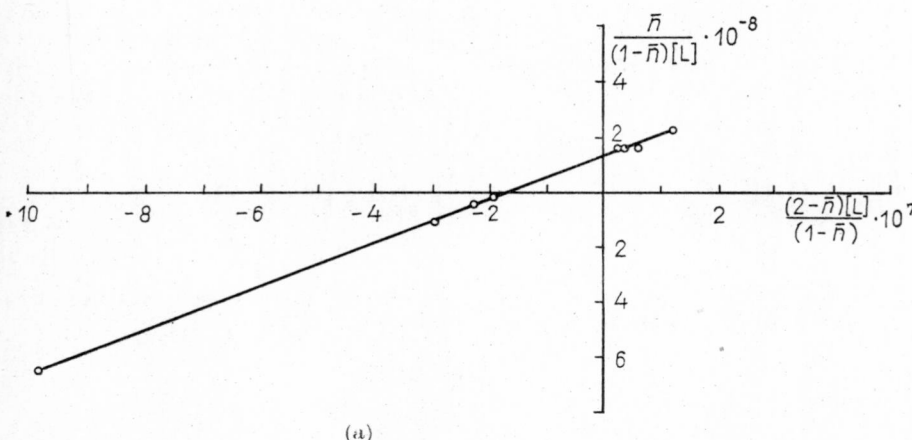

(a)

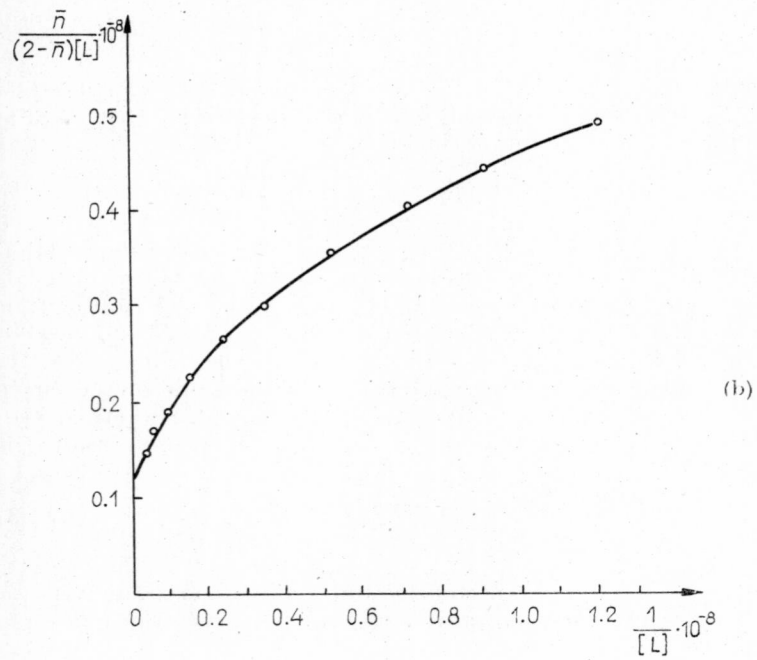

(b)

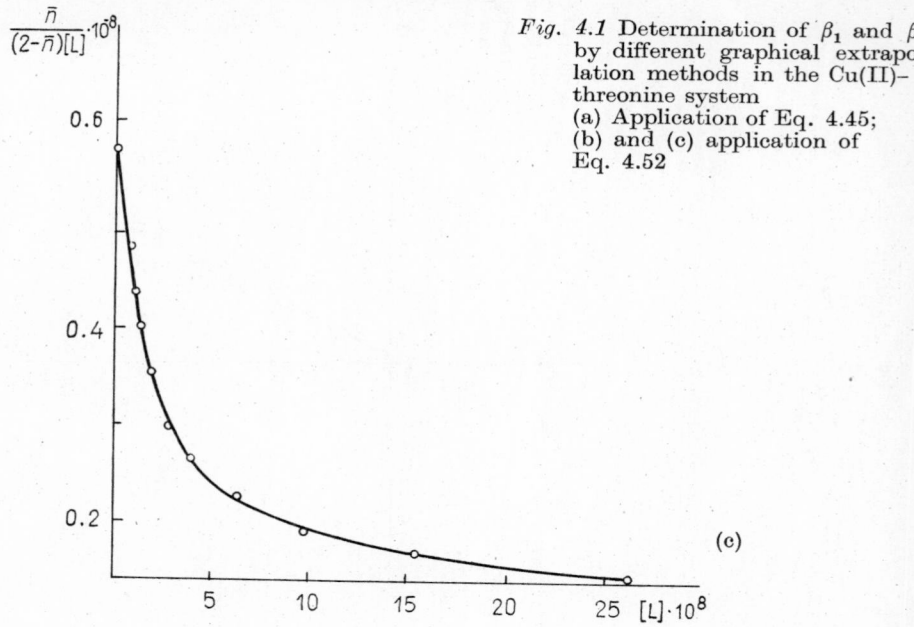

Fig. 4.1 Determination of β_1 and β_2 by different graphical extrapolation methods in the Cu(II)–threonine system
(a) Application of Eq. 4.45;
(b) and (c) application of Eq. 4.52

4.3.2 ELIMINATION METHOD

This method can be applied when only two simultaneous equilibria have to be considered. In such cases the complex formation function is

$$\bar{n} = \frac{\beta_1[L] + 2\beta_2[L]^2}{1 + \beta_1[L] + \beta_2[L]^2} \tag{4.54}$$

while the partial mole fraction of the complexes can be given as

$$\alpha_n = \frac{\beta_n[L]^n}{1 + \beta_1[L] + \beta_2[L]^2} \tag{4.55}$$

where $n = 0$, 1 or 2.

Both of these equations can be written in the form

$$xp_1 + yp_2 = 1 \tag{4.56}$$

where the variables x and y are functions of $\bar{n}$ and $[L]$ or of α_n and $[L]$, and the parameters p_1 and p_2 are related to the stability constants. There are six possible transformations for each of the functions $\bar{n}([L])$ and $\alpha_n([L])$. The possible values of x and y are tabulated in Table 4.4 [9]. (see p. 64).

To obtain the stability constants, each pair of experimental values $(\bar{n}, [L]; \alpha_0, [L]; \alpha_1, [L]; \alpha_2, [L])$ is used to calculate pairs of values (x, y) which are then plotted in a co-ordinate system of axes $1/x$ and $1/y$. All

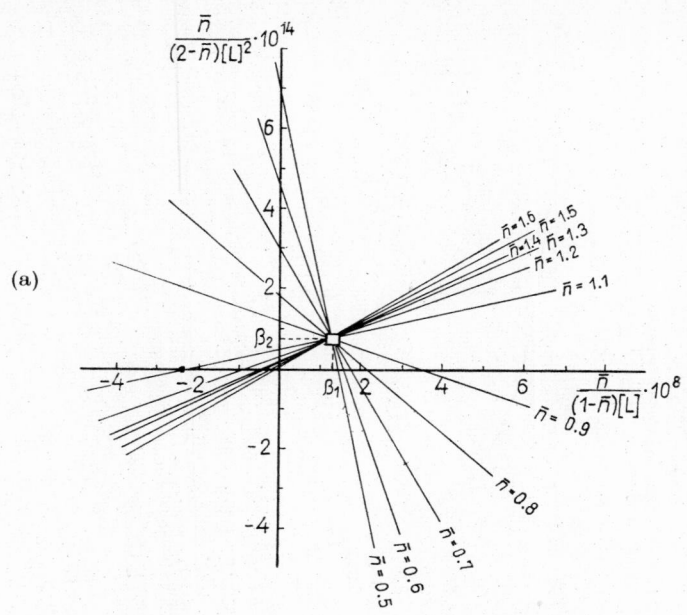

(a)

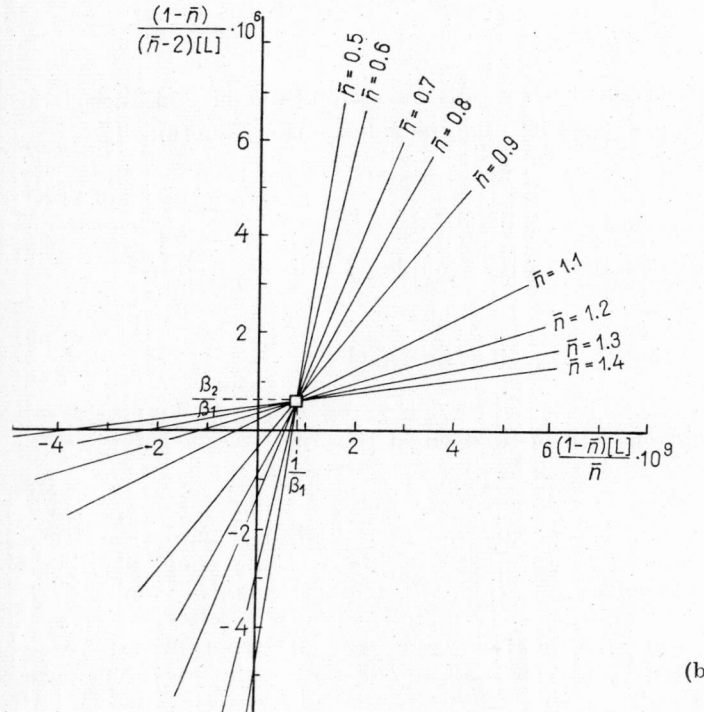

(b)

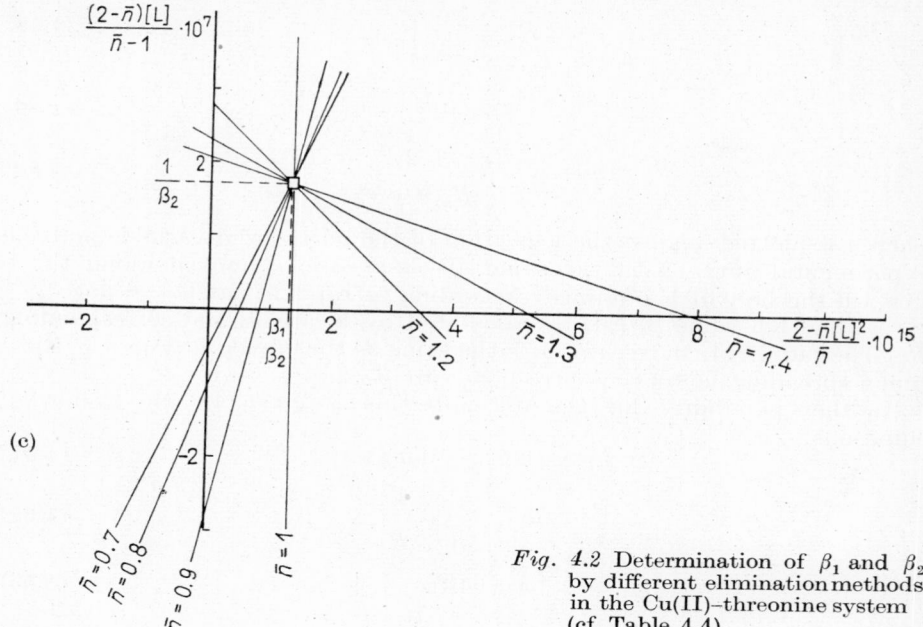

Fig. 4.2 Determination of β_1 and β_2 by different elimination methods in the Cu(II)–threonine system (cf. Table 4.4).

the straight lines have one common point, the co-ordinates of which give the correct values of p_1 and p_2, i.e. the desired constants. The elimination method was first applied by Prytz [20] to evaluate the stability constants of the chloro-tin(II) complexes from a function $\alpha_0(L)$, and later widely used by Schwarzenbach's school [22, 23] to calculate both acidic dissociation and complex stability constants. A generalization of the method was given by Rossotti and Rossotti [9]. Figure 4.2 illustrates the application of the elimination method to the copper(II)-threonine system [20a].

4.3.3 Curve-fitting methods

In simple systems ($N = 1$ or 2) the so-called curve-fitting methods developed by Sillén [24] can be conveniently applied to obtain the equilibrium constants. In these cases the shape and the position of the formation curve (or the curves of α_n *vs.* [L]) are unequivocally determined by one or two constants which can be obtained by comparing the experimental curve with a series of calculated ones. The data corresponding to the best fitting curve are taken as the constants of the system. For this comparison the formation function (Eq. 4.54) is normalized, and a new variable is introduced:

$$\mathrm{L} = \frac{1}{2} \log \beta_2 + \log [\mathrm{L}] \tag{4.57}$$

$$R_K = \left(\frac{K_1}{K_2}\right)^{\frac{1}{2}} \tag{4.58}$$

$$v = 10^{\mathrm{L}} \tag{4.59}$$

$$\bar{n} = \frac{R_K v + 2 v^2}{1 + R_K v + v^2}. \tag{4.60}$$

Then $\bar{n}$ is plotted against the logarithm of the normalized variable and the experimental curve with the same abscissa scale is moved along the L axis till the best fit is obtained. According to Eq. 4.57, at L $= 0$, $\log \beta_2 = = -2 \log [\mathrm{L}]$, while β_1 can be calculated by means of the corresponding R_K. The family of curves $\bar{n}$ vs. L is the same as that given in Fig. 3.1. Bjerrum's spreading factor (x) corresponds to $R_K/2$.

Another possibility for the normalization is given by the following equations:

$$\mathrm{L}' = \log \beta_1 + \log [\mathrm{L}] \tag{4.61}$$

$$R'_K = \frac{\beta_2}{\beta_1^2} \equiv \frac{K_2}{K_1} \tag{4.62}$$

$$v' = 10^{\mathrm{L}'} \tag{4.63}$$

$$\bar{n} = \frac{v' + 2R'_K v'^2}{1 + v' + R'_K v'^2}. \tag{4.64}$$

As before, the experimental curve is compared with the family of the calculated ones, and the value of $-\log [\mathrm{L}]$ at $\mathrm{L}' = 0$ on the best fitting curve gives $\log \beta_1$, while β_2 can be calculated from the corresponding R_K.

Sillén [25] developed curve-fitting methods even for $N = 3$. The so-called projection strip method was developed by Rossotti, Rossotti and Sillén [26]. This also applies to normalized functions, but in this case the function $R_K = = f(\mathrm{L})_{\bar{n}}$ is calculated, while in curve-fitting the family of curves corresponding to the function $\bar{n} = f(\mathrm{L})_{R_K}$ is used.

TABLE 4.4

Transformations of Eqs. 4.54 and 4.55 to the form $xp_1 + yp_2 = 1$

Parameter		Function			
		$\alpha_0 [\mathrm{L}]$		$\alpha_1 [\mathrm{L}]$	
p_1	p_2	y	x	y	x
β_1	β_2	$\dfrac{\alpha_0 [\mathrm{L}]}{1 - \alpha_0}$	$\dfrac{\alpha_0 [\mathrm{L}]^2}{1 - \alpha_0}$	$\dfrac{(1 - \alpha_1) [\mathrm{L}]}{\alpha_1}$	$-[\mathrm{L}]^2$
$\beta_1 - 1$	$\beta_1 - 1\beta_2$	$\dfrac{1 - \alpha_0}{\alpha_0 [\mathrm{L}]}$	$-[\mathrm{L}]$	$\dfrac{\alpha_1}{(1 - \alpha_1) [\mathrm{L}]}$	$\dfrac{\alpha_1 [\mathrm{L}]}{1 - \alpha_1}$
$\beta_2 - 1$	$\beta_1\beta_2 - 1$	$\dfrac{1 - \alpha_0}{\alpha_0 [\mathrm{L}]^2}$	$-\dfrac{1}{[\mathrm{L}]}$	$-\dfrac{1}{[\mathrm{L}]^2}$	$\dfrac{1 - \alpha_1}{\alpha_1 [\mathrm{L}]}$

TABLE 4.4 *(continued)*

Parameter		Function			
		$\alpha_2\,[L]$		$n\,[L]$	
p_1	p_2	y	x	y	x
α_1	β_2	$-[L]$	$\dfrac{(1-\alpha_2)\,[L]}{a_2}$	$\dfrac{(1-\bar{n})\,[L]}{\bar{n}}$	$\dfrac{(2-\bar{n})\,[L]^2}{\bar{n}}$
β_1-1	$\beta_1-1\beta_2$	$-\dfrac{1}{[L]}$	$\dfrac{(1-\alpha_2)\,[L]}{\alpha_2}$	$\dfrac{\bar{n}}{(1-\bar{n})\,[L]}$	$\dfrac{(\bar{n}-2\,[L])}{1-\bar{n}}$
β_2-1	$\beta_1\beta_2-1$	$\dfrac{\alpha_2}{(1-\alpha_2)\,[L]^2}$	$\dfrac{\alpha_2}{(1-\alpha_2)[L]}$	$\dfrac{\bar{n}}{(2-\bar{n})[L]^2}$	$\dfrac{\bar{n}-1}{(2-\bar{n}}$

From Eq. 4.60

$$R_K = \frac{(2-\bar{n})\,v^2 - \bar{n}}{(\bar{n}-1)\,v}. \tag{4.65}$$

The family of normalized functions $\log R_K(\text{L})\bar{n}$ is given in Table 4.5 [27] for the range $0.1 \leqslant n \leqslant 0.9$, and those for the range $1.1 < \bar{n} < 1.9$ may be calculated from the symmetry relationship

$$L\bar{n}_{=1-d} = L\bar{n}_{=1+d}. \tag{4.66}$$

TABLE 4.5

The family of normalized functions $\log R_K(\text{L})\bar{n}$

L	$\log R_K$								
	$\bar{n}=0.1$	$\bar{n}=0.2$	$\bar{n}=0.3$	$=0.4$	$\bar{n}=0.5$	$\bar{n}=0.6$	$\bar{n}=0.7$	$n=0.8$	$\bar{n}=0.9$
1.60	0.641	0.996	1.230	1.423	1.599	1.776	1.968	2.202	2.554
1.40	.433	.792	1.028	1.221	1.398	1.575	1.767	2.001	2.353
1.20	.212	.582	0.818	1.017	1.195	1.372	1.565	1.800	2.152
1.00	— .046	.357	.607	0.806	0.987	1.166	1.360	1.596	1.949
0.90	— .210	.231	.491	.695	.879	1.059	1.255	1.492	1.848
.80	— .436	.087	.365	.578	.766	0.950	1.147	1.385	1.741
.70	— .868	— .092	.221	.449	.645	.834	1.035	1.275	1.632
.60	— ∞	— .367	.040	.298	.509	.707	0.914	1.159	1.519
.50		—1.102	— .231	.102	.345	.561	.779	1.032	1.398
.40		— ∞	— .961	— .213	.120	.380	.617	0.884	1.261
.35			— ∞	— .521	— .032	.254	.517	.798	1.184
.30				— ∞	— .308	.093	.395	.697	1.095
.25					—1.039	— .155	.234	.584	0.996
.20					— ∞	— .771	— ,016	.407	.865
.15						— ∞	— .641	— .147	.693
.10							— ∞	— .569	.414
.08								— ∞	.223
.05									— .532
.045									—1.184
.04									— ∞

The application of the projection strip method is illustrated in Fig. 4.3 which refers to the copper (II)-α-amino-isobutyric acid system [21]. The experimental formation curve is plotted with the same abscissa scale used for the plot of R_K against L. For each value of $\bar{n}$ used to calculate the theoretical curves, a distance corresponding to the experimental uncertainty is marked off on the log [L] axis. These distances, the projection strips, are superimposed on the family of theoretical curves, parallel to the absciss, until the best fit is obtained for the series of $\bar{n}$ used. Then log R_K is obtained as the ordinate value corresponding to this position of the series of strips, and log $\beta_2 = -2$ log [L] at the point of intersection of the log R_K axis with the projection strip (L = 0). On comparing Eqs. 4.18 and 4.57 it is evident that the normalized variable L is identical with pL_d of the correction term method. A disadvantage of the projection strip method is that many experimental points are required for the determination of the length of the 'strips'.

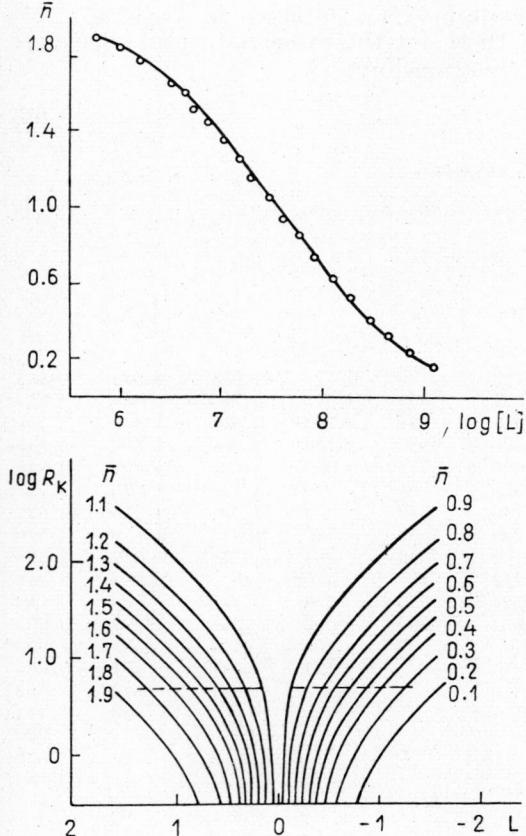

Fig. 4.3 Determination of the stability constants in the Cu(II)-aminobutyric acid system by the projection strip method

4.4 Numerical methods

The larger the maximum co-ordination number (N), the more difficult it is to calculate the stability constants directly. Nevertheless, if N does not exceed 3, it is convenient to obtain the stability constants by the direct solution of the basic equation (Eq. 3.3). The corresponding formulae were given by Block and McIntyre [12] who introduced the following symbols to derive easily manageable expressions:

$$J_n = (n - \bar{n})\,[L]^n$$

$$J'_n = (n - \bar{n}')\,[L']^n$$

$$J''_n = (n - \bar{n}'')\,[L'']^n$$

$$L_n = \bar{n}\,J'_n - \bar{n}'\,J_n$$

$$L'_n = \bar{n}\,J''_n - \bar{n}''\,J_n$$

$$L''_n = \bar{n}'\,J''_n - \bar{n}''\,J'_n$$

$$M_{np} = J_n\,J''_p - J'_n\,J_p$$

$$M'_{np} = J_n\,J''_p - J''_n\,J_p$$

$$M''_{np} = J'_n\,J''_p - J''_n\,J'_p\,.$$

If these abbreviations are used, the stability constants can be directly calculated by means of the following equations:

$N = 1$

$$K_1 = \frac{\bar{n}}{J_1} \tag{4.67}$$

$N = 2$

$$K_1 = \frac{(\bar{n}\,J'_2 - \bar{n}\,J_2)}{(J_1\,J'_2 - J'_1\,J_2)} \tag{4.68a}$$

$$K_2 = \frac{(\bar{n}'\,J_1 - \bar{n}\,J'_1)}{(\bar{n}\,J'_2 - \bar{n}'\,J_2)} \tag{4.68b}$$

$N = 3$

$$K_1 = \frac{L''_3\,M_{23} - L'_3\,M''_{23}}{M''_{13}\,M'_{23} - M'_{13}\,M''_{23}} \tag{4.69a}$$

$$K_2 = \frac{L'_1\,L''_3 - L''_1\,L'_3}{L''_2\,L'_3 - L'\,L''_3} \tag{4.69b}$$

$$K_3 = \frac{L''_1\,L'_2 - L'_1\,L''_2}{L'_1\,L''_3 - L''_1\,L'_3} \tag{4.69c}$$

$\bar{n}$, [L]; $\bar{n}'$, [L'] and $\bar{n}''$, [L''] refer to corresponding pairs. In the case of $N = 3$, the equilibrium constant can be obtained by means of other equations too, but the expressions given above are the simplest. Equation 4.67 is the same as the one used by Rossotti and Rossotti [17] (Eq. 4.44), while Eqs. 4.68a and 4.68b for $\bar{n} = 0.5$ and $\bar{n}' = 1.5$ can be reduced to Eqs. 4.15 and 4.16. It is advantageous if the values of the complex formation function used in the calculations are in the ranges

$$0.2 < \bar{n} < 0.8$$

$$1.2 < \bar{n} < 1.8$$

$$2.2 < \bar{n} < 2.8.$$

The method requires very good experimental data. Even relatively small errors result in negative values for K_n. As appears from Table 4.6, it is imperative to use a correct value for N.

TABLE 4.6

Effect of using incorrect N

System	N		$\log K_1$	$\log K_2$	$\log K_3$
	True	Used			
Cu^{2+}-bapa^a	1	1	15.81		
	1	2	15.80	13.39	
	1	3	15.81	Neg.	Neg.
Cu^{2+}-en^b	2	1	10.36^c		
	2	2	10.35	8.94	
	2	3	10.30	3.93	6.05
Ag$^+$−NH$_3$	2	1	3.69^c		
	2	2	3.18	3.84	
	2	2	3.19	3.84	0.90
Ni^{2+}-en	3	1	7.75^c		
	3	3	7.66^c	6.50^d	
	3	3	7.68	6.39	4.56

a bapa = (H$_2$NCH$_2$CH$_2$CH$_2$)$_2$NH; *b* en = H$_2$NCH$_2$CH$_2$NH$_2$; *c* calculated with $\bar{n} = 0.5$; for greater values of $\bar{n}$ the values calculated for K are not constant; *d* no $\bar{n} > 2$ was used;

4.5 Application of high-speed computers

It is self-evident and appears also from the former discussion that an increase in the number of complex species to be considered makes the evaluation of equilibrium constants more and more tedious. This tendency is particularly pronounced when besides the parent complexes, polynuclear, mixed-ligand and protonated complexes also occur in the system. By means of high-speed computers even these problems are easy to treat. Due to the structure of mathematical calculations, there is no possibility of simple modelling for analogous computers, and therefore digital computers are always used in the calculation of equilibrium constants.

McMasters and Schaap [28] were the first to use electronic computers for calculating stability constants. The basis of their approach was a least-squares treatment of polarographic data. The method was thoroughly studied and greatly developed by Rydberg and Sullivan [29–33]. They dealt mostly with the application of high-speed computers for the treatment of solvent extraction [31] and potentiometric titration [30, 31] data. The least-squares approach to the calculation of equilibrium constants by using electronic computers was applied and adapted in a number of important papers [e.g. 34–36]. The problem was most thoroughly studied by Sillén, Ingri and Dyrssen [37–39]. These authors developed a generalized least-squares treatment which can be applied where the measured quantity cannot be expressed as a linear or not even as an explicit function of the unknown parameters. The principle of this LETAGROP method (the Swedish word means 'search for the pit') is simple. By means of graphical or other methods a set of equilibrium constants is used as a first approximation. Then the computer systematically varies each constant in turn, keeping the others constant, until it finds the value that gives the least value for U, the error-square sum. With complicated systems the experimentally found curve can be described by assuming the formation of several sets of different species. These alternative sets have different minimum U values. The set giving the least U value may be considered as the real one. There is an increasing ambiguity in deducing the possible mechanisms from experimental data burdened with increasing error. Beyond a certain U value there is no reason to prefer one set of constants or one set of species to other sets. In these cases the careful consideration of all the chemical evidence may decide between the alternatives.

Ingri and Sillén elaborated several programmes [39], written in the Ferranti Mercury Autocode system, for special purposes, particularly for treating complicated hydrolytic equilibria involving polynuclear complexes. Most recently, computer programmes have been published [40] for the rigorous analysis of potentiometric complex formation and proton dissociation data. These programmes perform the following calculations: (1) Complex formation function; (2) Solution of simultaneous equations for stability constants; (3) Successive approximations of stability constants; (4) Theoretical complex formation function; (5) Formation function for protonation of ligand; (6) Solution of simultaneous equations for proton dissociation constants; (7) Successive approximation of proton dissociation constants; (8) Theoretical formation function for protonation of ligand.

The greatest advantage of the application of high-speed computers is the dramatic decrease in the time necessary to evaluate the stability constants from a great number of experimental data. This gain in time makes it more attractive for chemists to look for several sets of species and corresponding constants, which in turn permits a better approximation of

experimental data and gives a more quantitative and less subjective basis
for rejecting possible explanations. In the case of simple systems the
constants obtained by graphical methods or by desk machines agree well
with those calculated by electronic computers.

4.6 Calculation of stability constants based on the relative concentration changes of the different species

The calculation methods discussed in this Chapter are based on relationships between the stability constants and the absolute concentrations of
one or more species involved in the equilibria. It is worth mentioning that
there are some further possibilities for evaluation of equilibrium constants
from relative concentration data [41]. The partial mole fraction of the
uncomplexed metal ion and that of the complex species with maximum
co-ordination number are monotonic functions of the free ligand concentration, while the partial mole fractions of all the other complexes show
a maximum. Evidently the ligand concentration $[L]_n$ where the partial
mole fraction of the nth complex is at its maximum is the only positive
root of Eq. 4.70

$$\sum_{i=0}^{N} (n-i)_i \, [L]_n^i = 0 \quad (n = 1, 2 \ldots N-1),\tag{4.70}$$

which can be obtained by differentiating Eq. 3.8. As the number of the
intermediate complexes is $(N-1)$ and the number of stability constants
is N, Eq. 4.70 itself does not allow the calculation of the stability constants from the free ligand concentrations at the maximum. However,
if any of the stability products is also known, Eq. 4.70 renders possible
this calculation. For β_1 it is relatively easy to get information from measurements in the presence of metal ion excess, while β_N can be obtained from
experiments performed at high ligand concentrations. The equations suitable
for calculation can be derived by the determinant method. For $N = 4$
the following equations are obtained:

1.* $[L]_1$, $[L]_2$, $[L]_3$ are measured, β_1 is known:

$$\beta_n = \frac{D_{1,n0} + D_{1,n1} \times \beta_1}{D_1}\tag{4.71}$$

2. $[L]_1$, $[L]_2$, $[L]_3$ are measured, β_4 is known:

$$\beta_n = \frac{D_{4,n0} + D_{4,n4} \times \beta_4}{D_4}.\tag{4.72}$$

---

* At the maxima the complex formation function has an integral value. Thus it
is not necessary to determine the free ligand concentrations at these points because
these quantities can be obtained from the relationship:

$$[L]_n = T_L - n \times T_{Me}.$$

In Eqs. 4.71 and 4.72 the following symbols are used:

$$D_1 = [L]_1^2[L]_2^3[L]_3^4 - 4[L]_1^3[L]_2^4[L]_3 + 3[L]_1^4[L]_2^3[L]_3$$

$$D_{1,20} = [L]_2^3[L]_4 + 12[L]_1^3[L]_2^4 - 9[L]_1^4[L]_2^3 - 4[L]_1^3[L]_3^4$$

$$D_{1,21} = 8[L]_1^3[L]_2^4[L]_3 - 6[L]_1^4[L]_2^3[L]_3 - 2[L]_1^3[L]_2[L]_3^4$$

$$D_{1,30} = 2[L]_1^2[L]_3^4 - 2[L]_2^4[L]_3^2 + 6[L]_1^4[L]_3^2 - 6[L]_1^2[L]_2^4$$

$$D_{1,31} = [L]_1^2[L]_2[L]_3^4 + 3[L]_1^4[L]_2[L]_3^2 - 4[L]_1^2[L]_2^4[L]_3$$

$$D_{1,40} = 3[L]_1^2[L]_2^3 - 4[L]_1^3[L]_3^2 + [L]_2^3[L]_3^2$$

$$D_{1,41} = 2[L]_1^2[L]_2^3[L]_3 - 2[L]_1^3[L]_2[L]_3^2$$

$$D_4 = 2[L]_1^2[L]_2[L]_3([L]_1[L]_3 - [L]_2^2)$$

$$D_{4,10} = 3[L]_1^2[L]_2^3 - 4[L]_1^3[L]_3^2 + [L]_2^2[L]_3^3$$

$$D_{4,14} = [L]_1^2[L]_2^2[L]_3^2(4[L]_1[L]_2^2 - 3[L]_1^2[L]_2 - [L]_2[L]_3^2)$$

$$D_{4,20} = 8[L]_1^3[L]_3 - 6[L]_1^3[L]_2 - 2[L]_2^3[L]_3$$

$$D_{4,24} = 6[L]_1^4[L]_2^3[L]_3 + 2[L]_1^3[L]_2[L]_3^4 - 8[L]_1^3[L]_2^4[L]_3$$

$$D_{4,30} = 3[L]_1^2[L]_2 - 4[L]_1^2[L]_3 + [L]_2[L]_3^2$$

$$D_{4,34} = 4[L]_1^2[L]_2^4[L]_3 - 3[L]_1^4[L]_2[L]_3^2 - [L]_1^2[L]_2[L]_3^4 .$$

From the expression for D_4 it is directly seen that D_4 becomes zero when the ligand concentrations corresponding to the maxima of the partial mole fraction curves give a geometrical series. Evidently this is a limiting case which is never reached, but can well be approached. Assuming that $[L]_1$, $[L]_2$ and $[L]_3$ form a geometrical series, that is

$$q = \frac{[L]_2}{[L]_1} = \frac{[L]_3}{[L]_2}, \tag{4.73}$$

Eq. 4.71 is simplified to

$$\beta_2 = [L]_1^{-2}(1 - 3q^{-4}) - 2[L]_1^{-1}q^{-2}\beta_1 \tag{4.74a}$$

$$\beta_3 = [L]_1^{-2}q^{-2}\beta_1 \tag{4.74b}$$

$$\beta_4 = [L]_1^{-4}q^{-4}. \tag{4.74c}$$

A comparison of Eqs. 4.74b and 4.74c shows that, independently of the value of q, $K_1/K_2 = K_3/K_4$. If the value of q is large enough, $\beta_2 = [L]_1^{-2}$ and an even more general relationship is obtained:

$$K_1K_2 : K_2K_3 : K_3K_4 = 1 : q^{-2} : q^{-4}. \tag{4.75}$$

Thus a symmetry in the ligand concentrations belonging to the maximum values of the partial mole fractions indicates a symmetry in the successive constants.

Among the experimental techniques applied for the study of complex equilibria there are some methods [42] for the determination of the integer points of the complex formation function, i.e. the free ligand concentrations at which the partial mole fraction of a particular complex is of maximum value. These methods will be discussed in detail in Chapter 5, but some features should be mentioned here. The extractability or the solubility exhibits a maximum when the system $Me^{m+} - L^{k-}$ is at the isoelectric point [43], that is when

$$\sum_{i=0}^{i=\frac{m-1}{k}} (m - ik) [MeL_i^{(m-ik)+}] = \sum_{j=\frac{m+1}{k}}^{j=N} (jk - m) [MeL_j^{(jk-n)-}] . \qquad (4.76)$$

At the isoelectric point the partial mole fraction of the electrically neutral complex MeL_n $(n = m/k)$ is of maximum value. This point of the complex formation function can also be determined from sorption experiments with anion-exchange resins [44]. If only one of the successive complexes is positively charged, sorption experiments with cation-exchange resins provide directly the maximum for this particular complex, since the ligand anion is absorbed on the cation-exchanger only in the form of this complex. The analysis of the curve of solubility $vs.$ ligand concentration provides information on all the integer points of the complex formation function [45]. The most generally applicable method — at least in principle — is Raman spectroscopy [46], because each Raman-active species gives a characteristic line, which reaches maximum intensity when its partial mole fraction reaches its maximum value.

If a parameter proportional to the partial mole fraction of any intermediate complex is known as a function of the free ligand concentration, all of the successive equilibrium constants can be obtained. A rearrangement of the logarithmic form of Eq. 3.8 leads to

$$\log([L]^n/\alpha_n) = \log(1 + \beta_1[L] + \ldots + \beta_N[L]^N) - \log\beta_n . \qquad (4.77)$$

By differentiating Eq. 4.77 we get

$$\frac{d\log([L]^n/\alpha_n)}{d\log[L]} = \bar{n} . \qquad (4.78)$$

This equation provides the complex formation function if α_n is known as a function of $[L]$. Equation 4.78 is also obtained if not α_n itself but a quantity which is proportional to α_n is measured because the proportionality factor is eliminated by the logarithmic differentiation. However, this procedure increases the uncertainty of the measured quantity, so another possibility is also considered. To any fraction of α_n (c) there belong two values of $[L]$ ($\overrightarrow{[L]}_{n,c} > [L]_n$ and $\overleftarrow{[L]}_{n,c} < [L]_n$) from which

$$\alpha_n([L]) = c\, \alpha_{n,\,max}.\tag{4.79}$$

These concentrations are evidently the roots of the equation

$$\frac{(1 + \beta_1[L]_n + \dots + \beta_N[L]_n^N)}{\beta_n[L]_n^n} = c\,\frac{(1 + \beta_1[L]_{n,c} + \dots + \beta_N[L]_{n,c}^N)}{\beta_n[L]_{n,c}^n}.\tag{4.80}$$

Equation 4.80 provides linear equations for the β_i values. This procedure requires knowledge of $[L]_n$. However, a combination of equations such as Eq. 4.80 referring to both $\overleftarrow{[L]}_{n,c}$ and $\overrightarrow{[L]}_{n,c}$ leads to the equation

$$\sum_0^N \beta_i\, \overleftarrow{[L]}_{n,c}^{i-n} - \sum_0^N \beta_i\, \overrightarrow{[L]}_{n,c}^{i-n} = 0.\tag{4.81}$$

The solution of a set of equations of this type — consisting of at least N equations — gives the whole set of stability products even if $[L]_n$ itself is unknown.

4.7 Calculation of the concentration of different species from the stability constants and total concentrations

One of the primary aims of equilibrium studies is to calculate the concentration of each constituent of complex systems. As appears from Eq. 3.8, the partial mole fraction of each complex species is independent of the total concentrations of both the central ion and the ligand and can easily be calculated if the stability constants are known. If $N = 2$ the complex system consists of the ions Me, MeL and MeL_2. Obviously, the partial mole fractions of Me and MeL_2 change monotonically as a function of the free ligand concentration; $\alpha_0 = 1$ and $\alpha_2 = 0$, when there is no ligand in the system ($[L] = 0$), while $\alpha_0 = $ zero and $\alpha_2 = 1$ if the free ligand concentration exceeds a certain limit. Theoretically this is the case only if $[L] = \infty$, which condition of course has no chemical reality, but unless the complex is extremely weak or the solubility of the salt supplying the ligand is too small, this limiting free ligand concentration can be achieved. α_1 shows a maximum at $\bar{n} = 1$, i.e. when

$$[L]_1 = \frac{1}{\sqrt{K_1 K_2}}.\tag{4.82}$$

Inserting Eq. 4.82 in the expression for α_1

$$\alpha_1 = \frac{K_1[L]}{1 + K_1[L] + K_1 K_2[L]^2}\tag{4.83}$$

we get that the maximum value of the partial mole fraction is given by

$$\alpha_{1,\,max} = \frac{1}{2\sqrt{\dfrac{K_2}{K_1} + 1}}.\tag{4.84}$$

Besides $[L]_1$ and $\alpha_{1,\,max}$ there is another characteristic quantity of the distribution curve (α_1 $vs.$ log $[L]$), namely the half-width, i.e. the logarithm of the ratio of the two ligand concentrations at which the value of α_1 is $1/2\,\alpha_{1,\,max}$. An equation for this can be obtained by combining Eqs. 4.83 and 4.84:

$$\frac{2\,K_1\,[L]}{1 + K_1\,[L] + K_1\,K_2\,[L]^2} = \frac{1}{2\,\sqrt{\dfrac{K_2}{K_1} + 1}}.\qquad(4.85)$$

After rearranging Eq. 4.85 and solving for $[L]$, we get

$$\Delta\,\log\,[\overset{\leftrightarrow}{L}] = \log\,\frac{[\overrightarrow{L}]}{[\overleftarrow{L}]} = \log\left[2 + \frac{1}{2}\sqrt{\frac{K_1}{K_2}} + \sqrt{\left(2 + \frac{1}{2}\sqrt{\frac{K_1}{K_2}}\right)^2 - 1}\right]^2.$$

$$(4.86)$$

The meaning of $[\overleftarrow{L}]$ and $[\overrightarrow{L}]$ is self-explanatory. If $K_1 \gg K_2$, the value of the half-width is equal to log (K_1/K_2), while if $K_2 \gg K_1$, $\Delta \log\,[\overset{\leftrightarrow}{L}]$ is log $(2 + \sqrt{3})^2 = 1.144$.

In more complicated systems, ($N > 3$), neither the ligand concentration at the positions of maximum partial mole fractions, nor the half-widths, can be explicitly expressed as a function of the stability constants. However, the distribution of the complexes can be calculated and the complex systems well characterized by means of various distribution diagrams. The construction and use of these diagrams are shown in Figs. 4.4–4.6 [47]. The advantages of the different diagrams can be judged by comparing them.

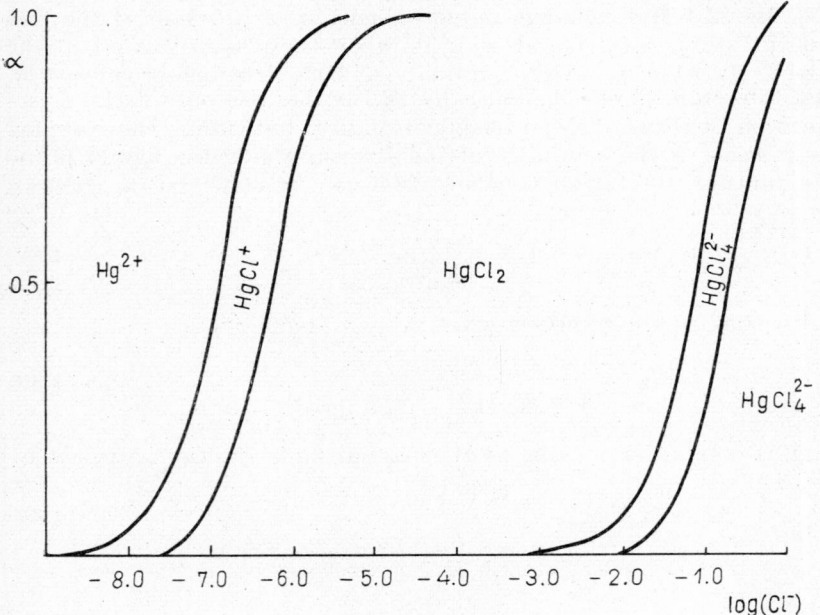

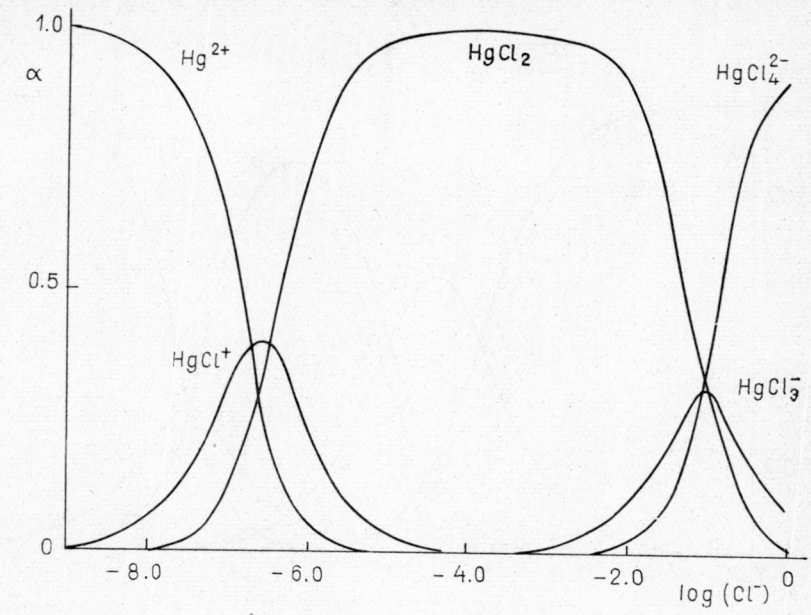

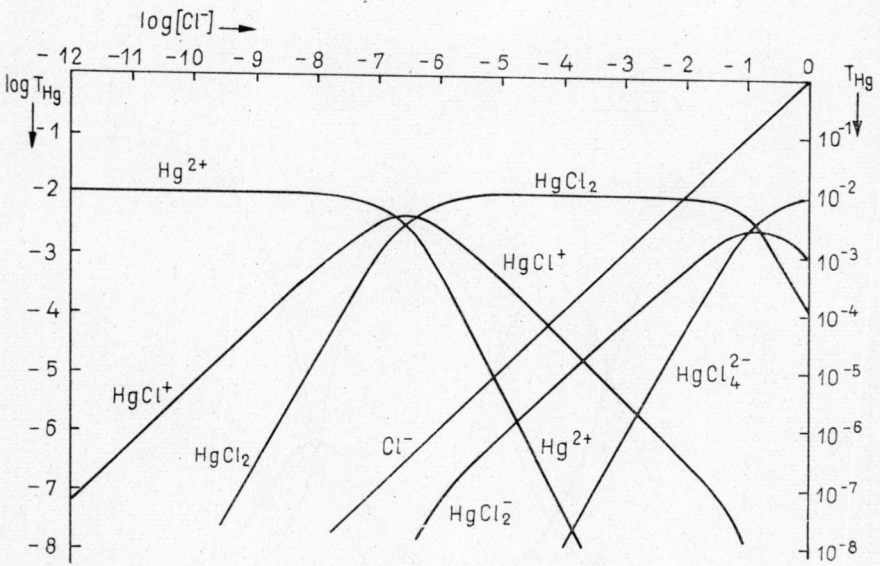

Fig. 4.4 Distribution of complexes $HgCl_n^{2-n}$ ($n = 0,1,2,3,4$) as a function of log [Cl$^-$] if for a given value of [Cl$^-$] a vertical line is drawn at the corresponding log [Cl$^-$], the segment of this line falling in a certain area represents the partial mole fraction of the corresponding complex. (Reproduced with permission from *Acta Chem. Scand. 3*, 539 (1949))

Fig. 4.5 Distribution of complexes $HgCl_n^{2-n}$ as a function of log [Cl$^-$]

Fig. 4.6 Logarithmic diagram for $T_{Hg} = 0.01$ M and varying [Cl$^-$], showing the concentrations of the different species. (Reproduced with permission from *Acta Chem. Scand. 3*, 539 (1949))

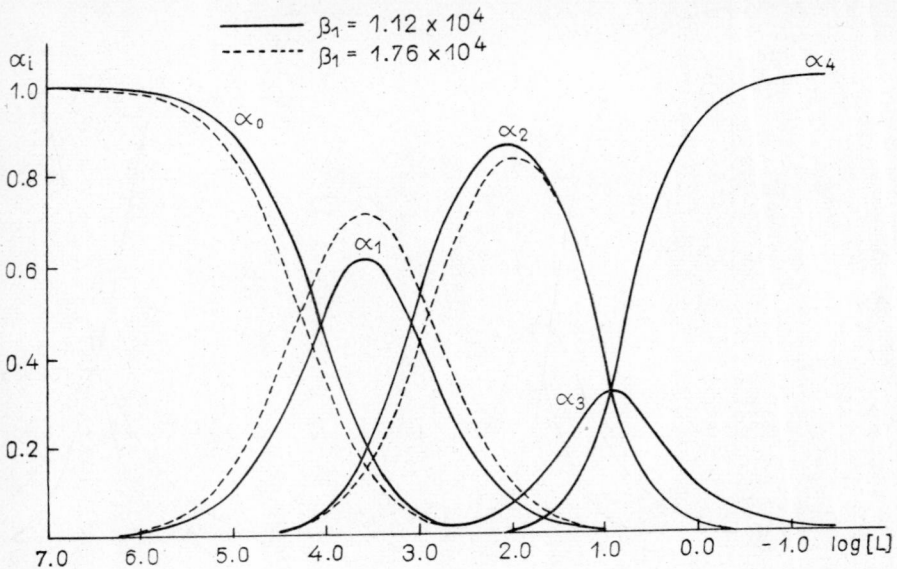

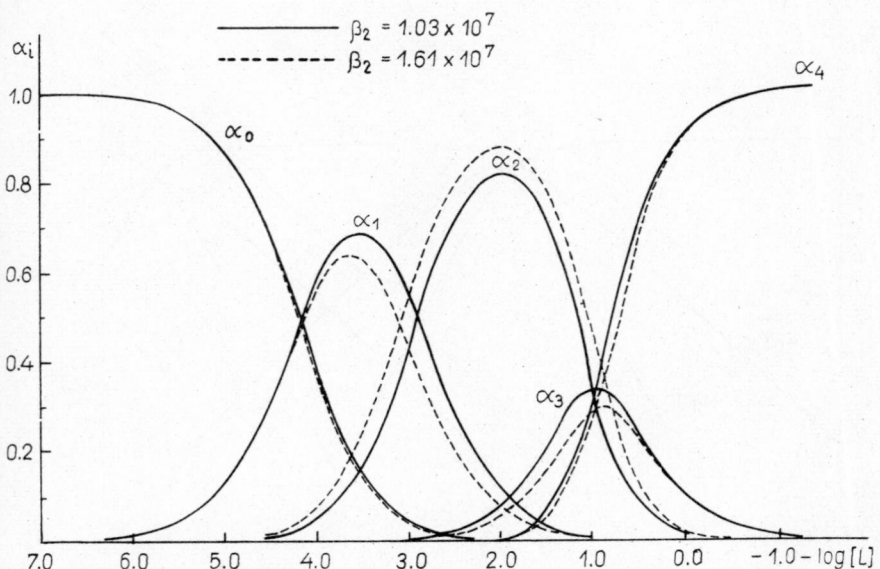

Fig. 4.7 Effect of change in successive stability products on the distribution of $AgBr_n^{1-n}$ complexes as a function of free bromide concentration. The original set of stability products [50] is: $\beta_1 = 1.40 \times 10^4$; $\beta_2 = 1.28 \times 10^7$; $\beta_3 = 8.9 \times 10^7$; $\beta_4 = 7.76 \times 10^8$. In each of the four distribution diagrams one of the successive

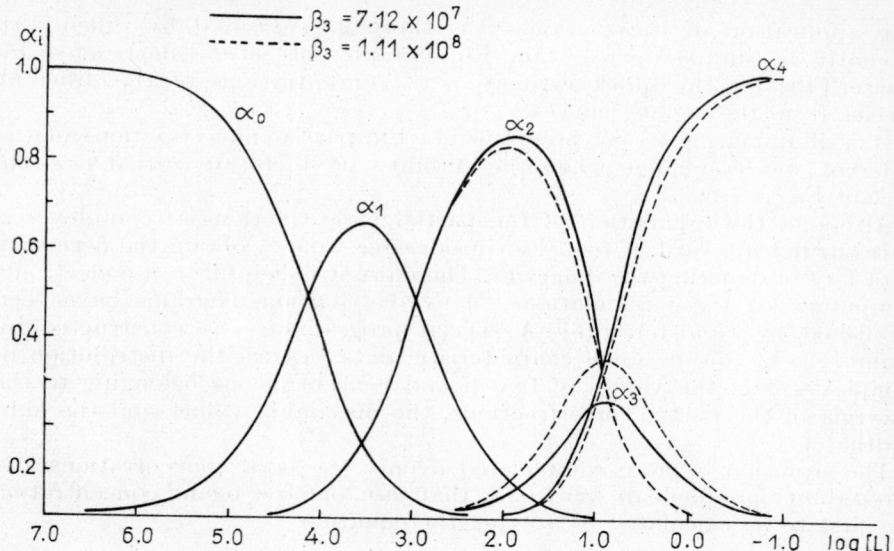

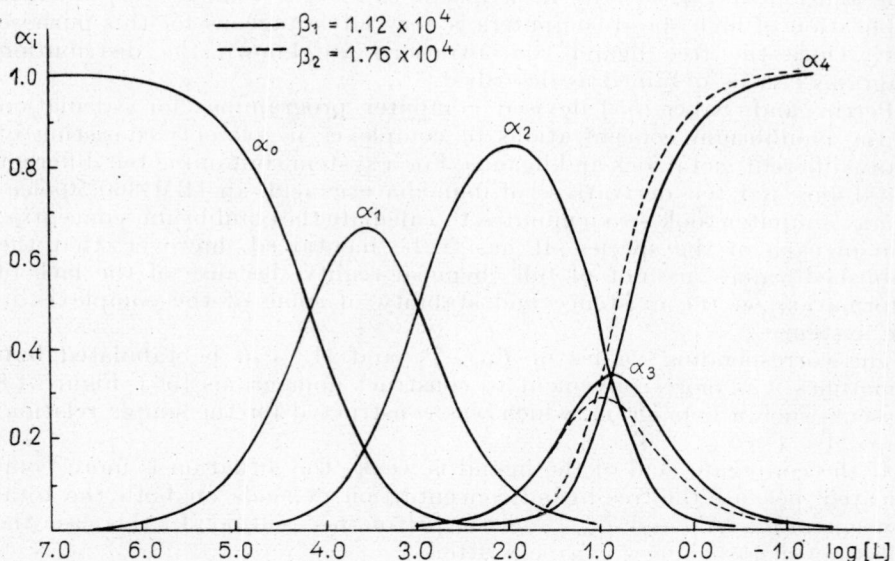

products is varied by $\pm 20\%$. (The varying values only are given on each relevant diagram.) The dotted curves are obtained from the higher, the full curves from the lower values of the stability products. The effect on the shape and the position of peaks should be noted

The application of these graphs was thoroughly treated by Sillén [48]. Recently, a simple device, the 'Equiligraph' has been constructed by Freiser [49] for the quick estimation of concentrations of the different species from the log–log curves.

It is illuminating to see how sensibly the partial mole fractions of the different species change when the stability products are varied (see the text of Fig. 4.7).

Although the calculation of the partial mole fractions of complexes is quite straightforward, it requires considerable time to obtain the necessary data for constructing the diagrams. Therefore it is helpful to use electronic computers for these calculations. Many distribution diagrams have been published by Goldstein [51]. A special programme was constructed by Huhn [52] to obtain some characteristic data besides the distribution of complexes, e.g. the values of free ligand concentrations belonging to the maxima of the partial mole fractions, the maximum values and the half-widths.

The situation is more complicated if only the total concentrations and the stability products are known. In that case, one free ligand concentration has first to be calculated by solving the equation

$$\sum_0^N \beta_i \, [\mathrm{L}]^{i+1} - \sum_0^N (\mathrm{T_L} - i \, \mathrm{T_{Me}}) \, \beta_i \, [\mathrm{L}]^i = 0. \tag{4.87}$$

The solution of Eq. 4.87 by desk machines is very tedious [53] and the application of high speed computers is very advantageous for this purpose [54]. Once the free ligand concentrations are known the distribution diagrams can be obtained as described.

Perrin and Sayce [55] devised computer programmes for calculation of the equilibrium concentrations of complexes in systems consisting of many different metal ions and ligands. For a system containing ten different metal ions and ten derivatives of iminodiacetic acid, an IBM 360/50 electronic computer took seven minutes to calculate the equilibrium concentration of each of the species. It has to be mentioned, however, that the published values are not of full chemical reality, because of the lack of information on the existence and stability of some of the complexes in the system.

The corresponding values of T_{Me}, T_L and [L] can be tabulated but sometimes it is more convenient to construct nomograms [56]. Figure 4.8 presents such a nomogram, which was constructed for the simple relationship: $[\mathrm{L}] + \bar{n} \, T_{\mathrm{Me}} = T_L$.

If the conjugate acid of the ligand is weak, the situation is more complicated, because the free ligand concentration depends on both the total concentrations (T_L and T_{Me}) and the pH of the solution. In this case the following mass-balances can be written

$$T_{\mathrm{Me}} = [\mathrm{Me}] + [\mathrm{MeL}] + [\mathrm{MeL_2}] + \ldots + [\mathrm{MeL}_N] = \sum_0^N [\mathrm{MeL}_i] \tag{4.88}$$

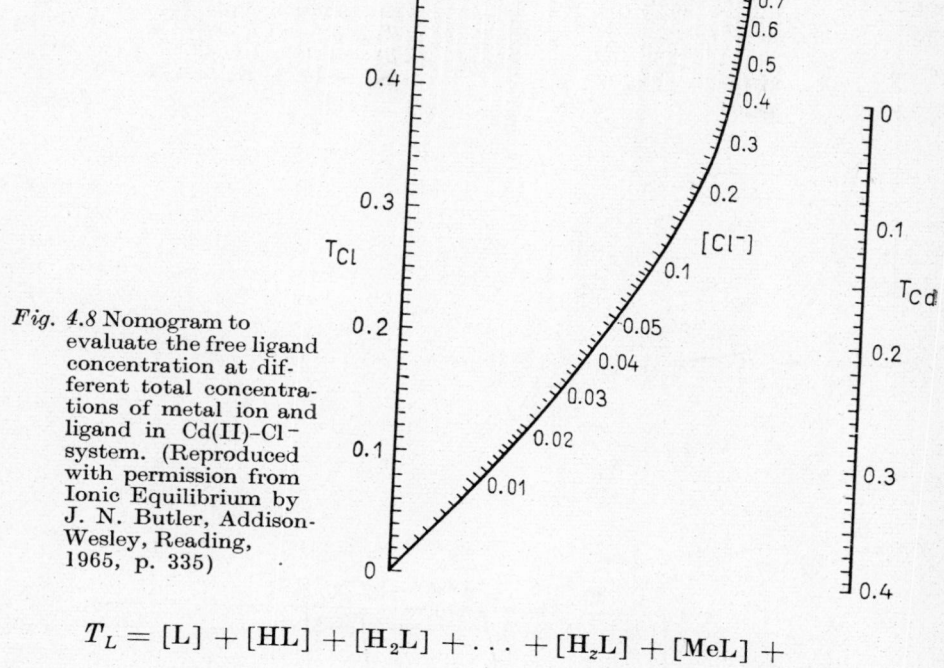

Fig. 4.8 Nomogram to evaluate the free ligand concentration at different total concentrations of metal ion and ligand in Cd(II)–Cl$^-$ system. (Reproduced with permission from Ionic Equilibrium by J. N. Butler, Addison-Wesley, Reading, 1965, p. 335)

$$T_L = [L] + [HL] + [H_2L] + \ldots + [H_zL] + [MeL] +$$

$$+ 2[MeL_2] + \ldots + N[MeL_N] = \sum_0^z H_jL + \sum_1^N i\,[MeL_i]. \quad (4.89)$$

Taking into account the acidic dissociation constants of the conjugate acid of the ligand, the first summation term in Eq. 4.89 can be written as

$$\sum_0^z [H_jL] = [L] \left(1 + \frac{[H^+]}{K_z^A} + \frac{[H^+]^2}{K_z^A \cdot K_{z-1}^A} + \ldots + \frac{[H^+]^z}{K_z^A \ldots K_1^A} \right). \quad (4.90)$$

The calculation of the free ligand concentration requires the solution of the following equation

$$\sum_0^N \frac{\beta_i}{\alpha_L} [L]^{i+1} - \sum_0^N (T_L - iT_{Me})\,\beta_i[L]^i = 0 \quad (4.91)$$

where α_L denotes the ratio $\sum_0^z \frac{[H_jL]}{[L]}$. Besides constructing the common distribution diagrams (α_i *vs.* log [L]) it is helpful to construct graphs

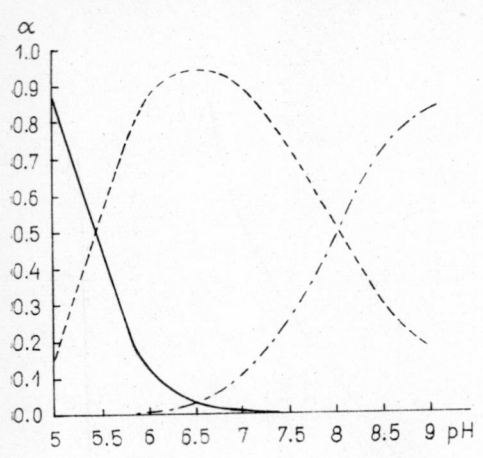

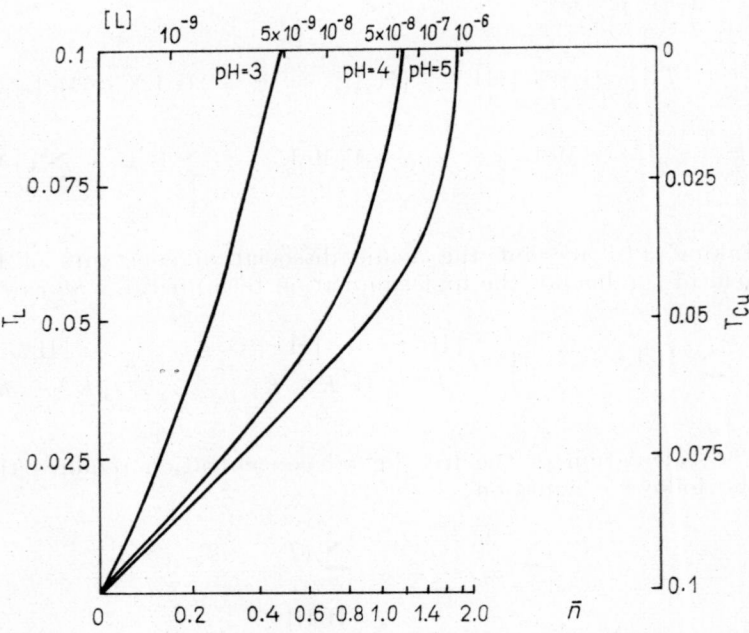

Fig. 4.9 Distribution of complex species in the Co(II)–histidine system
as a function of pH. $T_{Co} = 5 \times 10^{-4}$ M;
$T_{histidine} = 1.5 \times 10^{-3}$ M;
$\beta^1 = 1.98 \times 10^7$; $\beta_2 = 4.11 \times 10^{11}$;
$K_1'' = 10^{-6}$; $K_2'' = 6.76 \times 10^{-10}$;
—— α_{Co}^{+2} - - - - α_{CoL}^+ —·— α_{CoL_2}

Fig. 4.10 Nomogram to evaluate the free ligand concentration at different total
concentrations of metal ion and ligand in the Cu(II)–α-aminobutyric acid system
at different pH values

where the partial mole fractions of the complexes are plotted against pH at constant T_L and T_{Me}. Figure 4.9 shows such a distribution diagram for the Co(II)–histidine system [57]. The nomographic presentation of data is exemplified by Fig. 4.10 where the corresponding values of T_{Cu}, T_L and [L] are plotted at different pH values for the Cu(II)–aminoisobutyric acid system.

REFERENCES

1. LEDEN, I., *Z. phys. Chem. Leipzig A 188*, 160 (1941).
2. FRONAEUS, S., *Acta Chem. Scand. 4*, 72 (1950).
3. SULLIVAN, J. G. and HINDMAN, J. C., *J. Am. Chem. Soc. 74*, 6091 (1952).
4. BJERRUM, J., *Metal Ammine Formation in Aqueous Solution*, Haase, Copenhagen, 1941. Reprinted 1957.
5. SCHRØDER, K. H., *Acta Chem. Scand. 20*, 1401 (1966).
6. IRVING, H. and ROSSOTTI, H. S., *J. Chem. Soc.* 3397 (1953).
7. GERGELY, A., NAGYPÁL, I. and MOJZES, J., *Acta Chim. Acad. Sci. Hung. 51*, 381 (1967).
8. ROSSOTTI, F. J. C. and ROSSOTTI, H. S., *Acta Chem. Scand. 9*, 1166 (1956).
9. ROSSOTTI, F. J. C. and ROSSOTTI, H. S., *The Determination of Stability Constants*, McGraw Hill, New York, 1961, p. 108.
10. HENDERSON, L. J., *J. Am. Chem. Soc. 30*, 954 (1908).
11. HASSELBACH, K. A., *Biochem. Z. 78*, 116 (1917).
12. BLOCK, R. P. and MCINTYRE, G. H., *J. Am. Chem. Soc. 75*, 5667 (1953).
13. SPEAKMAN, J. C., *J. Chem. Soc.* 397 (1940).
14. OLERUP, H., *Svensk Kem. Tidskr. 85*, 324 (1943).
15. FOMIN, V. K. and MAJROVA, E. P., *Zh. Neorgan. Khim. 1*, 1703 (1956).
16. IRVING, H., *J. Chem. Soc.* 4056 (1962).
17. SCATCHARD, G., unpublished work, quoted in ref. [18].
18. EDSALL, J. T., FELSENFELD, G., GOODMAN, D. S. and GURD, F. R. N., *J. Am. Chem. Soc. 76*, 3054 (1954).
19. POE, A. J., *J. Phys. Chem. 67*, 1070 (1963).
20. PRYTZ, M., *Z. anorg. allgem. Chem. 172*, 147 (1928).
20a. GERGELY, A., *Thesis*, Debrecen, 1967
21. IRVING, H. and PETTIT, L. D., *J. Chem. Soc.* 1546 (1963).
22. SCHWARZENBACH, G., WILLI, A. and BACH, R. O., *Helv. Chim. Acta 30*, 1303 (1947).
23. SCHWARZENBACH, G. and ACKERMANN, H., *Helv. Chim. Acta 31*, 1029 (1948).
24. SILLÉN, L. G., *Acta Chem. Scand. 10*, 186 (1956).
25. SILLÉN, L. G., *Acta Chem. Scand. 10*, 803 (1956).
26. ROSSOTTI, F. J. C., ROSSOTTI, H. S. and SILLÉN, L. G., *Acta Chem. Scand. 10*, 203 (1956).
27. ROSSOTTI, F. J. C. and ROSSOTTI, H. S., *J. Phys. Chem. 63*, 1041 (1959).
28. MCMASTERS, D. L. and SCHAAP, W. B., *Indiana Acad. Sci. 67*, 111 (1958).
29. RYDBERG, J. and SULLIVAN, J. C., *Acta Chem. Scand. 13*, 186 (1959).
30. SULLIVAN, J. C., RYDBERG, J. and MILLER, W. F., *Acta Chem. Scand. 13*, 2023 (1959).
31. RYDBERG, J. and SULLIVAN, J. C., *Acta Chem. Scand. 13*, 2057 (1959).
32. RYDBERG, J., *Acta Chem. Scand. 14*, 157 (1960).
33. RYDBERG, J., *Acta Chem. Scand. 15*, 1723 (1961).
34. CHOPOORIAN, J. A., CHOPPIN, G. R., GRIFFITH, H. C. and CHANDLER, R., *J. Inorg. Nucl. Chem. 21*, 21 (1961).
35. HUGUS, Jr., Z. Z., *Proc. 7th Intern. Conf. Coord. Chem. Stockholm, 1962*, p. 204.
36. LANSBURY, R. C., PRICE, V. E. and SMEETH, A. G., *J. Chem. Soc.* 1896 (1965).

37. DYRSSEN, D., INGRI, N. and SILLÉN, L. G., *Acta Chem. Scand. 15*, 964 (1961).
38. SILLÉN, L. G., *Acta Chem. Scand. 16*, 158 (1962).
39. INGRI, N. and SILLÉN, L. G., *Acta Chem. Scand. 16*, 173 (1962).
40. ROMARY, J. K., DONNELLY, D. L. and ANDREWS, A. C., *J. Inorg. Nucl. Chem. 29*, 1805 (1967).
41. HUHN, P. and BECK, M. T., *Kémiai Közlemények 26*, 401 (1966).
42. BECK, M. T. and HUHN, P., *Acta Chim. Acad. Sci. Hung. 20*, 285 (1959).
43. BECK, M. T., *Acta Chim. Acad. Sci. Hung. 4*, 227 (1954).
44. FRONAEUS, S., *Svensk Kem. Tidskr. 65*, 1 (1953).
45. HAIGHT, G. .P, JR., *Acta Chem. Scand. 16*, 209 (1962).
46. WOODWARD, L., *Proc. 8th Intern. Conf. Coord. Chem.*, Springer, Vienna, 1964, p. 15.
47. SILLÉN, L. G., *Acta Chem. Scand. 3*, 539 (1949).
48. SILLÉN, L. G., in *Treatise on Analytical Chemistry* (edited by I. M. Kolthoff and P. J. Elving) Part I, Vol. 1, Interscience, New York, 1959.
49. FREISER, H., *School Sci. Math.* 227 (1967).
50. BERNE, E. and LEDEN, I., *Z. Naturforsch. 8A*, 719 (1953).
51. GOLDSTEIN, G., *Equilibrium Distribution of Metal-Ion Complexes*, ORNL-3620, 1964.
52. HUHN, P., to be published.
53. HUHN, P. and BECK, M. T., *Acta Phys. Chem. Szeged, 4*, 45 (1958).
54. CURTHOYS, G. and BRISLEY, W., *Anal. Chim. Acta 28*, 577 (1963).
55. PERRIN, D. D. and SAYCE, I. G., *Talanta 14*, 833 (1967).
56. BUTLER, J. N., *Ionic Equilibrium*, Addison-Wesley, Reading, 1964, p. 335.
57. LEBERMAN, R. and RABIN, B. R., *Trans. Faraday Soc. 55*, 1660 (1959).

Chapter 5

EXPERIMENTAL METHODS FOR THE DETERMINATION OF STABILITY CONSTANTS

There are many solution properties which change measurably as a result of complex formation. In principle all these measurements can give information on the existence and stability of the different species. By a careful selection one may find one or more suitable experimental methods for the investigation of all kinds of complexes. It must be emphasized, however, that a careful consideration is necessary as to whether the effects of complex formation are quantitatively reflected in the measured property. To choose an adequate method is a prerequisite for obtaining reliable equilibrium constants. Besides the classical methods such as spectrophotometry and potentiometry which are available in every chemical laboratory, the armoury of experimental methods has been recently enriched by a number of new approaches, such as nuclear magnetic resonance and electron spin resonance spectroscopy. Sometimes a linear relationship is assumed between the price of the instrument and the value of information obtained by its means. Although there is no doubt that the new methods requiring very expensive instruments may make it possible to determine otherwise inaccessible equilibrium constants, this view is completely wrong. Research in the field of equilibrium chemistry is relatively inexpensive, and in most cases reliable stability constants can be obtained by using simple and easily available methods and instruments.

Generally, the relationship between composition and the value X of a given property of a system containing several complex species is

$$X = a_0 c_0 + a_1 c_1 + \ldots + a_N c_N = \sum a_i c_i \tag{5.1}$$

where c_i stands for the concentration of the ith complex, while a_i is the corresponding intensive factor. This equation is valid for example in the case of visible and ultraviolet spectrophotometry, calorimetric measurements, etc.

If the value of each intensive factor is the same, the relationship is described by

$$X = a \sum c_i. \tag{5.2}$$

Equation 5.2 is valid in the case of colligative properties.

If only one of the complexes contributes to the measured property, i.e. the value of all but one of the intensive factors is zero, the following equation holds:

$$X = a_n c_n. \tag{5.3}$$

This relationship is to be applied in the case of electrode potential measurements, Raman spectroscopy and — with some reservation — liquid–liquid distribution.

5.1 The measurement of colligative properties

The decrease in the number of solute particles is a striking consequence of most complex-forming reactions. Therefore in these cases all solution properties which depend on the number of dissolved particles are changed by complex formation. The measurement of the depression of freezing point was used at the beginning of this century to evaluate some acid dissociation constants. In modern experimental methods thermistors are used instead of Beckman thermometers. In the last decades the method has been used to study complex equilibria involving mononuclear and polynuclear complexes. In most cases only qualitative conclusions on the existence of certain species were reached. The quantitative aspects of the method have been treated by Souchay [1], Kenttämaa [2] and particularly by Rossotti and Rossotti [3]. Obviously, this method has some inherent disadvantages and limitations. The electrolyte concentration must necessarily be fairly high, so relatively few media are available (e.g. 1.15 M KNO_3; 0.258 M $KClO_3$; 0.048 M $KClO_4$ and 3.50 M Na_2SO_4 the latter being used most frequently), the operating temperature of the experiments is determined by the cryoscopic system, and so on. The present author agrees with Tobias [4], who on thorough consideration of the conditions of the salt cryoscopic method for quantitative study of complex equilibria stated that the results obtained by this method 'are best regarded as indicative rather than conclusive'.

Analogously, the same objections may be made to application of ebullioscopy in equilibrium studies. Furthermore the boiling point elevations can be measured less precisely than the freezing point depressions and the higher temperature needed is frequently inconvenient.

If the ligand is volatile, the determination of its partial pressure is possible. From the partial pressure the concentration (activity) in solution can be simply calculated. This method was used by Jannik Bjerrum [5] in the study of copper(II) ammines. It seems to be very promising to apply the method for studying systems containing several volatile ligands, for example ammonia and organic amines, to derive stability constants of mixed ligand complexes.

The equilibrium between halide ions and halogens gives a possibility of using vapour pressure measurements to determine the stability constants of halide complexes. From the equilibria

$$Hal_2 + Hal^- \rightleftharpoons Hal_3^-$$

$$Hal_3^- + Hal_2 \rightleftharpoons Hal_5^-$$

the free halide concentration can be calculated from the partial pressure of the halogen. Scaife and Tyrrell [6] determined the stability constants of the Br_3^- and Br_5^- species from the study of the $Br_2 - Br^-$ system, and then the study of the $HgBr_2 - Br^- - Br_2$ system furnished the stability constants of the $HgBr_3^-$ and $HgBr_4^{2-}$ complexes [7]. Careful analysis of the experimental data showed the existence of the interesting species $HgBr_6^{2-}$, proving the capability of the bound halide ions to form a donor–acceptor complex with the halogen.

5.2 Optical methods

One of the most spectacular effects of complex formation is the change of spectral properties. The reasons for light absorption by the complexes are as follows. The excitation of the electrons of both the metal ion and the ligand(s) is influenced by their interaction. The electrons of transition metal ions are easily excited and consequently absorb in the visible region, that is these ions give coloured compounds. The electron systems of non-transition metals ions and of the ligands are much more stable; excitation of the electrons requires much greater energy, so these species absorb radiation in the ultraviolet range of the spectrum. Owing to the interaction of the central ion and the ligand a charge transfer from the ligand to the metal ion (and vice versa) may occur on irradiation. This phenomenon is the reason for the so-called charge transfer spectra in the visible of the near-ultraviolet region. As a matter of course, complex formation results in changes in the vibrational–rotational characteristics of the ligands, leading to the shift of peaks in the infrared spectrum. All these changes are measures of complex formation. The relationship between the absorbance and the composition of a given solution is given in the ideal case by the Beer-Lambert law:

$$\log \frac{I_0}{I} = A = d \sum \varepsilon_i \, c_i \tag{5.4}$$

where d is the cell length, A is the absorbance of the solution, ε_i the molar absorptivity of the ith species (which has concentration c_i); A and ε_i refer to a given wavelength. The intensive factor, the molar absorptivity, depends on the medium and on the temperature. The molar absorptivities of the metal ion, of the ligand and – in most cases – of the co-ordinatively saturated complex, can be determined directly, but those of the intermediary complexes are just as unknown as their stability constants. If two species in the solution have at a certain wavelength the same molar absorptivity, the series of the spectra of solutions, when the sum of the concentrations is constant but the ratios of their concentrations are changed, exhibits a common crossing point, the so-called isosbestic point. Because it is very unlikely that a third species has the very same molar absorptivity at this wavelength, the occurrence of isosbestic point(s) in the formation of a series of complexes gives evidence for the simultaneous existence of two absorbing species. It is a great advantage of the spectrophotometric method that the measurements can be made at many different wavelengths. In selecting the wavelengths it is helpful to use different bands of the spectrum. The agreement of stability constants obtained from measurements at different wavelengths is strong evidence for their reliability.

Spectrophotometric measurements in the ultraviolet, visible and infrared regions can easily be made with commercial instruments. These instruments consist of a light source, a monochromator, the absorption cell assembly and the detector. Nowadays photocells and photomultipliers are used as detectors, because the use of photographic plates to measure light intensities is much more tedious and subject to more sources of error.

The error of spectrophotometric measurements is determined by chemical (and photochemical) and by instrumental factors [8]. By careful experimental work these factors can be minimized, but besides the apparent deviations from Beer's law there are some real ones. The molar absorptivities refer to a certain medium. In the study of very weak complexes, high ligand concentrations must be applied, which in fact means a change in the medium. Prue [9] has pointed out that even without any specific interaction between particles, purely topological reasons may lead to a change in the charge transfer spectra. From this change an apparent association constant of the order of 0.2 1. mole^{-1} can be calculated. Therefore this is the lower limit of the stability constants which can be obtained spectrophotometrically. Woldbye and Bagger [10] have found experimentally and explained theoretically that in the spectrophotometry of circularly dichroic substances there are deviations from the Beer–Lambert law, but they do not exceed 1% of the absorbance.

5.2.1 THE METHOD OF CONTINUOUS VARIATION

Though the principle of continuous variation was first applied by Ostromisslensky [11] and Denison [12], nevertheless the discovery of the method is frequently attributed to Job [13] who applied it to complex systems. It was very popular between the mid-forties and mid-fifties and is still frequently applied, although its limitations are quite clear. The principle of the method is simple. Suppose that the metal ion and the ligand form one complex

$$m \, \text{Me} + n \, \text{L} = \text{Me}_m \text{L}_n .$$ (5.5)

A series of solutions is prepared in which the sum of the total concentrations of Me and L is constant, but their proportions are continuously varied:

$$T_{\text{Me}} + T_{\text{L}} = C$$ (5.6)

$$x = C \, \frac{T_{\text{L}}}{T_{\text{Me}} + T_{\text{L}}} .$$ (5.7)

The equilibrium concentrations of the species Me, L and Me_mL_n are given by the equations

$$[Me] = C(1 - x) - m[Me_mL_n] \tag{5.8}$$

$$[L] = Cx - n[Me_mL_n] \tag{5.9}$$

$$[Me_mL_n] = \beta_{mn} [Me]^m[L]^n . \tag{5.10}$$

Evidently, the concentration of the complex reaches a maximum at a certain value of x. This value (x_{Max}) can be obtained by differentiating Eqs. 5.8–5.10 with respect to x, setting $d[Me_mL_n]/dx = 0$ and eliminating [Me] and [L]:

$$m(Cx - n[Me_mL_n]) = n(C(1 - x) - m[Me_mL_n]). \tag{5.11}$$

That is

$$mx_{max} = n(1 - x_{max}) \tag{5.12}$$

and

$$\frac{n}{m} = \frac{x_{max}}{1 - x_{max}} . \tag{5.13}$$

If any property of the solution which varies linearly with $[Me_mL_n]$ is measured and plotted as a function of x, the ratio of the stoichiometric coefficients (n/m) can be obtained. The method of continuous variation is most frequently applied in the evaluation of spectrophotometric data, but in principle it can be applied in connection with polarimetry, NMR, magnetic susceptibility measurements, etc.

If only one complex is formed, the method can furnish some further information. Klausen and Langmyhr [14] have shown that if $x_{max} = 0.5$, i.e. $m = n$, and the curve exhibits inflections and parabolic portions for values of x near zero and one, this is an indication that $m = n > 1$. On the other hand the absence of inflections and parabolic portions indicates the formation of a complex with $m = n = 1$. It must be borne in mind, however, that in the case of complexes of high stability the inflections may not be appreciable even when $m = n > 1$ (see Fig. 5.1).

If the complex is very stable, the curve of absorbance (or other suitable measured property) versus x consists of two strictly linear intersecting portions. In the case of moderately stable complexes there is no such linearity, and from the deviation between the curve and that constructed by extrapolation from the slopes near the extreme values of x, the value of the stability constant can be calculated.

The simplicity of this technique makes it very attractive, but before it is applied the following limitations have to be considered.

(i) The method is not suitable for determining the composition of complexes if n/m is greater than 3. For complexes of mole ratios of 4 : 1, 5 : 1 and 6 : 1 the abscissa values are 0.800, 0.833 and 0.857, respectively.

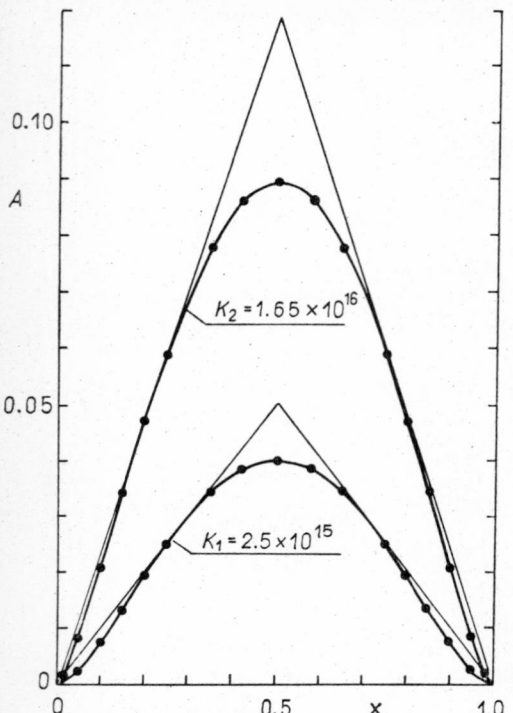

Fig. 5.1 Theoretical curves of continuous variation when MLe and Me_2L_2 complexes are formed. $K_1 = 2.5 \times 10^{15}$ and $K_2 = 1.65 \times 10^{16}$. $T_{Me} + T_L = 0.8 \times 10^{-5}$ M (Reproduced with permission from *Anal. Chim. Acta 28*, 335 (1963))

This means that an error of 2% in the measurements may result in a unit change in the observed mole ratio.

(*ii*) The change in the activity coefficients with changes in concentration may influence the molar ratio at which the concentration of the complex is of maximum value. To eliminate this effect the measurements have to be carried out at constant and sufficiently high ionic strength [15, 16].

(*iii*) The measured property must be a strictly linear function of the concentration of the complex [15].

(*iv*) If the conjugate acid of the ligand is weak, the position of the extremum depends on the pH [15, 17].

(*v*) If more than one complex species is formed in the system, the method gives unreliable results [17]. Although Vosburgh and Cooper [18] as well as Katzin and Gebert [19] applied it even in the case of stepwise complex formation, it is quite plausible that in these cases the method can give information on the system only if the conditions are exceptionally favourable.. If several complexes MeL_i are formed, and only one of them (MeL_n) is coloured. the position of the extremum can be obtained from the general equation

$$\frac{x_{\max}}{1 - x_{\max}} = n + \frac{1}{C(1 - x_{\max})} \sum_{1}^{N} (n - 1) \, i[MeL_i]. \qquad (5.14)$$

The relationships are even more complicated if all the complex species contribute to the measured property. Figures 5.2 and 5.3 illustrate the limitations of the methlod of continuous variation in these cases, since the position of the maximum is influenced by the total concentration and the wavelength used.

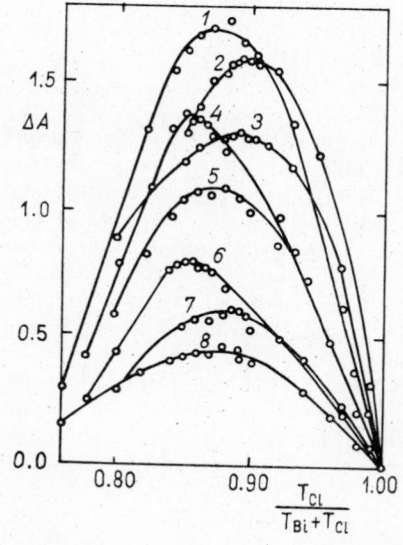

Fig. 5.2 Continuous variation curves for the Fə(III)–SCN⁻ system. $T_{Fe} + T_{SCN} = 0.1$ M (Curve A), 0.01 M (Curve B), 0.004 M (Curve C), 0.002 M (Curve D), 0.001 M (Curve E) [20] (Reproduced with permission from *Zhurnal Obshchei Khim.* *16*, 1551 (1946))

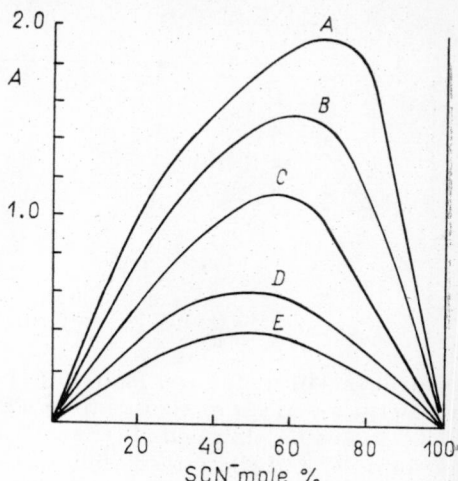

Fig. 5.3 Continuous variation curves for the Bi(III)–Cl⁻ system [20a]

No	λ	$T_{Bi}+T_{Cl}$, M	$HClO_4$, M	$NaClO_4$, M	I
1	366	0.20	8.0	1.0	9.2
2	348	0.10	7.0	0.5	7.6
3	330	0.04	4.0	0.2	4.2
4	363	0.50	9.0	0.0	9.5—10
5	369	0.20	8.0	1.0	9.2
6	366	0.50	9.0	0.0	9.5—10
7	360	0.20	8.0	1.0	9.2
8	375	0.20	8.0	1.0	9.2

(Reproduced with permission from *J. Am. Chem. Soc. 79*, 4576 (1957))

5.2.2 THE MOLE-RATIO METHODS

In the case of successive complex formation, more reliable information can be obtained by the mole-ratio methods [21]. Here the total concentration of the metal ion is kept constant while the total concentration of the ligand is increased. The absorbance is plotted as a function of the total ligand concentration. If only one complex of high stability is formed, the graph consists of two linear intersecting parts. The ratio of total con-

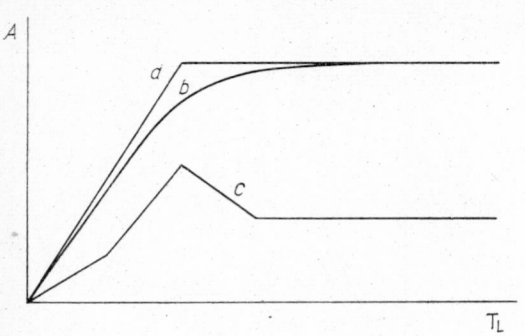

Fig. 5.4 Absorbance as a function of ligand concentration at constant metal ion concentration.
a — one stable complex is formed;
b — one weak complex is formed;
c — successive formation of stable complexes of different absorbance

centrations of metal and ligand at the intersection gives directly the ligand to metal ion ratio in the complex (Fig. 5.4). The stability constant in this case cannot be obtained. If the complex is moderately stable a steep curve is obtained and the intersection of the slopes of the two linear parts of the curve gives the ratio. However, there is an increasing uncertainty with decreasing stability. In favourable cases the stability constant can be simply derived from the difference between the measured curve and that constructed from the slopes. In the case of weak complexes this simple method fails to give any reliable information.

If N complexes are formed and each of them is stable enough to avoid overlapping equilibria, the curve consists of N sections. The slopes of the sections are determined by the molar absorptivities of the complexes (Fig. 5.4). According to Meyer and Ayres [22] the mole ratios of successive complexes can be determined if

$$K_n \geqslant 600\,K_{n+1}. \tag{5.15}$$

Harvey and Manning [23] have pointed out that the composition can also be obtained from the slopes of the curves of absorbance versus total ligand concentration. They have found that the break-points on the curves become more pronounced at higher ionic strength. This means that in these cases the stability of the complexes increases with increasing ionic strength. This, however, is not a general phenomenon. The method of Bent and French [24] has the advantage that not only the ratio n/m but the coefficients themselves can be obtained. Taking the logarithm of the equilibrium constant in reaction 5.5 we get

$$\log[\mathrm{Me}_m\mathrm{L}_n] = m\log[\mathrm{Me}] + n\log[\mathrm{L}] + \log\beta_{mn}. \tag{5.16}$$

If the complex is weak enough $[\mathrm{L}] = T_\mathrm{L}$, $[\mathrm{Me}] = T_\mathrm{Me}$. The absorbance is proportional to the concentration of the complex, so that plotting $\log A$ as a function of T_L at constant T_Me gives a straight line with slope n. Analogously, the value of m can be obtained from experiments at constant T_L and varying T_Me. Kingery and Hume [25] have pointed out that a clever combination of Job's and Bent and French's methods gives much more information than either of them alone.

Although the problem of determining the composition of the stability constant in the case of the formation of one complex is fairly simple, papers offering new methods and modifications are continually being published [14, 26, 27].

5.2.3 THE DILUTION METHOD

For weak complexes where only one complex MeL_n is formed and the absorbance of the metal ion and of the ligand is negligible the dilution method is convenient [28]. Assuming that the concentration of the ligand is much higher than that of the metal ion, and expressing concentrations in mole/l (m_i/v), the stability constant is given by

$$\beta_n = \frac{\dfrac{m_n}{v}}{\dfrac{m_0 - m_n}{v}\left(\dfrac{m_L}{v}\right)^n} \tag{5.17}$$

and the absorbance

$$A = \frac{\varepsilon_n m_n d}{v}. \tag{5.18}$$

Combining Eqs. 5.17 and 5.18 and multiplying both sides by $\varepsilon_n d\, v^n/A$ leads to

$$v^n = \frac{\beta_n \varepsilon_n m_0 m_L^n}{vA} - \beta_n m_L^n. \tag{5.19}$$

Therefore a plot of v^n against $1/vA$ should yield a straight line, with an intercept ($\beta_n m_L^n$) and slope ($\beta_n \varepsilon_n dm_0 m_L^n$) from which the two unknown constants can be calculated.

5.2.4 DETERMINATION OF THE STABILITY CONSTANT OF COMPLEXES OF COMPOSITION MeL

If $N = 1$, only one complex is formed and the absorbance of the solution is given by

$$A = (\varepsilon_0[Me] + \varepsilon_1[MeL] + \varepsilon_L[L])\, d. \tag{5.20}$$

Usually it is possible to find a wavelength at which $\varepsilon_L = 0$, so that Eq. 5.20 reduces to

$$A = (\varepsilon_0[Me] + \varepsilon_1[MeL])\, d. \tag{5.21}$$

Considering that $T_{Me} = [Me] + [MeL]$ and $\beta_1 = [MeL]/[Me][L]$, Eq. 5.21 can be written

$$A = T_{Me}\, \frac{\varepsilon_0 + \varepsilon_1 \beta_1[L]}{1 + \beta_1[L]}\, d. \tag{5.22}$$

Writing $\bar{\varepsilon}$ instead of A/dT_{Me}, a rearrangement of Eq. 5.22 leads to

$$\frac{\bar{\varepsilon} - \varepsilon_0}{\varepsilon_1 - \bar{\varepsilon}} = \beta_1[L]. \tag{5.23}$$

The value of ε_0 can be directly determined, and if the complex is stable β_1 can be obtained from measurements at high ligand concentration where practically the total amount of metal ion introduced is converted into MeL. The value of β_1 can be obtained most conveniently by plotting the left-hand side of Eq. 5.23 against [L], β_1 then being equal to the slope [29].

If the complex is not stable enough its molar absorptivity cannot be directly determined and Eq. 5.23 cannot be applied. In these cases a high ligand excess must evidently be applied to achieve formation of the complex in considerable amount, and therefore $[L] = T_L$, and hence,

$$A = d\left[\varepsilon_1 T_{Me} T_L \frac{\beta_1}{1 + \beta_1 T_L} + \varepsilon_0\left(T_{Me} - T_{Me} T_L \frac{\beta_1}{1 + \beta_1 T_L}\right)\right]. \qquad (5.24)$$

A rearrangement of this equation leads to

$$\frac{A/d - \varepsilon_0 T_{Me}}{T_{Me} T_L} = A^* = (\varepsilon_1 - \varepsilon_0)\frac{\beta_1}{1 + \beta_1 T_L}. \qquad (5.25)$$

To evaluate the constants ε_1 and β_1, the reciprocal of Eq. 5.25 can be used:

$$\frac{1}{A^*} = \frac{T_L}{\varepsilon_1 - \varepsilon_0} + \frac{1}{\beta_1(\varepsilon_1 - \varepsilon_0)}. \qquad (5.26)$$

On plotting $1/A^*$ as a function of T_L a straight line should be obtained, and from the slope and intercept the unknown constants can be calculated. Equation 5.26 was first used by Stearns and Wheland [30] to determine the acid dissociation constants of alcohols. The method was also applied in studying the Cu^{2+}–Cl^- [31] and the $Ni(CN)_4^{2-}$–CN^- [32] systems. This and analogous expressions were used most frequently in studying weak donor–acceptor complexes [33–38]. The work in this field cleared up the applicability of the spectrophotometric method to the obtaining of reliable equilibrium constants for weak complexes. If the complex is weak enough, the concentration of the complex is strictly proportional to the total ligand concentration up to its highest value. It means that the value of the intercept is zero and only the product $\beta_1\varepsilon_1$ can be obtained. As Person [39] has pointed out, independently of the molar absorptivity of the complex the necessary requirement that the constant be determinable is

$$T_{L,\,max} = \frac{0.1}{\beta_1} \qquad (5.27)$$

where $T_{L,max}$ is the highest possible total ligand concentration. For most anions the value of $T_{L,max}$ is less than 5 M, and for organic liquids (pyridine, ethylenediamine) it is about 10 M. This means that the lower limit of β_1 is about 0.01–0.02. As already pointed out, however, even larger apparent equilibrium constants may be produced by effects associated with the medium [9].

The solvation in such concentrated solutions must also be considered. That is, the formation of the complex is represented by the reaction

$$MeS_n + LS_m \rightleftharpoons MeLS_p + qS \qquad (5.28)$$

where S stands for the solvent molecule and $q = n + m - p$. Therefore the equilibrium constant has to be written as

$$\beta_1 = \frac{[MeLS_p][S]^q}{[MeS_n][LS_m]} . \qquad (5.29)$$

Carter, Murrell and Rosch [40] have pointed out that if the solvation of the species is not considered it leads to an overestimate of the molar absorptivity of the complex and an underestimate of the stability constant.

Finally, the method of Nakagura [41] should be mentioned. The stability constant is calculated from pairs of results. If the absorbances measured at constant T_{Me} and at zero and two different total ligand concentrations are A_0, A and A', the stability constant is given by

$$\beta_1 = \frac{T_L(A_0 - A') + T_L'(A - A_0)}{T_L T_L'(A' - A)} . \qquad (5.30)$$

Rediscovery is a fairly frequent phenomenon nowadays. This method for the more general case of formation of a single complex of the composition MeL_n was suggested much earlier by Edmonds and Birnbaum [42].

5.2.5 DETERMINATION OF THE STABILITY CONSTANTS IN THE CASE OF SUCCESSIVE COMPLEX FORMATION

If the ratio of the successive constants is large enough, the formation of each complex occurs in a separate step, and the equilibrium constants can be simply obtained by applying Eq. 4.23 for each step. A good example of the application of this principle is given by Irving, Rossotti and Harris [43] in determining the acid dissociation constants of a number of dibasic organic acids.

The problem is much more complicated in the case of overlapping complex formation steps. A number of methods are given for $N = 2$. The basic relationship is

$$A = \frac{\varepsilon_0 + \varepsilon_1 K_1[L] + \varepsilon_2 K_1 K_2[L]^2}{1 + K_1[L] + K_1 K_2[L]^2} d T_{Me}. \qquad (5.31)$$

As was shown by Thamer and Voigt [44] and by Ang [45], the calculation is much easier if the molar absorptivity of the complex MeL is larger (or smaller) than the molar absorptivities of Me and MeL_2. In these cases the change of the absorbance as a function of ligand concentration is not

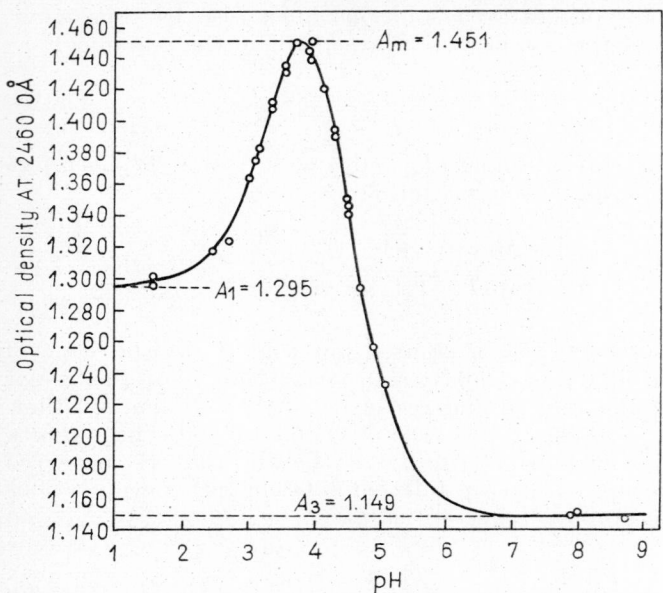

Fig. 5.5 Absorbance of isophthalic acid at 246 nm as a function of pH. Circles are experimental points, solid curve is calculated by Thamer and Voigt's method. (Reproduced with permission from *J. Phys. Chem. 56*, 225 (1952))

monotonic (Fig. 5.5). The stability constants can be conveniently obtained from the pairs of ligand concentrations belonging to the same absorbance. Let $\bar{\varepsilon}$ be the observed molar absorptivity $(A/d\, T_{\mathrm{Me}})$ at the two free ligand concentrations $[L]_1$ and $[L]_2$. Then, it follows from Eq. 5.31 that

$$[L]_1^2(\bar{\varepsilon} - \varepsilon_2) + [L]_1 \frac{1}{K_2}(\bar{\varepsilon} - \varepsilon_1) + \frac{1}{K_1 K_2}(\bar{\varepsilon} - \varepsilon_0) = 0 \qquad (5.32)$$

and

$$[L]_2^2(\bar{\varepsilon} - \varepsilon_2) + [L]_1 \frac{1}{K_2}(\bar{\varepsilon} - \varepsilon_1) + \frac{1}{K_1 K_2}(\bar{\varepsilon} - \varepsilon_0) = 0 \qquad (5.33)$$

from which

$$\bar{\varepsilon} = \varepsilon_1 - \frac{P}{K_2} \qquad (5.34)$$

where

$$P = \frac{(\varepsilon - \varepsilon_0)\,([L]_1 + [L]_2)}{[L]_1\,[L]_2} \qquad (5.35)$$

and

$$\bar{\varepsilon} = \varepsilon_2 + \frac{Q}{K_1} \qquad (5.36)$$

where

$$Q = \frac{(\varepsilon_1 - \bar{\varepsilon})}{[L]_1 + [L]_2} \, . \tag{5.37}$$

The unknown stability constants and molar absorptivity (ε_1) can be obtained from the slopes and intercepts of straight lines by plotting ε first as a function of P and then of Q.

The numerical method of Thamer [46] does not require any particular relationship between the molar absorptivities of the three species containing the central ion. To derive the stability constants one must measure the absorbance at five different ligand concentrations. The constants can be explicitly calculated by means of Eqs. 5.38 and 5.39

$$K_1 =$$

$$= \frac{-\lambda_1([L]_2 + [L]_3) + \lambda_2([L]_2 + [L]_4) - \lambda_3([L]_2 + [L]_5) - \lambda_4([L]_3 + [L]_4)}{\lambda_1[L]_2[L]_3 - \lambda_2[L]_2[L]_4 + \lambda_3[L]_2[L]_5 + \lambda_4[L]_3[L]_4 - \lambda_5[L]_3[L]_5 + \lambda_6[L]_4[L]_5} +$$

$$+ \frac{\lambda_5([L]_3 + [L]_5) - \lambda_6([L]_4 + [L]_5)}{\lambda_1[L]_2[L]_3 - \lambda_2[L]_2[L]_4 + \lambda_3[L]_2[L]_5 + \lambda_4[L]_3[L]_4 - \lambda_5[L]_3[L]_5 + \lambda_6[L]_4[L]_5} \tag{5.38}$$

and

$$K_2 = \frac{\lambda_1 - \lambda_2 - \lambda_3 + \lambda_4 - \lambda_5 + \lambda_6}{\{-\lambda_1([L]_2 + [L]_3) + \lambda_2([L]_2 + [L]_4) - \lambda_3([L]_2 + [L]_5) - \lambda_4([L]_3 + [L]_4) + \lambda_5([L]_3 + [L]_5) - \lambda_6([L]_4 + [L]_5)\}} \tag{5.39}$$

where

$$\lambda_1 = (A_2 - A_1)(A_3 - A_1)([L]_1 - [L]_4)([L]_1 - [L]_5)([L]_2 - [L]_3)([L]_4 - [L]_5) \tag{5.40}$$

$$\lambda_2 = (A_2 - A_1)(A_4 - A_1)([L]_1 - [L]_3)([L]_1 - [L]_5)([L]_2 - [L]_3)([L]_3 - [L]_5) \tag{5.41}$$

$$\lambda_3 = (A_2 - A_1)(A_5 - A_1)([L]_1 - [L]_3)([L]_1 - [L]_4)([L]_2 - [L]_5)([L]_3 - [L]_4) \tag{5.42}$$

$$\lambda_4 = (A_3 - A_1)(A_4 - A_1)([L]_1 - [L]_2)([L]_1 - [L]_5)([L]_3 - [L]_4)([L]_2 - [L]_5) \tag{5.43}$$

$$\lambda_5 = (A_3 - A_1)(A_5 - A_1)([L]_1 - [L]_2)([L]_1 - [L]_4)([L]_3 - [L]_5)([L]_2 - [L]_4) \tag{5.44}$$

$$\lambda_6 = (A_4 - A_1)(A_5 - A_1)([L]_1 - [L]_2)([L]_1 - [L]_3)([L]_4 - [L]_5)([L]_2 - [L]_3). \tag{5.45}$$

As a matter of course the situation is much more complicated if $N > 2$. Although many authors [47–53] have dealt with this problem the different successive approximation methods cannot give unambiguous information on the systems if more than three species are present simultaneously in the system. The most detailed treatment of these systems was given by Newman and Hume [47]. In the following, Yatsimirskii's extrapolation method [50] is briefly treated. The general relationship between the absorbance and stability products is given by

$$\frac{A}{T_{Me} d} \equiv \bar{\varepsilon} \left\{ \frac{\varepsilon_0 + \varepsilon_1 \beta_1 [L] + \varepsilon_2 \beta_2 [L]^2 + \ldots + \varepsilon_N \beta_N [L]^N}{1 + \beta_1 [L] + \beta_2 [L]^2 + \ldots + \beta_N [L]^N} \right\}. \tag{5.46}$$

Subtracting ε_0 from both sides of Eq. 5.46 and writing $\Delta\varepsilon_i$ for $(\varepsilon_i - \varepsilon_0)$ we get

$$\Delta\bar{\varepsilon} = \frac{\Delta\varepsilon_1 \beta_1 [L] + \Delta\varepsilon_2 \beta_2 [L]^2 + \ldots + \Delta\varepsilon_N \beta_N [L]^N}{1 + \beta_1 [L] + \beta_2 [L]^2 + \ldots + \beta_N [L]^N}. \tag{5.47}$$

By use of the function

$$F \equiv \frac{\Delta\bar{\varepsilon}}{[L]}, \tag{5.48}$$

Eq. 5.47 can be written as

$$F = \frac{\Delta\varepsilon_1 \beta_1 + \Delta\varepsilon_2 \beta_2 [L] + \ldots + \Delta\varepsilon_N \beta_N [L]^{N-1}}{1 + \beta_1 [L] + \beta_2 [L]^2 + \ldots + \beta_N [L]^N}. \tag{5.49}$$

Plotting F as a function of $[L]$ and extrapolating to $[L] = 0$ yields

$$\lim_{[L] \to 0} F = F_1 = \Delta\varepsilon_1 \beta_1. \tag{5.50}$$

Let another function be defined:

$$F' \equiv \frac{F - F_1}{[L]}. \tag{5.51}$$

By plotting F' against $[L]$ and extrapolating to $[L] = 0$, we get

$$\lim_{[L] \to 0} F' = F_2 = \Delta\varepsilon_2 \beta_2 - \Delta\varepsilon_1 \beta_1^2. \tag{5.52}$$

Further composite constants can be obtained by analogous procedures. Another set of composite constants can be obtained by introducing new functions. Let both the nominator and denominator of the right-hand side of Eq. 5.47 be divided by $[L]^N$:

$$\Delta\bar{\varepsilon} = \frac{\Delta\varepsilon_N \beta_N + \Delta\varepsilon_{N-1} \beta_{N-1} [L]^{-1} + \ldots + \Delta\varepsilon_1 \beta_1 [L]^{1-N}}{\beta_N + \beta_{N-1} [L]^{-1} + \ldots + \beta_1 [L]^{1-N} + [L]^{-N}}. \tag{5.53}$$

By p lotting $\Delta\bar{\varepsilon}$ as a function of $1/[L]$ and extrapolating to $1/[L] = 0$ we get

$$\lim_{\frac{1}{[L]} \to 0} \equiv G_1 = \Delta\varepsilon_N. \tag{5.54}$$

Introducing the further function

$$G' = (\Delta\bar{\varepsilon} - G_1)[[L] \tag{5.55}$$

we obtain

$$\lim_{\frac{1}{[L]} \to 0} G' \equiv G_2 = (\Delta\varepsilon_{N-1} - \Delta\varepsilon_N) \frac{\beta_{N-1}}{\beta_N} . \tag{5.56}$$

Further composite constants of this series can be analogously obtained. The values of the single constants can be simply calculated from the two series of composite constants (F and G). In the case of $N = 2$ the corresponding formulae are:

$$\beta_1 = \frac{F_1 G_1 - F_2 G_2}{F_1 G_2 + G^2} \tag{5.57}$$

$$\beta_2 = \frac{F_1^2 + F_2 G_1}{F_1 G_2 + G_1^2} \tag{5.58}$$

$$\Delta\varepsilon_1 = \frac{F_1^2 G_2 + F_1 G_1^2}{F_1 G_1 - F_2 G_2} \tag{5.59}$$

$$\Delta\varepsilon_2 = G_1. \tag{5.60}$$

5.2.6 THE METHOD OF CORRESPONDING SOLUTIONS

As already shown, in the case of the formation of mononuclear binary complexes the distribution of complexes is unambiguously determined by the free ligand concentration. Those solutions in which the total concentrations of the metal ion and the ligand are different, but the free ligand concentrations — and consequently the value of the complex formation function, as well as the mole fractions of the complexes — are the same, are termed corresponding solutions. There are many possibilities for deciding the correspondence of two solutions. The spectral identity of two solutions, manifested by the equality of the apparent molar absorptivity at every wavelength, is a crucial proof that these solutions are corresponding.

Evidently, for two corresponding solutions the following relationship holds

$$\bar{n} = \frac{T_L - [L]}{T_{Me}} = \frac{T'_L - [L]}{T'_{Me}} . \tag{5.61}$$

Equation 5.61 can be solved for both $\bar{n}$ and $[L]$:

$$\bar{n} = \frac{T_L - T'_L}{T_{Me} - T'_{Me}} \tag{5.62}$$

$$[L] = \frac{T_{Me} T'_L - T'_{Me} T_L}{T_{Me} - T'_{Me}} . \tag{5.63}$$

From the pairs of values of $\bar{n}$ and $[L]$ the stability constants can be calculated by the methods discussed in Chapter 4. Experimentally, the method of corresponding solutions requires the study of two series of solutions of

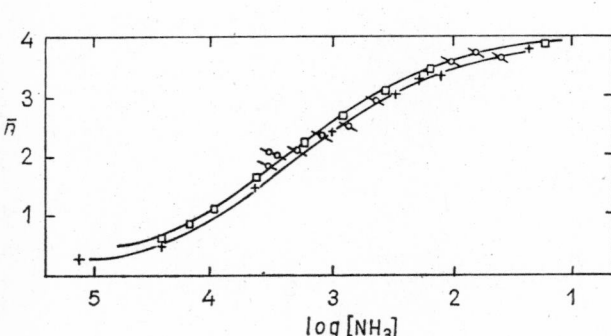

Fig. 5.6 Formation function of the Cu(II)–NH$_3$ system. o -values determined spectrophotometrically by the method of corresponding solutions; + -values determined potentiometrically. □ from NH$_3$ vapour pressure (Reproduced with permission from *Kgl. Dansk. Vid. Selskab Mat. fys. Medd.* *21*, No. 4. (1941))

total central ion concentrations T_{Me} and T'_{Me} and of varying total ligand concentrations. The spectrum of each solution is recorded and the composition of the strictly corresponding solutions is evaluated by interpolation. Figure 5.6 shows the complex formation function of the copper(II)–ammonia system determined by different methods. The method is applicable in the case of moderately stable complexes when the value of the bound ligand concentration $(T_L - [L])$ is significantly smaller than T_L; nevertheless $[L]$ is not near zero. This method does not require the knowledge of the molar absorptivities of the different complex species and does not provide explicit information on them. These constants can be obtained from the absorbance of the solutions after calculation of the distribution of the complexes.

In spite of the great potential of the method of corresponding solutions developed by J. Bjerrum [54], it has been applied to only a few systems. Jezowska–Trzebiatowska, Pajdowski and Starosta [55] have determined as many as six successive constants in studying the chromium(III)–glycine system. Although the reality of these constants is doubtful, owing to possible side-reactions and some improper features of the calculation, this is not related to the applicability of the method.

Bjerrum has already pointed out that the method may be suited to the study of non-coloured systems, when the coloured metal ion–ligand system plays the role of the indicator, and in fact, a competitive reaction is

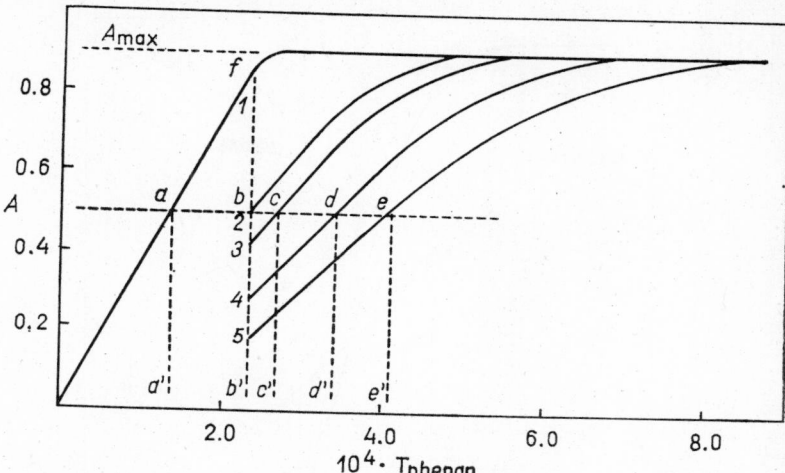

Fig. 5.7 Absorbance of Fe(II) and Co(II) solutions at 510 nm as a function of 1,10-phenanthroline concentration. $T_{Fe} = 8 \times 10^{-5}$ M; $T_{Co} = 0$ M (Curve *1*); 1.2×10^{-4} M (Curve *2*); 1.6×10^{-4} M (Curve *3*); 2.4×10^{-4} M (Curve *4*); 3.2×10^{-4} M (Curve *5*). Points *a, b, c, d, e* and *f* refer to corresponding solutions. (Reproduced with permission from *J. Chem. Soc.* 3957 (1955))

studied. This application of the principle of corresponding solutions was thoroughly studied by Irving and Mellor [56]. The method of applying the theory of corresponding solutions to systems consisting of two different central ions and one ligand is illustrated by Fig. 5.7. Curve *1* shows the absorbance of 8×10^{-5} M iron(II) at 510 nm as a function of the 1,10-phenanthroline concentration. Curves *2–5* refer to solutions of the same iron(II) concentration but containing also cobalt(II) ions in varying concentration (see the legend of the Figure). The absorbance at 510 nm is due essentially to the tris(1,10-phenanthroline)iron(II) complex. The reduction in absorbance on increasing the cobalt(II) concentration is the consequence of the formation of different phenanthroline complexes of cobalt(II). Solutions of the same absorbance are corresponding. As the total iron concentration is the same in each solution, the concentrations of the iron complexes and the free phenanthroline in solutions of the same absorbance are identical. Therefore the whole of the phenanthroline represented by the concentration difference *a'b'* must have formed complexes with the cobalt present in solution 2; the whole of the phenanthroline represented by concentration difference *c'd'* must have formed complexes with the amount of cobalt equivalent to the difference in cobalt concentration between solutions 3 and 4, etc. In this way the necessary pairs of $\bar{n}$ and [L] can be obtained. This procedure can furnish, however, only a part of the formation curve, because the accuracy of the measurements rapidly decreases when the ratio of the actual absorbance A_s to the maximum absorbance A_{max} (that is when the total amount of the indicator metal is complexed)

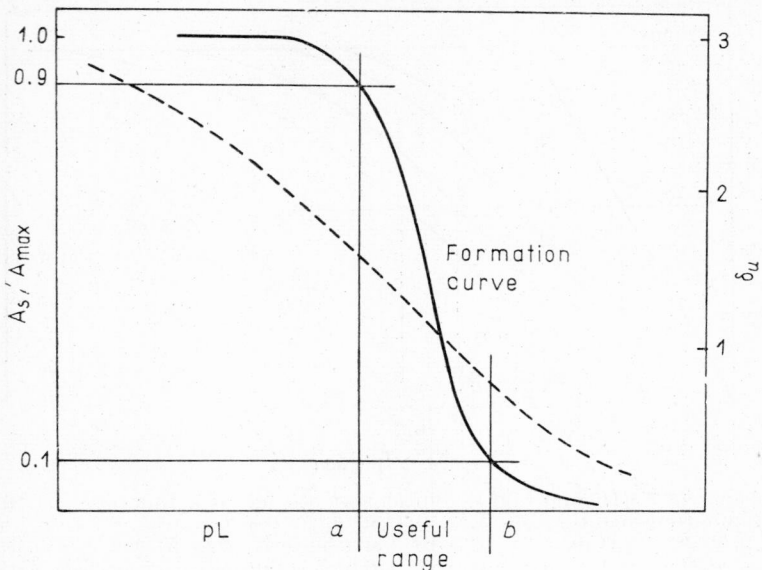

Fig. 5.8 Schematic representation of the formation curve for an arbitrary metal–1,10-phenanthroline system (broken line) superimposed on a graph representing the variation of concentration of tris(phenanthroline)iron(II) ion present, as a function of the logarithm of free ligand concentration. (Reproduced with permission from *J. Chem.Soc.* 3957 (1955))

approaches unity or zero. For practical purposes this ratio must be in the range $0.1 < A_s/A_{max} < 0.9$. This range is related to a range of the free ligand concentration, and therefore the portion of the formation curve accessible to study is limited (see Fig. 5.8).

5.2.7 THE STUDY OF COMPETING REACTIONS

As a matter of course, competitive reactions can be studied spectrophotometrically by other means than the principle of corresponding solutions. It is important to bear in mind the possibility of the formation of binuclear complexes in case of systems containing two different metal ions and one ligand, and of the formation of mixed ligand complexes in systems containing two different ligands and one kind of central ion. Hughes and Martell [57] have determined the stability constants of a number of complexes of ethylenediaminetetra-acetic acid, by using competitive reactions. Musgrave and Keller [58] have determined the stability constants of mercury(II)–azide complexes based on the competition between mercury(II) and iron(III) for the azide ion. The competition of 8-hydroxyquinoline and diethyldithiocarbamate for copper(II) was studied by Janssen [59]. Kleiner [60] has obtained the stability constant of the AlF^{2+} ion by studying the iron(III)–aluminium(III)–thiocyanate–fluoride system.

Exchange reactions involving dithizone and diethyldithiocarbamate as ligands and zinc(II), silver(I) and arsenic(III) as central ions were recently studied in non-aqueous solutions by Růžička and Starý [61]. Most frequently the competition between a metal ion and proton is studied. In these systems the formation of protonated species must be considered (see Chap. 6).

5.2.8 INFRARED SPECTROPHOTOMETRY

From the rapid development of instrumentation in infrared spectroscopy it is expected that this method of great potential value will be very widely used in the future. There are two basic experimental problems. First, in general fairly concentrated solutions must be studied because of the great self-absorbance of the solvents; secondly, special cell windows are required for work with aqueous solutions. Earlier cell windows were made from silver chloride and calcium fluoride, but the introduction of IRTRAN windows made the experimental work more convenient. In evaluating the results either the absorbance at a particular frequency is considered, or the integrated absorption, i.e. the total area of an absorption band $\int_{v_1}^{v_2} \log \dfrac{I_0}{I} \, dv$.

To a first approximation the latter quantity is equal to the product of the maximum absorbance and the half-width of the absorption band.

The greatest advantage of infrared studies is that they make it possible to decide whether the co-ordination occurs in the inner or the outer sphere, and to identify the donor atom of the ligand. This latter possibility is very important in the case of such ligands as SCN^-, OCN^-, NO_2^-, etc.

Infrared spectrophotometry has been applied in the study of copper(I)–cyanide [62], silver(I)–cyanide [63], tetracyanonickelate–cyanide [64], mercury(II)–cyanide–halide [65], cadmium(II)–cyanide, transition metal ion–thiocyanate [66] and transition metal ion–sulphate systems [67]. The data can be treated by the calculation methods discussed earlier.

5.2.9 RAMAN SPECTROSCOPY

When a beam of light traverses a transparent medium free from dust particles, two kinds of scattered light can be observed. The elastic collisions between photons and molecules lead to Rayleigh scattering; the frequency of the scattered light is identical with that of the incident light. On the other hand the inelastic collisions result in the appearance of new line(s), the Raman spectrum. The Raman spectrum is characteristic of the species present and makes it possible to get information on the concentration and structure of the different complexes. Some of the vibrations are infrared-active, others are Raman-active and in many cases both techniques can be applied. Just as in the field of infrared spectroscopy, owing to the rapid instrumental development the popularity of this method is continually increasing. Recently Tobias published an excellent review on the theory and application of Raman spectroscopy in inorganic chemistry [68].

It is hoped that the construction of laser Raman spectrophotometers would make it possible to study dilute solutions too.

Just after the discovery of the phenomenon by Raman, Rao [69] tried to determine the dissociation constant of nitric acid by the method. Later this problem was studied more exactly by Redlich [70]. Delwaulle was the pioneer of the application of the Raman technique to complex equilibria. He studied the formation of mixed ligand complexes in the $AsCl_3$–$AsBr_3$ [71], $SnCl_4$–$SnBr_4$ [72], PCl_3–PBr_3 [73], $Cd(II)$–Cl^-–$Br^-(I^-)$ [74], $Hg(II)$–Br^-–I^- [75], $Zn(II)$–Br^-–Cl^- [76] systems, etc. Recently the following systems were studied by this method: BCl_3–BBr_3 [76, 77], $Cd(II)$–$Cl^-(Br^-, I^-)$ [78], $Zn(II)$–Br^- [79, 80], $Tl(III)$–Cl^- [81, 82], $Ta(V)$–F^- [83], $(CH_3)_2Sn^{2+}$–Cl^- [84], $Ga(III)$–Br^- [85], $In(III)$–SO_4^{2-} [86], $Ca(II)$–NO_3^- [87], $Ga(III)$–I^-–$Br^-(Cl^-)$ [88], $Pb(II)$–OH^- [89], $Bi(III)$–OH^- [90], $Ce(IV)$–NO_3^- [91], $Hg(II)$–ethylenediamine [92]. The method was used to detect weak interactions between perchlorate ion and different metal ions.

5.2.10 OPTICAL ROTATORY POWER

A molecule is optically active if it has neither a centre nor a plane of symmetry, i.e. if its structure is not superimposable upon its mirror image. A metal complex is optically active if either the whole complex molecule is asymmetric or one of the ligands is optically active. Optical rotatory power measurements make it possible to study two kinds of equilibrium reactions. If a metal complex can be resolved and the racemization is a very slow reaction, the uptake of different ligands in the outer co-ordination sphere can be followed by measuring the rotation as a function of the ligand concentration. These investigations will be treated in Chapter 9. On the other hand, if the ligand is optically active the formation of inner complexes can be studied by rotation measurements.

This approach is very important, particularly in comparison with the results of spectrophotometric studies. A spectrophotometric study is based on the effect of the ligand(s) on the electronic system of the central ion, whilst the data derived from a polarimetric investigation reflect the action of the central ion on the ligand. Another advantage of polarimetry is that it gives a possibility for the study of systems which have no characteristic absorption even in the far ultraviolet.

The rotation of a solution is described by an expression strictly analogous to the Beer–Lambert law, the corresponding intensive factor being the molar optical rotation. Earlier the measurements were made at the Na_D line, but nowadays commercial instruments are available both for the visible and the ultraviolet region. For the details of the theory and instrumentation the reader may consult some excellent reviews [93, 94].

Polarimetry was used to determine the dissociation constants of both the carboxylic groups [95], and the alcoholic hydroxyl groups of tartaric acid [96, 97]. Nebbia has determined the dissociation constants of ascorbic acid, glutamic acid and lysine [98–100]. Figure 5.9 shows the change of

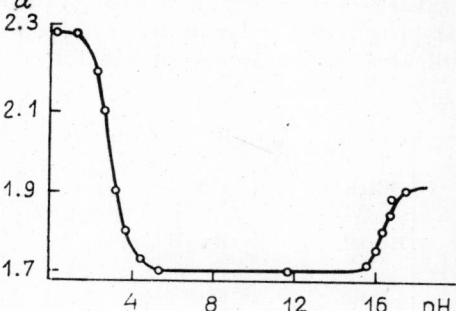

Fig. 5.9 Optical rotatory power of mandelic acid at the sodium D line as a function of pH. (Reproduced with permission from *Naturwiss. 48*, 571 (1961))

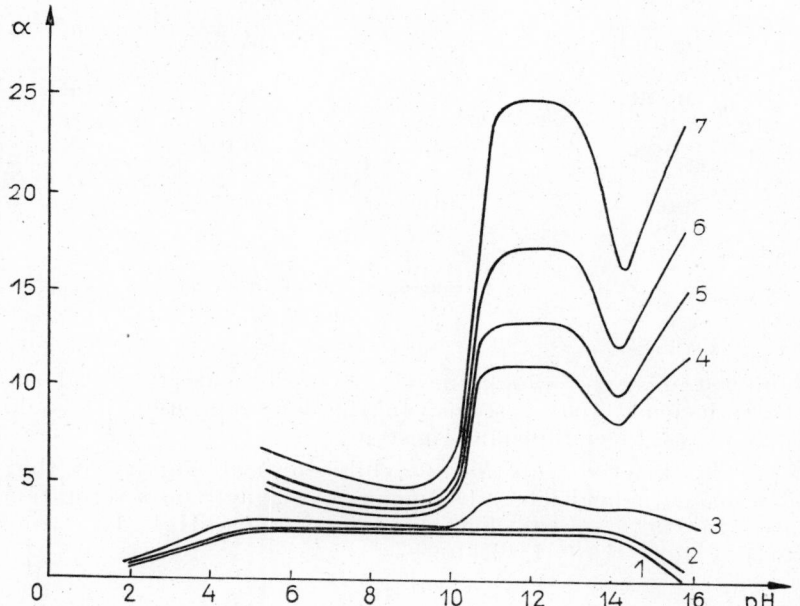

Fig. 5.10 Optical rotatory power of Cu(II) — tartaric acid system at 435.8 nm as a function of pH. $T_{tartrate} = 0.1$ M $T_{Cu} = 0$ M (Curve 1); 10^{-3} M (Curve 2); 5×10^{-3} M (Curve 3); 2×10^{-2} (Curve 4); 2.5×10^{-2} M (Curve 5); 3.33×10^{-2} M (Curve 6); 5×10^{-2} M (Curve 7)

the optical rotatory power (sodium D line) of mandelic acid as a function of pH [101]. The decrease in the rotatory power in acidic solution is due to the ionization of the carboxylic group, and the increase in strongly alkaline medium is caused by the dissociation of the alcoholic hydroxyl group. Earlier, information obtained from polarimetric studies [102, 103] was mostly qualitative, but some recent quantitative investigations illustrate well the potential of this method. Figure 5.10 shows the optical rotatory power of the copper(II)–tartaric acid system as a function of pH [104]. The polarimetric study, together with some other information, indicates the existence of the following species:

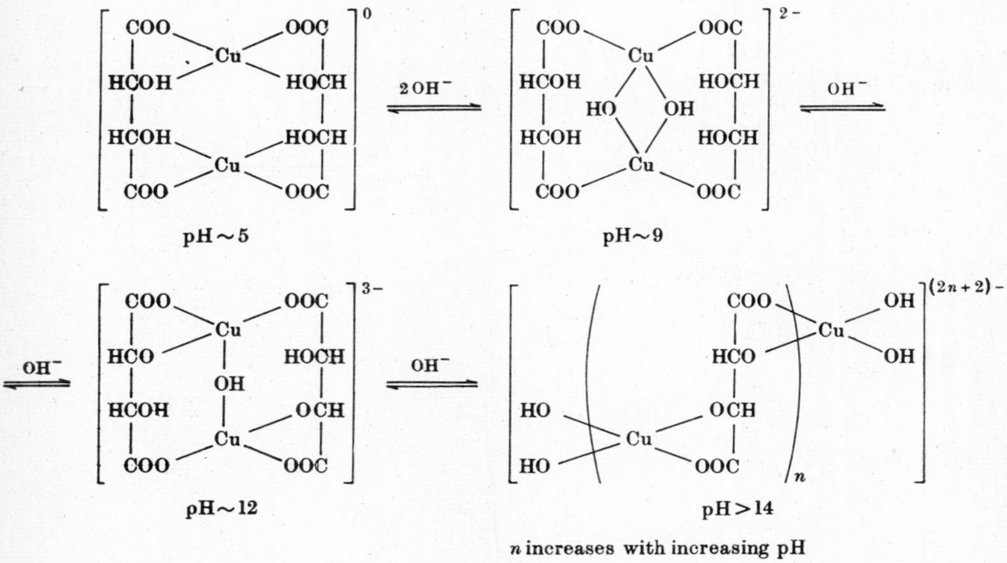

pH ~ 5   pH ~ 9

pH ~ 12   pH > 14

n increases with increasing pH

Frei carefully studied the aluminium(III)–tartaric acid [105] and the boric acid–tartaric acid [106] systems. In the former system he could calculate as many as 18 equilibrium constants.

Certain optically inactive substances exhibit optical activity if they are placed in a magnetic field. The phenomenon of magneto-optical rotation has been used to study only a few complex systems: HgI_2–I^- [107], $Hg(SCN)_2$–SCN^- [108], $AlCl_3$–LiCl [109].

5.3 Calorimetry

The heat evolved when a ligand is added to a solution of a metal ion is given by

$$Q = V \sum_{1}^{N} [MeL_i] \cdot \Delta H_i + \Delta H_{dil} \qquad (5.64)$$

where Q is the total heat evolved (or absorbed), V is the volume, ΔH_i is the molar formation enthalpy of the ith species, and ΔH_{dil} is the heat of dilution. There is a striking similarity between Eqs. 5.4 and 5.64 which suggests that all the methods for the mathematical treatment of spectrophotometric data should be applicable to calorimetric data. In principle, however, the calorimetric method is much more general than the spectro-photometric or any other method, because with every chemical reaction is connected the evolution or absorption of heat. The absorbance of light is a constant property of a system, but the heat evolved (or absorbed) must be measured at the moment of the mixing of the reactants. If the

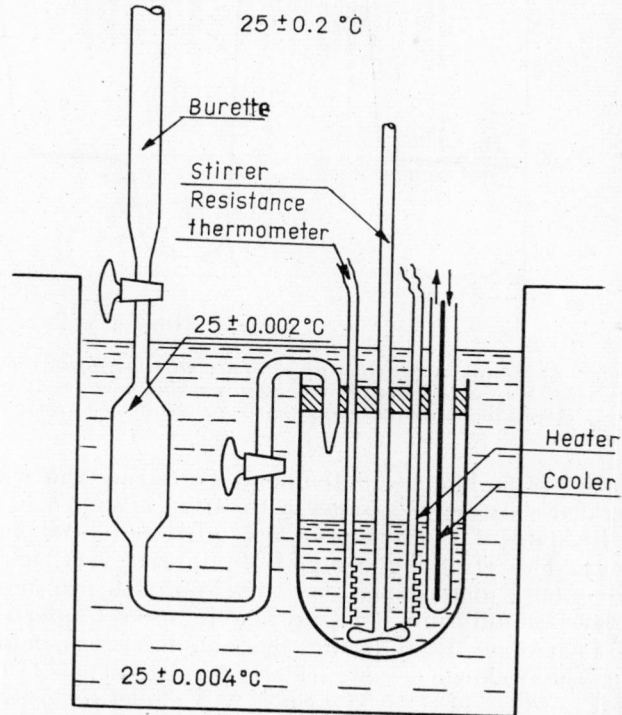

Fig. 5.11 Principle of Schlyter and Sillén's calorimeter. (Reproduced with permission from *Acta Chem. Scand.* 385 (1959))

measurements are made under adiabatic conditions, the change of tem-perature is proportional to the enthalpy change. The introduction of thermistors to follow minute changes of temperature has resulted in the construction of a number of convenient types of apparatus [110–116]. The apparatus constructed by Schlyter and Sillén [112, 113] and developed by Johansson [115] is particularly suitable for the study of complex equilib-ria. The schematic arrangement of the calorimeter is shown by Fig. 5.11. The temperatures of both solutions to be mixed are exactly the same. The heater and the cooler make it possible to restore the temperature of the solution to its original value after addition of each increment of the titrant (usually the solution of the ligand). This method eliminates any

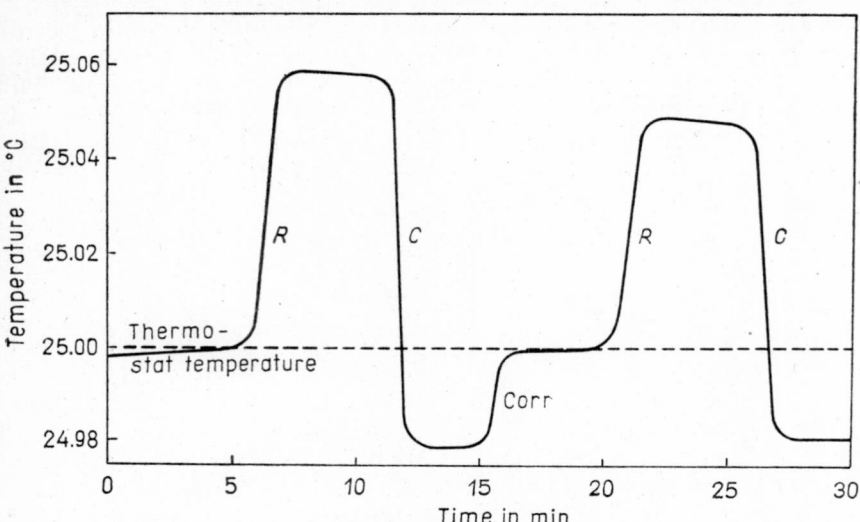

Fig. 5.12 Temperature history of part of a titration: R — reaction heat evolved; C — cooling applied; Corr — temperature corrected by means of a heater. (Reproduced with permission from *Acta Chem. Scand. 13*, 385 (1959))

considerable change of temperature during the whole titration. The temperature history of part of a titration is shown in Fig. 5.12.

In spite of the exceptionally great potential value of the calorimetric method in studying complex equilibria, very few equilibrium constants have been determined by this approach. In most calorimetric studies known equilibrium constants were used to derive formation enthalpies and entropies. Equilibrium constants have been calorimetrically determined for the following systems: $HgBr_2$–Br^- [117–119], Ag^+-pyridine [119], $HgCl_2$–Cl^- [118], $(CH_3)_2SnCl$–N,N-dimethylacetamide [120], $Cu(CN)_2^-$–CN^- [121]. Christensen *et al.* [122] have determined the pK values for proton ionization from HSO_4^- and HPO_4^{2-}. Arnek [118] has made a comparative study of the potentiometric and calorimetric methods in the case of the $HgCl_2$–Cl^- and $HgBr_2$–Br^- systems. As appears from the data of Table 5.1 calorimetric study provided nearly as exact data as potentiometry did.

TABLE 5.1

System	Method	$\log K_3$	$\log K_4$	$\log K_3 K_4$
$HgCl_2$–Cl^-	Potentiometry	0.75 ± 0.12	1.38 ± 0.12	2.13 ± 0.06
	Calorimetry	1.08 ± 0.21	1.09 ± 0.24	2.17 ± 0.09
$HgBr_2$–Br^-	Potentiometry	2.76 ± 0.30	1.49 ± 0.04	4.25 ± 0.04
	Calorimetry	2.32 ± 0.30	2.00 ± 0.42	4.30 ± 0.27

Considering that calorimetric studies also provide values of the formation enthalpy and entropy of each complex species, it is reasonable to expect that this method will soon come into general use. Tyrrell aud Beezer have written a useful monograph on calorimetric titrations [118a].

5.4 Measurements based on reaction kinetics

The rate of chemical reactions is proportional to the concentrations of the reactants, raised to a certain power. Kinetic and equilibrium studies of complex systems are complementary. Equilibrium data can be derived from kinetic measurements, and independent equilibrium data provide an invaluable help in deducing sound reaction mechanisms. Every chemical reaction is evidently a time process: such adjectives as instantaneous or immediate reflect only that the reaction in question is so fast that it cannot be followed by the method applied. However, modern instrumental methods [123], especially those developed by Eigen, make possible the study of the fastest chemical reactions. These methods enable us to distinguish between co-ordination in the inner and in the outer sphere [124]. This aspect will be treated in Chapter 9. Many reactions are catalysed by ligands, metal ions and metal complexes. The study of these catalytic reactions can also provide information on the stability constants of the complex species [125].

If the reaction is first-order for the species of which the stability constant is to be determined, the rate can be described by an expression strictly analogous to Eq. 5.4, the corresponding intensive factors being the elementary rate constants, so for the mathematical treatment of data the methods already discussed can be applied. However, some important differences have to be borne in mind. The rate constants are usually much more sensitive to temperature variations than are the molar absorptivities. In spectrophotometry the measurement can be made at different wavelengths but in kinetics a species — or more correctly a reaction — can be characterized by a single intensive factor. Especially in the case of catalytic reactions must it be remembered that thermodynamically less stable species may contribute considerably to the observed rate, because it can be expected that the value of the intensive factor will exhibit a large value in just that case. This makes possible the detection of such species and in favourable cases even the determination of their stability constants. Evidently, the more complicated the system, the greater the ambiguity in the interpretation of the results. In such cases the stability constants must not be estimated from the kinetic measurements alone, since only a careful consideration of the data provided by several approaches may give reliable information on the composition, structure and stability of the species.

5.4.1 STOICHIOMETRIC REACTIONS

The equilibrium constant being the ratio of the rate constants of the forward and the reverse reactions, kinetic study of the formation and decomposition of the complexes can furnish stability constants. This principle was used by Postmus and King [126] for the determination of the stability constant of $CrSCN^{2+}$. The constant found agrees excellently with the data obtained from the study of equilibrated solutions [127].

The stability of the tris(phenanthroline)–iron(II) and tris(dipyridyl)–iron(II) complexes is favoured, that is $\beta_3 \gg \beta_2, \beta_1$. The value of β_3 was determined in the case of phenanthroline [128] and its derivatives [129] and of dipyridyl [130] from kinetic studies of the formation and decomposition reactions.

The different relaxation methods were applied in the determination of the stability constants of a number of ion-pairs [124, 131–135].

The rate of the formation of the ethylenediaminetetra-acetate–chromium(III) complex depends on the pH. At the same pH rate is a function of the age of the chromium(III) solution; in slightly acidic solution a dimerization occurs, and the hydroxide-bridged dimeric chromium(III) complex reacts much more slowly with EDTA than do the mononuclear complexes. From this behaviour the dimerization constant can be determined [136].

The inhibition of stoichiometric reactions can also be used to obtain thermodynamic information. The oxidation of thiosulphate by iron(III) is hindered by cadmium ions because of formation of the CdS_2O_3 complex. Its stability could be determined from the retarding effect [137]. Similarly, the retarding effect of chloride and bromide ions on the $BrCN–I^-$ reaction furnished data on the stability of different interhalogen complexes [138]. Sykes has determined the stability of $FeSO_4^+$ and $FeNO_3^{2+}$ complexes from the retarding effect of the sulphate and nitrate ions on the iron(III)–iodide reaction [139].

5.4.2 CATALYTIC REACTIONS

The number of catalytic reactions is extremely high. In fact, it is fairly difficult to find a reaction without any catalytic effect. If the catalytic activity of the ligand or the central metal ion exceeds that of the complex, the formation of complexes results in a decrease of the rate, and vice versa.

5.4.2.1 *Catalysis by ligand*

In connection with the study of the equilibria in the mercury(II)–iodide system, Sherrill [140], as early as 1903, determined the free iodide concentration from the rate of decomposition of hydrogen peroxide.

There are many reactions catalysed by hydroxide ions. Rate studies for the determination of hydroxide ion concentration were first applied by Koelichen [141], who studied the hydrolysis of diacetone alcohol. Because of the volume change in the reaction the rate can be determined dilatometrically. The reaction leads to an equilibrium, and it is possible to study the reverse reaction too, that is the formation of diacetone alcohol from acetone. This reaction was used by Austerweil [142] to determine the solubility of lead hydroxide and the composition of the species in the solution. It must be considered, however, that owing to the high concentration of acetone, a change of the medium occurred.

The effect of different metal ions on the reactions catalysed by hydroxide ion was most thoroughly studied by Bell et al. [143, 144]. From the retarding effect of Ca^{2+}, Ba^{2+} and Tl^+ on the hydrolysis of diacetone alcohol [143], the stability constants of the monohydroxo-complexes of these ions could be determined. Similar results were achieved with the hydrolysis of carbethoxymethyltriethylammonium iodide [144]. However, it was found that the hydrolysis of ethyl acetate was not retarded by metal ions. This means that the hydrolysis of ethyl acetate is catalysed by hydroxo-complexes as efficiently as by hydroxide ions.

The decomposition of nitramide is catalysed by many anions such as acetate, mandelate, salicylate, malate and fumarate. From the decrease in the rate on addition of Ca^{2+}, Ba^{2+} and Zn^{2+} ions, the stabilities of the corresponding ion-pairs were estimated [145]. The complexes retain some of the original activity of the ligands.

The hydrolysis of the p-nitrophenyl acetate is catalysed by imidazole and carbobenzoxy-L-prolyl-L-histidylglycinamide. Complex-forming metal ions decrease the catalytic effect and the stability constants were obtained from the kinetic data [146, 147].

The iodine–azide reaction is specifically catalysed by ligands containing S(II). The metal complexes of these ligands are less active, and it seems to be promising to study this effect for the estimation of the stability of these complexes.

5.4.2.2 *Catalysis by metal ions and metal complexes*

The catalytic effect of transition metal ions on many inorganic and organic redox reactions is well known. The effect of salts on these catalytic reactions has also been very thoroughly studied [148]. The so-called primary salt effect is the consequence of the effect of electrolytes on the activity coefficients of the reactants. The secondary or specific salt effect, however, is the consequence of short range interactions, i.e. complex formation. Complex formation may decrease or increase the original effect of the metal ion, and if there is successive complex formation one of the intermediary complexes frequently has the highest catalytic efficiency.

The effect of ligands on some isotope exchange reactions has been particularly well studied and only a few examples are given here. Hudis and Wahl [149] have determined the stability constants of some iron(III) fluoride complexes from the effect of fluoride on the Fe(III)–Fe(II) reaction. Duke and Parcher [150] have determined the stability constants of the species $Ce(OH)_2^{2+}$, $Ce(OH)_3^+$ and $(CeOCeOH)^{5+}$ by studying the Ce(IV)–Ce(III) exchange reaction at different acidities. Brubaker and Michel [151] estimated the stability constants of the species TlOH, $TlSO_4^+$ $[Tl(OH)_2^+]$ and $TlSO_4^-$ from the dependence of the Tl(III)–Tl(I) reaction on pH and sulphate concentration. Cohen, Sullivan and Hindman [152] found that the rate of the Np(VI)–Np(V) exchange reaction is not a monotonic function of the chloride ion concentration, and could interpret the observed rate by means of the following rate and equilibrium constants (Table 5.2):

TABLE 5.2

Species	Rate constant $1. mole^{-1}sec^{-1}$	Stability constant $1.^{-1} mole$
NpO_2^{2+}	174.8	—
NpO_2Cl^+	573	0.81
NpO_2Cl_2	229	0.20

Stability constants were determined from the effect of different ligands on the autoxidation of iron(II) [153] and tin(II) [154]. The catalytic reactions of hydrogen peroxide were frequently used for obtaining equilibrium data. Yatsimirskii and Alekseeva [155] have determined the stability constants of several molybdenum(VI) complexes from the inhibiting effect of the ligands on the Mo(VI)-catalysed hydrogen peroxide–iodide reaction. From the inhibition of the iodide–hydrogen peroxide reaction hydrogen peroxide decomposition reaction by fluoride [156] and EDTA [157] the corresponding stability constants have been determined. It must be mentioned, however, that the rate of decomposition of hydrogen peroxide at certain acidities is not a monotonic function of the EDTA concentration [158].

Equilibrium data have also been obtained from the reactions of molecular hydrogen. From the effect of fluoride on the silver(I)–hydrogen reaction the stability constant of AgF was obtained [159]. Halpern [160] observed that the rate of the chromium(VI)–hydrogen reaction catalysed by the copper(II)–glycine complex shows a maximum at a certain pH. He concluded that a diprotonated bis(glycinato)–copper(II) complex is the most effective catalyst. It is more probable, and equally consistent with the data at hand, that the high activity is attached to the monoglycinato complex, the mole fraction of which is of maximum value at this pH according to independent data.

Taube [161] derived, from the pH-dependence of the Mn(III)–catalysed oxalate–bromine reaction, the stability constants of the first three oxalato complexes of manganese(III), and from the effect of chloride and fluoride ions he also estimated the stability of the manganese(III) complexes of these ligands.

Lister [162] has determined the stability constant of tellurato and periodato copper(III) complexes from the inhibiting effect of the anions on the copper-catalysed decomposition of hypochlorite.

5.4.2.3 *Central ion exchange reactions*

If a metal ion forms a catalytically active complex with ligand A and an inactive one with ligand B, then from the rate data the equilibrium constant of the reaction

$$MeA + B \rightleftharpoons MeB + A$$

can be determined. If the stability constant of the MeB complex is known, that of the MeA complex can be obtained simply. This principle was applied [163] in determining the stability constant of the triethylenetetramine–iron(III) complex from the retarding effect of EDTA on hydrogen peroxide decomposition catalysed by this complex.

Quite analogous considerations hold for the inhibition of enzyme-catalysed reactions, which was used to obtain information on the stability of metal–enzyme complexes [164, 165].

5.4.3 THE KINETIC APPROACH TO STRUCTURAL PROBLEMS

The following examples illustrate that kinetic experiments also give an insight into structural problems. The rate of reaction of co-ordinated nitrite ions with different reactants depends on the configuration of the complex [166]. Therefore the spatial arrangements of the ligands in a complex can be judged from a comparative study.

The iodine–azide reaction is catalysed by thiocyanate ions. Even the co-ordinated thiocyanate ion has this catalytic effect. However, the catalytic efficiency of the co-ordinated ion depends on the nature of the donor atom, that is, on whether a thiocyanato or an isothiocyanato complex is formed [167].

As these examples show, no general rule can be given for the approach to structural problems from kinetic studies. It is clear, however, that careful study can provide important information.

5.5 Liquid–liquid partition

The partition of a solute between two immiscible liquids may be used to investigate complex equilibria. The distributed solute may be the ligand or a species containing metal ion. The condition that the two solvents – one of which is usually water or an aqueous solution – should be immiscible, is often not completely fulfilled. If the mutual solubility of the two liquids is appreciable, the change of the medium and its effect on the complex equilibria have to be considered. In the study of complex equilibria in an aqueous solution, the other liquid used is an organic solvent. The extraction of the species can be followed by many methods. Most generally, radioactive isotopes are applied. This method makes it possible to work at very low metal ion concentration, permitting the study of mononuclear complexes even though the tendency to form polynuclear complexes is strong.

Technically, extraction experiments are very simple. It is important that both liquids be saturated with each other prior to the extraction. To evaluate the equilibrium constants the distribution of a solute must be determined as a function of concentration changes. Each value of this function, that is each point of the distribution curve, is determined in a separate experiment. Very recently Reinhardt and Rydberg [168] devised 'an apparatus for continuous measurement of distribution factors in solvent extraction'. This instrument is now commercially available under the name AKUFVE, which is the abbreviation of the Swedish equivalent of that part of the last sentence that is in inverted commas. This apparatus consists of five parts: mixer, separator (a centrifuge with rotation speed of up to 18000 rpm), external liquid system, detectors and a measuring

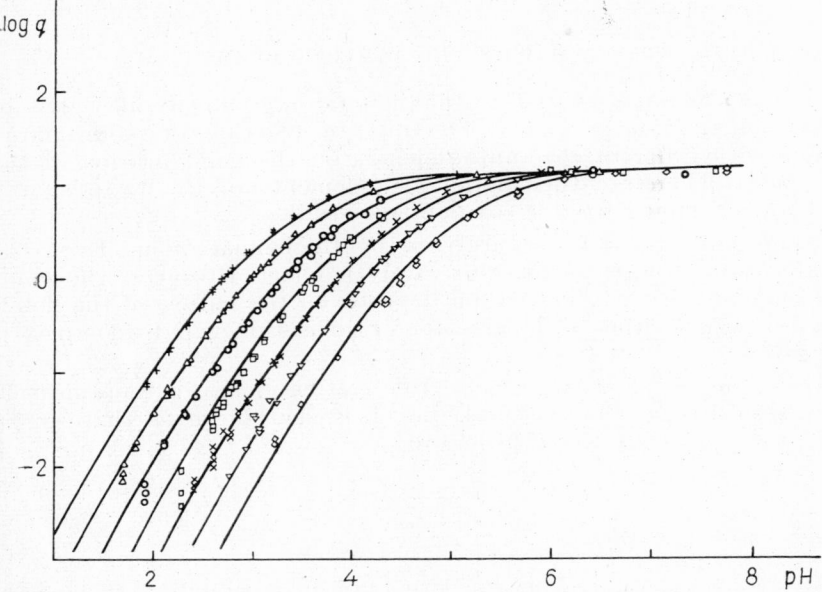

Fig. 5.13 Log q as a function of pH in the Cu(II)–acetylacetone–benzene–0.1 M NaClO$_4$ system obtained by *AKUFVE*; $+$ 0.0462 M acetylacetone; Δ 0.0223 M acetylacetone; ○ 0.0125 M acetylacetone; □ 0.00645 M acetylacetone; $\times$ 0.00329 M acetylacetone; ∇ 0.00169 M acetylacetone; ◇ 0.000852 M acetylacetone. All data were obtained and the curves automatically drawn within 25 hours. The computing time was only 10 minutes. (Unpublished figure kindly provided by Prof. J. Rydberg)

system. The value of this technique can be judged from the experiments on the copper(II)–acetylacetone (HAA) system (Fig. 5.13). Considering that the liquid–liquid extraction method is the one with the greatest potential in the study of complex equilibria, this new achievement seems to be very important.

The number of papers dealing with the different aspects and applications of liquid–liquid partition in solution chemistry is enormous and here we mention only some classical and some very recent papers. A recent book provides an account of the present state of the whole field [169].

5.5.1 Distribution of the ligand

If a neutral ligand or its conjugate acid can be extracted, while the metal complexes are not extractable, the formation of complexes can be followed by measuring the ligand distribution as a function of the metal ion (or ligand) concentration. The equations for the ligand distribution are as follows.

Distribution of the neutral ligand

$$q_\mathrm{L} = \frac{[\mathrm{L}]_\mathrm{org}}{[\mathrm{L}]_\mathrm{aq} + \overset{N}{\underset{1}{\sum}} i[\mathrm{MeL}_i]_\mathrm{aq}} . \tag{5.65}$$

Distribution of the electrically neutral conjugate acid

$$q_\mathrm{L} = \frac{[\mathrm{H}_n\mathrm{L}]_\mathrm{org}}{\overset{Z}{\underset{0}{\sum}} [\mathrm{H}_j\mathrm{L}]_\mathrm{aq} + \overset{N}{\underset{1}{\sum}} i[\mathrm{MeL}_i]_\mathrm{aq}} . \tag{5.66}$$

The principle was first used for determination of the dimerization constants of organic acids in benzene and in chloroform [170], and recently to determine acidic dissociation constants [171, 172]. Dawson and McCrae [173], as early as 1900, studied the formation of copper(II)–ammonia complexes by measuring the distribution of ammonia between the aqueous solution and organic solvents. More recently the complexes of hydrazine [174], pyridine [175], aniline [176], unsaturated hydrocarbons [177] and $(\mathrm{C}_2\mathrm{H}_5\mathrm{O})_2\mathrm{POOH}$ [178] were studied by this method.

5.5.2 Distribution of metal complexes

There may be many reasons for the extractability of metal ions and metal complexes by different organic solvents. If the ligand is charged and during successive complex formation an electrically neutral complex is formed, this may be extracted by an organic solvent of lower dielectric constant. Sometimes this uncharged complex is the last member of the series of successive complexes, but there are cases where further uptake of the ligand results in the formation of complex anions:

$$\mathrm{Me}^{m+} \overset{\mathrm{L}^-}{\longrightarrow} \mathrm{MeL}^{(m-1)+} \rightarrow \mathrm{MeL}_{m-1}^+ \rightarrow \mathrm{MeL}_m \rightarrow \mathrm{MeL}_{m+1}^- \rightarrow \mathrm{MeL}_N^{(m-N)}$$
$$\tag{5.67}$$

In most cases the anionic species are not extracted, so the curves of distribution ratio *vs.* ligand concentration will indicate whether anionic complexes are formed or not (Fig. 5.14). The distribution ratio is the ratio of the total concentrations of the metal ion species in the two phases:

$$q = \frac{\overset{N}{\underset{0}{\sum}} [\mathrm{MeL}_i]_\mathrm{org}}{\overset{N}{\underset{0}{\sum}} [\mathrm{MeL}_i]_\mathrm{aq}} . \tag{5.68}$$

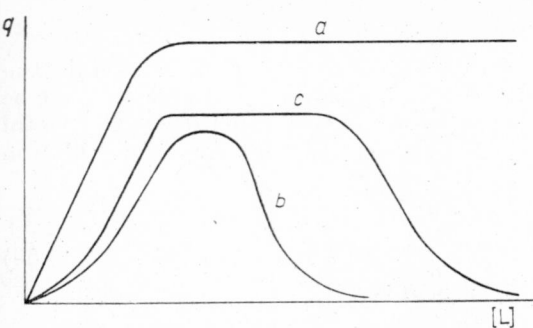

Fig. 5.14 q as a function of free ligand concentration. a — the co-ordinatively saturated complex is electrically neutral; b — complex anions are formed; c — complex anions are formed, the electrically neutral complex is of high stability

If MeL_m is the only complex present in appreciable concentration in the organic phase,

$$q = \frac{[MeL_m]_{org}}{\sum\limits_{0}^{N} [MeL_i]_{aq}} = \frac{[MeL_m]_{org}}{T_{Me_{aq}}} . \tag{5.69}$$

Then if $m = N$ the $q([L])$ function is monotonic, while if $N > m$ it has a maximum. If the stability of the neutral complex is high enough, the q vs. $[L]$ curve has a definite plateau. Considering that the distribution constant of the species MeL_m is

$$D = \frac{[MeL_m]_{org}}{[MeL_m]_{aq}} \tag{5.70}$$

the concentration of MeL_m in the aqueous phase can be expressed by means of the stability constants:

$$[MeL_m]_{aq} = T_{Me} \frac{\beta_m [L]_{aq}^m}{\sum\limits_{0}^{N} \beta_i [L]_{aq}^i} . \tag{5.71}$$

Therefore

$$q = \frac{D \beta_m [L]_{aq}^m}{\sum\limits_{0}^{N} \beta_i [L]_{aq}} . \tag{5.72}$$

It appears from Eq. 5.72 that for calculation of N unknown stability constants at least $N + 1$ distribution experiments are needed. If this curve shows a long horizontal portion, the value of D can be directly obtained:

$$D = q_{max}. \tag{5.73}$$

The determination of the stability constants is particularly easy if a maximum occurs on the q vs. [L] curve, because the formation of cationic and anionic complexes can be treated separately.

The principle was first used by Morse [179] for studying halide complexes of mercury(II), and the fundamental problems of the determination and computation have been treated by Rydberg [180–183], Dryssen [184–186] and Irving [187, 188].

When the same complex system is studied but with different organic solvents, it follows from Eq. 5.72 that the same stability constants should be obtained. If the difference of the stability constants is larger than the experimental error, this is an indication of an interaction between the complexes and the solvent molecules. Such an interaction is indicated by the q values. According to Fig. 5.15, the ratio $q_{CHCl_3}/q_{C_6H_6}$ for the thorium (IV)–acetylacetone system is constant at the same free ligand concentration, but the ratio $q_{C_6H_6}/q_{C_4H_9COCH_3}$ varies considerably with it. This behaviour could be quantitatively explained by the formation of thorium(IV)–acetylacetone–isobutyl methyl ketone mixed-ligand complexes.

In the case of organic solvents of low dielectric constants, the assumption that only the neutral complex exists in the organic phase is a very good approximation. It does not necessarily mean, however, that only one complex species is present in the organic phase. The negative charge(s)

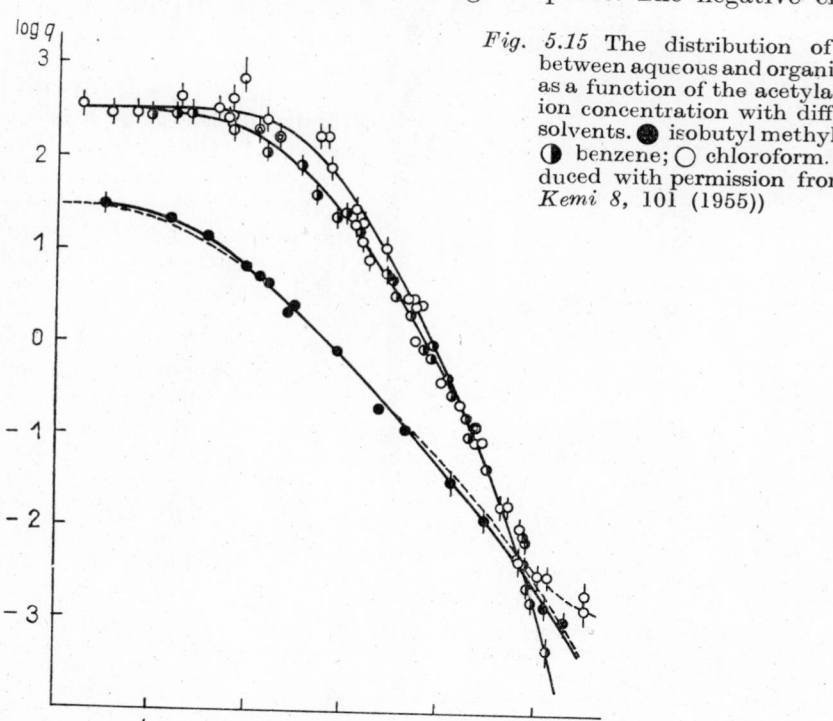

Fig. 5.15 The distribution of Th(IV) between aqueous and organic phases as a function of the acetylacetonate ion concentration with different solvents. ● isobutyl methyl ketone; ◑ benzene; ○ chloroform. (Reproduced with permission from *Arkiv Kemi 8*, 101 (1955))

on the anionic complexes may be compensated by the proton-uptake. It is well-known that a number of metal halides can be extracted in the protonated form, $HMeX_4$. The relationships are even more complicated if the dielectric constant of the organic solvent is not very low. These systems may find very important practical applications but are not suitable for the evaluation of the stability constants in aqueous solutions.

In recent years many extraction systems have been thoroughly studied [189]. The alkyl phosphate esters are particularly important organic solvents, and frequently extract the metal cation itself into the organic phase.

5.5.3 SYNERGISM

This term was originally applied to an enhancement of the extraction of metal ions by an organic solvent system containing different ligands, one being acidic and neutralizing the charge on the metal ion, and the other being a neutral donor molecule [190]. It was first thought that the phenomenon was restricted to very few metal ions and ligand systems, but later work showed it to be quite general. Although many factors must be considered, the work of Irving [191, 192], Healy [193, 194], Newman [195], Li [196], Taube [197] and others has clearly shown that the formation of mixed ligand complexes is decisive. The formation of mixed ligand complexes may increase the fraction of the metal ion concentration bound in the complex and may result in a favourable change in the extractability of the species. A characteristic synergic behaviour is shown in Fig. 5.16 for the cobalt(II)–thenoyltrifluoroacetone-pyridine system [198]; similar behaviour

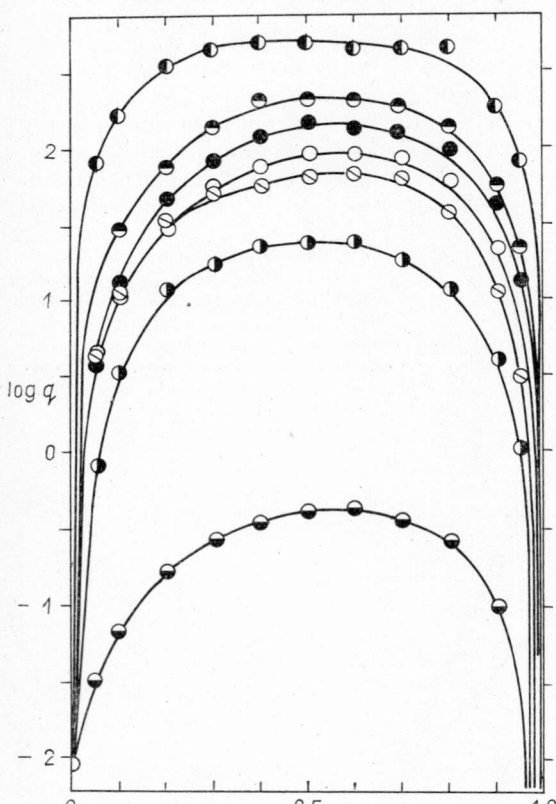

Fig. 5.16 The distribution of ^{60}Co between acetate buffer of pH 4.93 and mixtures of thenoyltrifluoroacetone (TTA) and a heterocyclic base (Q) in cyclohexane. $T_{TTA} + T_Q = 0.02$ M; ◗ isoquinoline; ◖ 3-methylpyridine; ● 4-methylpyridine; ○ pyridine; ◔ 3-chloropyridine; ◑ quinoline; ◓ 2-methylpyridine. (Reproduced with permission from *Solvent Extraction Chemistry* edited by D. Dyrssen, J. O. Liljenzin and J. Rydberg, North-Holland, Amsterdam, 1967, p. 91)

is exhibited if other bases like pyridine are used. The reason for the enhancement of the extraction by pyridine and its analogues is that two molecules of them are co-ordinated to the planar bis-chelate. A further synergistic effect was found by Irving [199] by applying two different organic bases. In this case the following equilibrium must be considered:

$$Co(TTA)_2A_2 + Co(TTA)_2B_2 = 2\,Co(TTA)_2AB,$$

characterized by the equilibrium constant

$$K = \frac{[Co(TTA)_2AB]^2}{[Co(TTA)_2A_2]\,[Co(TTA)_2B_2]} \tag{5.74}$$

Figure 5.17 shows the calculated distribution curves for the extraction of bis(thenoyltrifluoroacetone)–cobalt(II) into cyclohexane by mixtures of

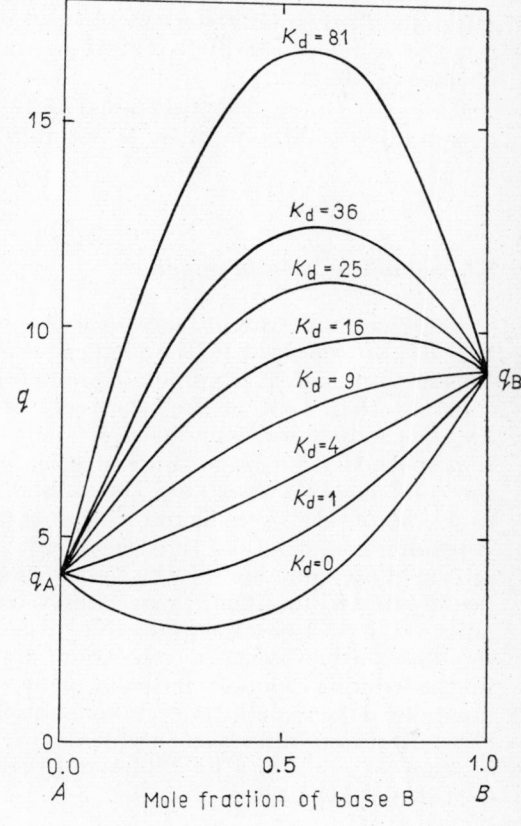

Fig. 5.17 Calculated distribution curves for the extraction of a chelated complex MeL_2 with synergic combination of two bases A and B such that q_A and q_B represent the distribution coefficients for MeL_2A_2 and MeL_2B_2 alone and K_d is the disproportionation constant. (Reproduced with permission from *Solvent Extraction Chemistry* edited by D. Dyrssen, J. O. Liljenzin and J. Rydberg, North-Holland, Amsterdam, p. 91)

two bases. It is interesting that if the value of the equilibrium constant is less than 4 (the statistical value), a minimum appears. This is the consequence of the fact that the concentration of $Co(TTA)_2A_2$ depends on the second power of the mole fraction of the ligand A. With increasing K_d, the synergic effect becomes more and more important (see Chapter 8).

5.5.4 COMPETITIVE REACTIONS

The method of liquid–liquid extraction is well suited to the study of competitive reactions. Although in principle it seems possible to follow the competition between two metals for a common ligand, so far only the competition between two ligands for the same central ion has been studied by solvent extraction. Connick and McVey [200] have determined the stability constants of a number of zirconium(IV) complexes, by measuring the hindering effect of some inorganic anions on the extraction of zirconium(IV)–thenoyltrifluoroacetone complex into benzene. Similar studies have been made with neptunium(IV) sulphate complexes by Sullivan and Hindman [201]. Many workers have studied hydrolytic reactions by measuring the influence of hydroxide ions on the extraction of different metal chelates [202–204].

Recently Starý [205] systematically studied competitive reactions in determining a large number of stability constants by liquid–liquid extraction.

5.6 Solubility measurements

The change of the charges of the species on successive complex formation is also reflected in the change of the solubility. The solubility of the electrically neutral complex is much less in solvents of high dielectric constant than that of the positively or negatively charged complex ions. The method shows some features analogous to those of the solvent extraction method. The basic difference is that the activity of the complex in the solid phase is constant. The solubility may reach a limiting value as a function of the free ligand concentration or may exhibit a minimum, depending on whether the electrically neutral complex is the last or an intermediate member of the series of successive complexes. For reliable results to be obtained, two criteria must be fulfilled. First, the system must be in a state of equilibrium. Reaching the equilibrium state usually requires longer than it does in liquid–liquid extraction. Second, any change in the composition of the solid phase at different ligand concentrations must be determined, and taken into account in the evaluation of the constants. Solubility measurements were very early applied to the study of complex equilibria by Bodländer [206], but the potential of the method is still not exhausted.

5.6.1 CALCULATION OF CONSTANTS FROM THE SOLUBILITY MINIMUM

If only two, oppositely-charged, species Me^{m+} and MeL_s^{m-sl} exist it is possible to calculate the stability constant from the position of the solubility minimum [207]. For the solubility minimum, that is at the isoelectric point [208], we can write

$$m[Me_{m+}] = (m - sl)\,[MeL_s^{m-sl}] \tag{5.75}$$

therefore

$$K = \frac{MeL_s^{m-sl}}{[Me^{m+}]\,[L^{l-}]^s} = \frac{m}{(m - sl)\,[L]_{iso}^s}. \tag{5.76}$$

It is possible to calculate several equilibrium constants from the position of the solubility minimum alone, if reasonable assumptions can be made about the ratios of successive constants. For example the symmetry of the EDTA solubility curve (solubility versus pH) near the isoelectric point suggested that the ratio of the dissociation constants of the positively-charged species H_5Y^+ and H_6Y^{2+} is equal to the ratio of the dissociation constants of the two anionic species H_3Y^- and H_2Y^{2-} [209]. The validity of this assumption was verified later on by independent experiments[210].

5.6.2 CALCULATION OF THE STABILITY CONSTANTS FROM THE SOLUBILITY CURVE

A method was developed by Sano [211] for the determination of the acidic dissociation constants of amino-acids. Essentially the same method was applied by Irving and Ewart [212] for the determination of dissociation constants of 8-hydroxyquinoline. The principles of the solubility method for the determination of successive complexes were established mostly by Leden [213–215], Ahrland [216] and Lieser [217], while Težak [218] gave an approximate graphical procedure. The mole fraction of the electrically neutral complex (MeL_s) is expressed by Eq. 5.71. The concentration of this complex in the solution is constant in the ligand concentration range for which the solid phase consists of this particular complex. The concentrations of the complexes can be expressed as functions of the free ligand concentration and this constant quantity:

$$[MeL_i^{m-il}] = [MeL_s]\,\frac{\beta_i}{\beta_s}\,[L]^{i-s}. \tag{5.77}$$

Therefore the solubility S is given by

$$S = \sum_0^N [MeL_i] = [MeL_s] \sum_0^N \frac{\beta_i}{\beta_s}\,[L]^{i-s}. \tag{5.78}$$

Equation 5.78 can be rearranged to

$$\frac{S}{[MeL_s]} = \sum_0^N \frac{\beta_i}{\beta_s}\,[L]^{i-s}, \tag{5.79}$$

which can easily be solved graphically if the concentration of MeL$_s$ is known. This is the case when the minimum exhibits a flat portion, which gives directly this constant concentration. The analysis of the solubility curve also gives this value if the minimum is sharp.

An interesting possibility for the determination of successive constants was pointed out by Haight [219]. He synthesized a number of salts of low solubility, of general formula $(R_4N^+)(MeL_{m+1}^-)$, where Me is Hg(II), Sn(II), Pb(II), Bi(III) and Sb(III), and L is a halide. If the tetra–alkyl-ammonium ion is not added to the solution in excess, the amount of the dissolved complex salt is

$$S = T_{Me} = T_{R_4N^+} = [MeL_{m+1}^-] \sum_0^N \frac{\beta_1}{\beta_{m+1}} [L]^{i-m+1} . \qquad (5.80)$$

If the solubility product of $(R_4N^+)(MeL_{m+1}^-)$ is

$$K_s = [R_4N^+][MeL_{m+1}^-] \qquad (5.81)$$

then Eq. 5.80 can be written as

$$S = \frac{K_s}{T_{R_4N^+}} \sum_0^N \frac{\beta_i[L^-]^i}{\beta_{m+1}[L^-]^{m+1}} \qquad (5.82)$$

or

$$S^2 = K_s \sum_0^N \frac{\beta_i[L^-]^i}{\beta_{m+1}[L^-]^{m+1}} . \qquad (5.83)$$

Since

$$\bar{n} = \frac{\sum_0^N i[MeL_i^{m-i}]}{S} \qquad (5.84)$$

Eqs. 5.82 and 5.83 can be written as

$$\bar{n}S = \frac{K_s}{S} \sum_0^N \frac{i\beta_i[L^-]}{\beta_{m+1}[L^-]^{m+1}} \qquad (5.85)$$

and

$$\bar{n}S^2 = K_s \sum_0^N \frac{i\beta_i[L^-]^i}{\beta_{m+1}[L^-]^{m+1}} . \qquad (5.86)$$

By differentiation we get

$$\frac{dS^2}{d[L^-]} = \frac{K_s}{[L^-]} \sum_0^N \frac{(i-m+1)[L^-]^i}{\beta_{m+1}[L^-]^{m+1}} . \qquad (5.87)$$

Combining Eqs 5.82–5.87 we get

$$\frac{[L^-]dS^2}{d[L^-]} = \bar{n}S^2 - (m+1)S^2 \qquad (5.88)$$

and

$$\frac{d \log S^2}{d \log [L^-]} = \bar{n} - m + 1 . \qquad (5.89)$$

It follows from similar reasoning that

$$\frac{d \log (S^2[L^-])}{d \log [L^-]} = \bar{n} - m + 2 \tag{5.90}$$

and

$$\frac{d \log \dfrac{S^2}{[L^-]}}{d \log [L^-]} = \bar{n} - m \tag{5.91}$$

or generally that

$$\frac{d \log (S^2[L^-]^{m+1-n})}{d \log [L^-]} = \bar{n} - n. \tag{5.92}$$

By this procedure the integer values of the complex formation function can be obtained. For these points the free ligand concentration can be directly calculated. For the other $\bar{n}$ values the corresponding free ligand concentrations can be estimated from

$$[L^-] = \frac{[L^-]_0}{1 + 2\, dS/d[L]}, \tag{5.93}$$

where $[L]_0$ stands for the ligand concentration before the complex salt is dissolved. Haight applied this method for several complex systems [220–222].

Solubility measurements are also very suitable for study of competitive reactions [223].

5.7 Ion-exchange measurements

The possibility of the application of ion-exchange to detect the formation of complex ions and to determine their stability arises from the change of the charge of the species during stepwise complex formation. Since complex species exist both in the solution and in the resin phase the relationships in most cases are too complicated for a complete and exact analysis of the results. Further, adsorption (chemisorption) effects are superimposed on the real ion-exchange reaction and the constant ionic medium principle cannot be applied. Therefore some more or less arbitrary assumptions must be introduced, which of course decrease the reliability of the stability constants obtained by these methods. Nevertheless ion-exchange investigations are extremely important because ion-exchange resins are widely applied for practical purposes ranging from analytical chemistry to large scale industrial processes.

For the details of the experimental technique and the description of the commercially available ion-exchange resins the reader should consult the appropriate handbooks [224–226]. A recent well-documented review by Marcus [227] gives a detailed treatment of ion-exchange studies of complex formation.

5.7.1 Application of Cation-Exchange Resins

The experimentally available characteristic quantity is the distribution Φ of the metal ion between the resin phase and the solution:

$$\Phi = \frac{\text{total amount of Me per g of air-dried resin}}{\text{total amount of Me per ml of solution}}, \quad (5.94)$$

and Φ is determined by the stability of the successive complexes and the equilibrium constant $(_rK_n)$ of the exchange reactions

$$\text{MeL}_n^{m-nl} + (m + nl)\,_rC = {_r}\text{MeL}_n^{m-nl} + (m - nl)\,C^+. \quad (5.95)$$

(The subscript r indicates the resin phase, C^+ is a non-complex forming univalent cation.) Therefore the distribution constant of the different species

$$\varphi_n = \frac{[_r\text{MeL}_n^{m-nl}]}{[\text{MeL}_n^{m-nl}]} \quad (5.96)$$

can be written as

$$\varphi_n = \beta_n \cdot {_rK_n} \left(\frac{_rC^+}{C^+}\right)^{m-nl} \quad (5.97)$$

and the gross distribution (Φ) can be expressed by the equation

$$\Phi = \frac{\sum\limits_0^N \varphi_i[\text{L}]^i}{\sum\limits_0^N \beta_i[\text{L}]^i}. \quad (5.98)$$

Because of the fairly large experimental uncertainty in the distribution values, the solution of Eq. 5.98 for the general case is not possible. However, if as Schubert assumed [228, 229] the only distributing species is the free metal ion, Eq. 5.98 can be written in a simple form

$$\frac{\varphi_0}{\Phi} = \sum\limits_0^N \beta_i[\text{L}]^i \quad (5.99)$$

which permits the evaluation of the stability constants by the well-known mathematical methods. Methods have been developed for the case where the adsorption of one or two cationic complex species is also considered [230–234], but the values obtained by these methods must be cautiously dealt with. Recently Loman and Van Dalen [234] suggested a comparative study with a number of cation-exchangers differing in the degree of cross linking. They claim that the accuracy of this method is comparable with that of potentiometric work.

If positively-charged complexes are formed and adsorbed on the cation-exchanger, the measurement of the distribution of the ligand between the resin and the solution gives semi-quantitative information on the stability of the complexes.

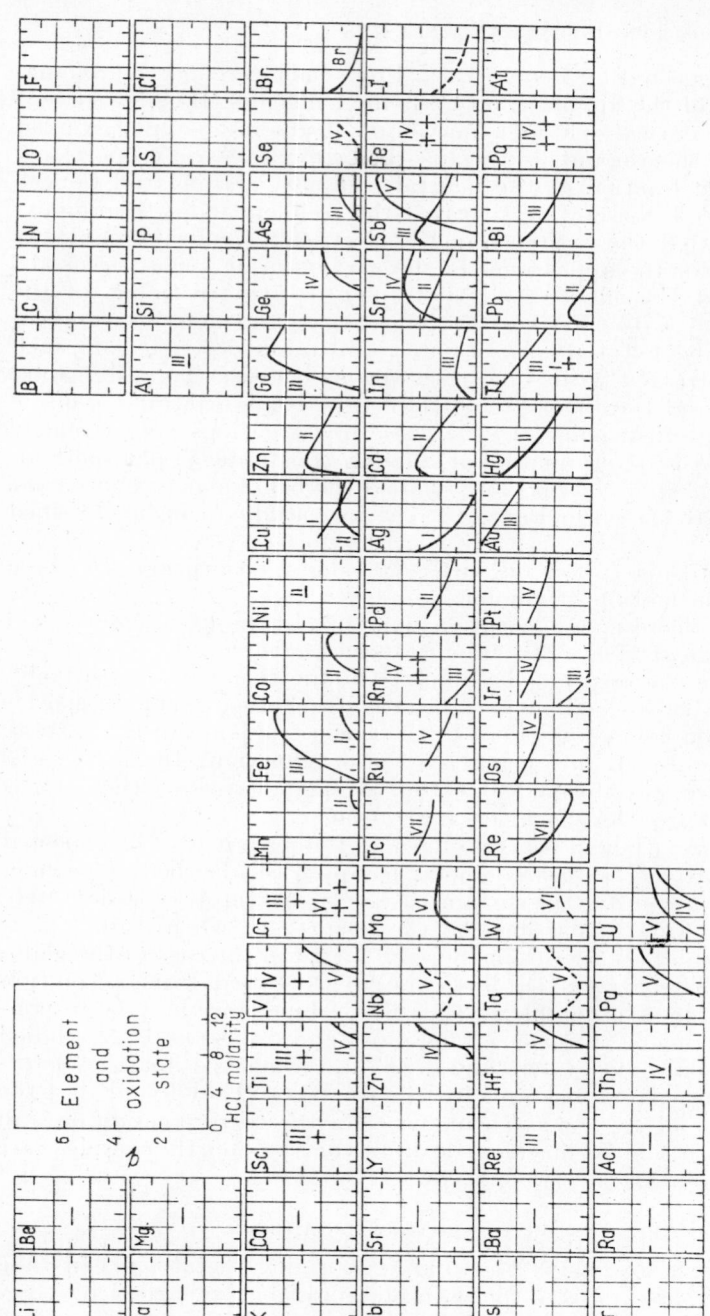

Fig. 5.18 Distribution of metal ions between the aqueous and anion-exchange resin phases as a function of HCl concentration. − no sorption; + weak sorption; ++ strong sorption. (Reproduced with permission from *Proc. Intern. Conf. Peaceful Uses Atomic Energy, Geneva,* 1955 Vol. 7 p. 113)

5.7.2 APPLICATION OF ANION-EXCHANGE RESINS

This was first suggested by Leden [235]. Fronaeus [236] laid the foundation of the theory of the application of anion-exchangers. He showed that the φ versus $[L^-]$ curves exhibit a maximum if anionic complexes exist and suggested that the amounts of the adsorbed complexes are proportional to the partial mole fraction of the neutral complex. Marcus and Coryell [237] and Kraus and Nelson [238] treated the problem more thoroughly. They pointed out that the existence of the electrically neutral complex is not a prerequisite for the appearance of the maximum on the φ versus $[L]$ curve. Marcus and Coryell claimed that quantitative treatment of the systems is possible on the basis of the effective ligand activity concept. However, it may be felt that the stability constants derived from such studies should be treated with some reservations because the systems are very complicated and the theory is not free from some arbitrary assumptions. From the practical point of view these investigations are extremely important, but for the determination of stability constants simpler methods can be found. Kraus and Nelson [239] studied the behaviour of many ions, and their results are shown in Fig. 5.18. The ions studied can be classified in four groups.

(*i*) Ions with no adsorption or with only slight adsorption. In these cases there is no formation of anionic complexes.

(*ii*) Ions with a decreasing adsorption function. In the hydrochloric acid concentration range studied only anionic species exist.

(*iii*) Ions with an adsorption maximum. On the rising part positively-charged, and on the descending part negatively charged complex species exist, the maximum being equivalent to the isoelectric point of the system.

(*iv*) Ions with only an increasing adsorption function. In these cases complex anions are also formed, but the average charge of the ions is positive at any hydrochloric acid concentration.

Because of the great ligand concentration on the resin it may be expected that the co-ordinatively saturated metal complex is adsorbed. This may hold for many systems, but in certain cases this assumption is definitely disproved: e.g. $Ag(CN)_3^{2-}$ is sorbed on the resin [240] when $Ag(CN)_4^{3-}$ is the predominant complex species in the solution, or in the case of the gadolinium(III)–glycollate system the bis complex is sorbed from a solution containing mostly the tris complex [241]. Steric hindrance may be responsible for the phenomena. Fronaeus, Lundquist and Sonesson [242] studied the cadmium(II)–bromide system and developed a method for the determination of the successive constants in the resin phase. They found that the values of K_1, K_2 and K_3 are much higher than the corresponding ones in the solution, while the formation constants of the fourth complex are practically the same in the two phases.

5.7.3 STUDY OF COMPLEXES WITH NEUTRAL LIGANDS

Cation-exchangers loaded with a complex-forming metal ion can take up neutral ligands, or cations of the resin phase can be exchanged by complex ions containing a neutral ligand. In fact the metal ions are sorbed

— if other complex forming ligands are absent — in the form of the aquo-complexes. Walton [243–245] has thoroughly studied the different metal-ammine systems. It was found that the stability of the ammine complexes is greater in the resin phase than in the solution, but in most cases the formation of higher complexes was hindered. This effect is shown in Fig. 5.19. It is assumed that polymeric complexes exist in the resin phase. A fairly general mathematical treatment of these systems was given by Helferich [246].

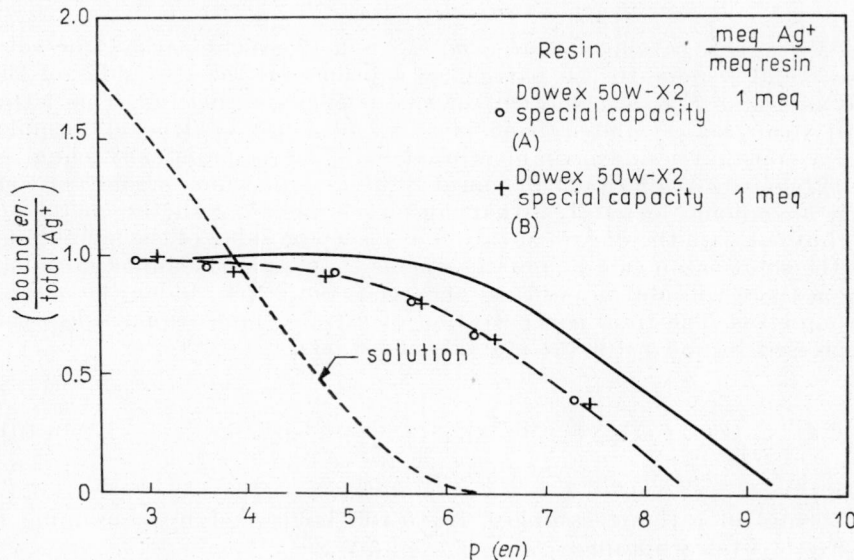

Fig. 5.19 Silver(I)–ethylenediamine complexes on special capacity partially sulphonated Dowex 50 resins. Resin (A) capacity 1.63 meq/g of dry H-resin; resin (B) capacity 2.32 meq/g of H-resin. (Reproduced with permission from *J. Phys. Chem. 66*, 78 (1962))

5.7.4 APPLICATION OF LIQUID EXCHANGERS

In the last decade much work has been devoted to the application of the so-called liquid exchangers. These are tertiary amines with long aliphatic chains, or various organic derivatives of phosphoric acids. Their application to obtain quantitative information on complex equilibria is much more convenient than that of ion-exchange resins and usually the uncertainty of the data obtained is much less. However, from the point of view of the theory and the methods of calculation, these systems may be regarded as liquid–liquid extraction systems, so the treatment given in Section 5.5 is valid for them also.

5.7.5 STUDY OF THE DONNAN MEMBRANE EQUILIBRIUM

If two electrolyte solutions are separated by a selective membrane that is ideally permeable to cations but impermeable to anions and solvent molecules, the anion concentrations on both sides must remain constant. The distribution of the cations is governed by the Donnan equilibrium. Therefore at equilibrium

$$\frac{[Me^{m+}]_s}{[Me^{m+}]_p} = \left(\frac{[C^+]_s}{[C^+]_p}\right)^m \cdot F \qquad (5.100)$$

where Me^{m+} is a cation of charge m, C^+ is a univalent cation, the subscripts s and p refer to the particular solutions on the two sides of the membrane, and F is the quotient of the activity coefficients. This latter quantity can be calculated from the extended Debye–Hückel formula, and if a constant ionic medium is used its value is practically equal to unity. Following Wallace [247], consider the case in which a solution containing an anionic ligand L^{l-} that forms complexes with the metal ion Me^{m+} but not with the univalent cation, is placed on side s of the membrane, while the solution on side p contains anions that are not complex-forming. If $l \geq m$ there will not be positively-charged complexes among the successive complexes. The total concentration of metal complexes on side s can be expressed by means of the stability constants:

$$T_{Me_s} = [Me^{m+}]_s \left(1 + \sum_0^N \beta i [L]^i\right). \qquad (5.101)$$

There being no cationic complex, Eq. 5.100 holds and by combining it with Eq. 5.101 we obtain

$$\frac{R-1}{[L]} = \sum_0^N \beta_i [L]^{i-1}, \qquad (5.102)$$

where R is defined as

$$R \equiv \frac{T_{Me_s}}{[Me^{m+}]_p} \cdot \left(\frac{[C^+]_p}{[C^+]_s}\right)^m. \qquad (5.103)$$

The concentration of uncomplexed Me^{m+} in solution s can be determined from its concentration in solution p by use of Eq. 5.100. This method has two advantages in comparison with the ion-exchange method. First, concentration determinations have to be made in solutions only; second, it is not necessary to consider complex formation in the resin phase.

Wallace studied the UO_2^{2+}–SO_4^{2-} system, applying an AMFion C-103 membrane and using tracers for the concentration determinations. A fair agreement was found between the constants evaluated from that study and those determined by other methods.

5.8 Potentiometric methods

The concentration changes caused by complex formation are reflected in the potential of well-chosen electrodes. The various potentiometric measurements are very frequently applied in the study of complex equilibria. In fact, most of the recorded stability data were determined by potentiometric methods. By potentiometric measurements both the metal ion and the ligand activities (concentrations) can be determined. It is a prerequisite of the application of potentiometric methods that the electrode reaction be reversible. In this case the activity (concentration) in question can be calculated from the measured potential by means of the Nernst equation.

There are two possibilities for the performance of the experiments: either a solution is prepared for each potential measurement or one solution is titrated with another and the potential is measured after each addition of titrant. The latter method is evidently much more rapid and with careful performance of the experiments is almost as exact as the more laborious first method, so it is much more popular.

For the potentiometric measurements an electrode must be selected, the potential of which is a well-defined function of the concentration of the ion to be determined, and also a reference electrode. The electrode, together with the solution in which it is immersed, comprises a half-cell. For the measurement of the potential this has to be combined with another half-cell containing the reference electrode. If the two electrodes are in contact with the same solution and respond to different constituents of the solution the electromotive force of the cell is the difference between the potentials of the two half-cells. If, however, the two electrodes are immersed in two different solutions, an electrolytic connection must be made between the two half-cells. This liquid junction has the result that the measured potential is not equal to the difference of the potentials of the two half-cells. The liquid junction potential is caused by the different mobilities of the different ions. Therefore potassium chloride solutions are frequently used to minimize the liquid junction potential, since the mobilities of the potassium and chloride ions are similar. It is advisable to avoid liquid junctions if possible. If this is impossible, it is necessary to use well-constructed salt bridges. The so-called 'Wilhelm' reference half-cell [248] with liquid junction is certainly one of the best. (The half-cell is named after the first name of the first author.)

5.8.1 DETERMINATION OF METAL ION CONCENTRATION

For the potentiometric determination of metal ions the following electrodes can be used:

(i) metal electrodes, including amalgam electrodes;
(ii) redox electrodes;
(iii) membrane electrodes;
(iv) electrodes of the third kind.

The order reflects the present frequency of their application. Metal electrodes are the most popular, but amalgam electrodes are sometimes more
convenient, especially because the time for reaching constant potential is
usually much shorter than in the case of metal electrodes. It must be taken
into account, however, that the potential is also influenced by the concentration of the amalgam.

If the metal in question is of variable oxidation state the following
equilibrium must be considered

$$Me(s) + Me^{m+} = 2 Me^{(m-1)+}.$$

In the case of copper, gold and indium this equilibrium is completely shifted
toward the left and therefore does not disturb the measurements. Just the
opposite is the case for thallium, tin and lead, and therefore the concentrations of the ions of the higher oxidation state cannot be measured conveniently. Mercury lies between these two extremes.

Some examples of the application of metal (and amalgam) electrodes
follow. A cadmium electrode was used by Flengas [249] for the study of
the $Cd^{2+}-CN^-$ system, and a cadmium amalgam electrode by Leden [250]
for the Cd^{2+}-halide system. A mercury electrode was used for the study
of the stability of bromide [251] and iodide [252] complexes of mercury(II)
at higher ligand concentrations. The $Ag^+-S_2O_3^{2-}$ system was studied by
Nilsson [253] with a silver electrode, the Cu^+-Cl^- system by Náray-Szabó
and Szabó [254] and the Cu(I)-diphenylphosphinobenzene-m-sulphonate by
George and Bjerrum [255], with a copper electrode. Amalgam electrodes
were used by Biedermann [256] in the investigation of the hydrolysis
of indium(III) and by Granér and Sillén [257] and by Olin [258] in that of
bismuth(III). A bismuth amalgam electrode was also used for the determination of the stability constants of bismuth(III)–halide complexes by
Ahrland and Grenthe [259] and for of the bismuth(III)–diphenylphosphino-
benzene-m-sulphonate complexes by Bjerrum and Wright [260]. Budevsky
and Platikanova [261] have recently studied the lead(II)–dithiocarbamino-
acetate system, using a lead amalgam electrode.

Redox systems are also frequently used in the investigation of complex
systems. Usually bright platinum is used as an inert electrode. The potential
of the Co^{2+}/Co^{3+} redox system was measured in studying the stability of
various cobalt(III)–amine complexes [262] and cobalt(III)–biguanide [263]
complexes.

By means of a suitable competitive reaction it is possible to use the
potentiometric method for the study of systems which themselves do not
give a reversible potential. For example the stability constants of alumi-
nium(III)–fluoride complexes were determined by Brosset and Orring [264]
from the effect of aluminium ions on the potential of the $Fe^{3+}-Fe^{2+}-F^-$
system. The equilibrium concentration of iron(III) and hence the potential
of the system is determined by the substitution reaction

$$FeF_n^{3-n} + Al^{3+} \rightleftharpoons AlF_n^{3-n} + Fe^{3+}$$

Schmid and Reilley [265] have determined the stability constants of the EDTA complexes of a number of metal ions by following the equilibrium of the reaction

$$HgY^{2-} + Me^{2+} \rightleftharpoons Hg^{2+} + MeY^{2-}$$

(Y^{4-} stands for ethylenediaminetetra-acetate), using a mercury electrode. The potential of the mercury electrode at 25°C is

$$E = E^0_{Hg} + 0.0269 \log [Hg^{2+}]. \tag{5.104}$$

Considering the stability constants of the HgY^{2-} and MeY^{2-} complexes, the potential of the $Hg^{2+}-Me^{2+}-Y^{4-}$ system is given by

$$E = E^0_{Hg} + 0.0296 \log \frac{[Me^{m+}][HgY^{2-}]}{[MeY^{2-}] K_{HgY}} + 0.0296 \log K_{MeY}. \tag{5.105}$$

For the calculation of K_{MeY} the value of K_{HgY} is necessary. This can be determined by measuring the potential of a mercury electrode in a solution containing the mercury(II) chelate and the free chelating agent. The same principle can be applied to different complex forming agents and different suitable metal electrodes.

There are many different types of membrane electrode. Although the theory of membrane electrodes is not quite clear, the origin of the potential of these electrodes is evidently the different mobilities of the ions in the membrane, these in turn being the consequence of the specific interactions between the ions and the functional groups of the membrane. A membrane electrode consists of a membrane and an internal reference solution in which a reference electrode, usually an Ag/AgCl electrode, is immersed. A detailed review of the earlier literature was given by Pungor and Havas [266]. Great progress has been achieved in recent years by using binding materials with favourable mechanical properties, e.g. silicone rubber. Glass electrodes made from special glasses respond to different univalent ions. This property was used by Palaty [267] in determining the stability constant of the weak sodium complex of EDTA. Rechnitz and his co-workers studied the sodium ion-malate [268] and sodium ion-citrate [269] systems, using a cation-sensitive glass electrode. Liquid ion-exchangers can also be applied as membrane electrodes. The liquid ion-exchanger is separated from the external solution by an extremely thin, porous, inert membrane disc. For a calcium-sensitive liquid-membrane electrode a calcium salt of an organophosphoric acid derivative is used [270]. Nakayama and Rasnick have recently studied [271] the dissociation and stability of calcium sulphate by means of such an electrode. Membrane electrodes are especially important in the investigation of complex equilibria involving alkali and alkaline earth metal ions since metal electrodes or redox electrodes cannot be applied in these cases.

Electrodes of the third kind [272] which can be represented by Me^{m+}, MeX(s), MX(s)/M, have been used to study of very few equilibria. Recently

Berecz and Szita [273] developed an electrode [Al^{3+}, $Al(OH)_3$, $Hg(OH)_2/Hg$] responding to aluminium ions, and studied the equilibria in the NaOH–$Al(OH)_3$ system.

From the potentiometric data, stability constants can be derived most conveniently by the Leden [274] or the Fronaeus [275] method. It is also possible [276, 277] to apply the principle of corresponding solutions to the evaluation of stability constants from potentiometric data.

5.8.2 DETERMINATION OF LIGAND CONCENTRATIONS

For the potentiometric determination of ligand concentrations the following electrode types can be used:

(*i*) electrodes of the second kind;
(*ii*) membrane electrodes;
(*iii*) redox electrodes.

If the electrode reaction

$$M_l L_z(s) + zl\,e^- = zM(s) + z\,L^{l-}$$

is reversible, the concentration of the ligand can be determined by means of an electrode of the second type, $M/M_l L_z$, L^{l-}. This type of electrode is frequently used as a reference electrode [278]. For the determination of halide ions the silver/silver halide electrodes can be conveniently applied. Anderegg [279] has studied the stability of the complexes of 1,10-phenanthroline by applying the 1,10-phenanthroline, $Hg_2(phen)_2(NO_3)_2(s)/Hg(l)$ half-cell.

Many membrane electrodes have been devised for the determination of ligands [280]. Anion-exchange resins and various precipitates embedded in suitable carrying materials are used for this purpose. Membrane electrodes can be used for the determination of Cl^-, Br^-, I^-, S^{2-}, CN^-, SO_4^{2-}, PO_4^{3-}, ClO_4^-. The iodide-sensitive membrane electrode was used by Burger and Pintér [281] for studying the formation of I_3^- and BiI^{2+} ions. According to our experience electrodes made with silicone rubber deteriorate if they are immersed in organic solvents. For the determination of fluoride, a unique electrode was developed by Frant and Ross [282], in which the membrane was a laser-type, rare-earth doped single crystal. It was used by Lingane [283] for analytical purposes and the excellent results indicate that its application in the study of the stability of fluoride complexes will be very fruitful. Some membrane electrodes may well be applicable as reference electrodes without liquid junction.

Shulman *et al.* [284] suggested the application of redox electrodes for the determination of ligand concentrations. They studied the complexes of various sulphur-containing ligands such as thiourea, diethyldithiophosphate and thiobiuret, as well as the stability of halide complexes. This principle can be used if the electrode reaction

$$L^{l-} = L^{k-} + (l - k)\,e^-$$

is reversible and only one form (the reduced one) can be co-ordinated to the metal ion.

TABLE 5.3

$C_{Cd} \times 10^3$	$C_I \times 10^3$	E	$[Cd^{2+}]_{exp} \times 10^3$	$[I^-]_{calc} \times 10^3$	$[I^-]_{exp} \times 10^3$
3.33	0	0	3.03	0	0
3.33	7.0	7.2	1.90	5.5	5.7
3.33	16.7	14.6	1.07	13.5	13.8
3.33	33.7	27.8	0.382	27.1	27.7
3.33	66.7	54.4	0.048 1	56.0	56.8
3.33	133.3	86.9	0.003 82	121.0	123.0
3.33	257.0	121.4	0.000 26	244.0	243.0
3.33	404.0	144.5	0.000 042 9	391.0	391.0

The determination of the free ligand concentration is very important because this quantity must be used in every calculation. The free ligand concentration can be calculated from the free metal ion concentration by a successive approximation method developed by Leden [250]. As appears from Table 5.3 the calculated and the directly determined free ligand concentrations are in good agreement in the case of the cadmium(II)–iodide system.

5.8.3 DETERMINATION OF HYDROGEN ION CONCENTRATION

The conjugate acids of most ligands are weak, consequently complex formation means a competition between protons and metal ions for the ligand. In such cases the stability constant can be derived from pH measurements. The determination of pH is the most common electrometric procedure and many excellent monographs [285, 286] are devoted to this topic; the experimental problems will therefore be dealt with very briefly.

For the potentiometric determination of pH, the hydrogen, quinhydrone and glass electrodes are most frequently used. A disadvantage of the hydrogen electrode and the H^+/H_2 (g) Pt half-cell, is that the measurement is fairly laborious and the equilibrium potential only slowly reached. Nevertheless, it permits the most accurate measurement of the hydrogen ion concentration.

When the quinhydrone electrode is used it must be remembered that it takes up protons in strongly acidic media [287] and with certain metal ions it forms complexes [288].

The glass electrode is the far most frequently applied for pH measurements. Owing to the high resistance of the glass electrode a valve potentiometer is necessary for the measurements. Many reliable instruments are commercially available. The accuracy of the measurements — depending on the experimental conditions and the instrument — is 0.002–0.1 pH unit. It is important to calibrate the instrument with buffer solutions of well defined pH. The accuracy can be increased if the difference between the measured pH and the pH of the buffer does not exceed one pH unit. In the case of aqueous solutions the scale-reading gives the pH directly. In mixed solvents, however, this is not true because of the change of the junction potential. Van Uitert and Haas [289] measured the 'pH' in dioxan–water mixtures with the glass electrode, assuming that the scale-reading is strictly proportional to the pH determined with the hydrogen electrode. The relationship between the scale-reading (B) and the pH is

$$\text{pH} = B - \log U_\text{H} \tag{5.106}$$

where U_H is a function of the composition of the solvent. If the value of U_H is known, the true pH can be simply obtained from measurements with a glass electrode. Van Uitert and Fernelius [290] determined the U_H values for aqueous dioxan solutions. When working with mixed solvents it is even more important to calibrate the instrument frequently with a buffer solution.

When the pH measurements are used for the determination of stability constants there are two ways of arranging the experiments: (*i*) the change of the pH is measured as a function of the ligand concentration; (*ii*) the pH is measured as a function of the concentration of acid (or base) added at constant metal ion and ligand concentration. In the second method the free ligand concentration, that is the extent of complex formation, can be varied much more widely.

The usual expression for $\bar{n}$

$$\bar{n} = \frac{T_\text{L} - [\text{L}]}{T_\text{Me}}$$

in the case of weakly basic ligands has to be written in the form

$$\bar{n} = \frac{T_\text{L} - [\text{L}^{i-}]}{T_\text{Me}} \tag{5.107}$$

where φ is defined by

$$\varphi = 1 + \frac{[\text{H}^+]}{K_{\text{H},Z}} + \frac{[\text{H}^+]^2}{K_{\text{H},Z} \cdot K_{\text{H},Z-1}} + \ldots + \frac{[\text{H}^+]^Z}{K_{\text{H},Z} \cdot K_{\text{H},Z-1} \ldots K_{\text{H},1}} \tag{5.108}$$

where Z is the maximum number of protons taken up by the ligand. The equation for the free ligand concentration can be derived by considering the mass balances

$$T_\text{L} = \sum_0^Z [\text{H}_j\text{L}^{j-l}] + \sum_1^N i[\text{MeL}_i] \tag{5.109}$$

$$T_\text{Me} = \sum_0^N [\text{MeL}_i], \tag{5.110}$$

the acidic dissociation constants of the ligand, and the electroneutrality principle.

$$[\text{L}^{l-}] = \frac{T_\text{L} - [\text{NaOH}] + [\text{HCl}] - [\text{H}^+] + [\text{OH}^-]}{\dfrac{Z[\text{H}^+]^Z}{K_{\text{H},1} \cdot K_{\text{H},2} \ldots K_{\text{H},Z}} + \dfrac{(Z-1)[\text{H}^+]^{Z-1}}{K_{\text{H},2} \cdot K_{\text{H},3} \ldots K_{\text{H},Z}} + \ldots + \dfrac{[\text{H}^+]}{K_{\text{H},Z}}} \tag{5.111}$$

All quantities in Eq. 5.111 are measurable or known: the total ligand concentration (T_L), the concentration of the added acid or base ([NaOH] and [HCl]), the pH and the acidic dissociation constant of the ligand. Therefore the stability constants can be calculated by using Eqs. 5.107 and 5.111.

Equations 5.107 and 5.111 provide the general solution of the problem of the calculation of stability constants from pH measurements. Another general treatment is given by Irving and Rossotti. In Table 5.4 a few examples are cited from the abundant literature of pH-measurement equilibrium studies. It must be mentioned that in the case of the formation of very stable complexes the method cannot be directly applied, but the stability constant can be evaluated from the study of suitable competitive reactions [291].

TABLE 5.4

Some examples of ligands the metal complexes of which have been studied by pH measurement

Ligand	Ref.
Ammonia	[292, 293]
Oxalic acid	[294]
Ethylenediamine	[292, 295]
1,3-Diaminopropane	[296]
Glycine	[297, 298]
Imidazole	[299]
Dimethylglyoxime	[300]
Glycylglycine	[301]
Acetylacetone	[302, 303]
Triethylenetetramine	[304]
Nitrilotriacetic acid and analogous compounds	[305, 306, 307]
α-Hydroxycyclohexanecarboxylic acid	[308]
8-Hydroxyquinoline	[302, 309]
2,2'-Dipyridyl	[310]
Ethylenediaminetetra-acetic acid	[311]
2-Hydroxy-1,3-diaminopropane-N,N,N',N'-tetra-acetic acid	[312]
Diethylenetriaminepenta-acetic acid	[313]
Triethylenetetraminehexa-acetic acid	[314]

5.9 Polarographic measurements

There are some similarities between the polarographic and the potentiometric determination of stability constants. However, the processes involved in the reduction on the surface of the dropping mercury electrode are much more complicated. The simplicity and versatility of the polarographic technique made this method very popular in the study of complex equilibria. The different possibilities will be dealt with briefly, as the reader may consult some excellent reviews [315–317].

5.9.1 MEASUREMENT OF THE SHIFT OF THE HALF-WAVE POTENTIAL

If the electrode process is reversible the potential of the dropping mercury electrode is given by the Heyrovský–Ilkovič equation [318]

$$E_{d.e.} = (E_{1/2})_s - \frac{RT}{zF} \ln \frac{i}{i_d - i} \qquad (5.112)$$

where $(E_{1/2})_s$ is the half-wave potential of the hydrated metal ion Me^{m+}, z is the number of electrons involved in the electrode reaction, i_d is the limiting diffusion current, i_d is the current corresponding to the potential $E_{d,e,}$ and R, T and F have their conventional meanings. Heyrovský and Ilkovič observed (and correctly explained) that the half-wave potential is shifted to a more negative value as a result of complex formation. Lingane [319] pointed out that a quantitative relationship exists between the shift of the half-wave potential and the stability constant of the complex formed

$$\Delta E_{1/2} = (E_{1/2})_s - (E_{1/2})_c = \frac{2.303 \, RT}{zF} \log \beta_n + \frac{2.303 \, RT}{zF} n \log [L] \qquad (5.113)$$

where $(E_{1/2})_c$ is the half-wave potential of the complex MeL_n. On plotting $\Delta E_{1/2}$ as a function of $\log [L]$ a straight line should be obtained and from the slope and the intercept the values of n and β_n can be calculated. In deriving 5.113 it was assumed that the diffusion coefficients of the simple ion and of the complex are the same and the activity coefficients are constant in the media used. The diffusion coefficients can be determined from the diffusion current and used for the correction of the value calculated from Eq. 5.113.

In the case of successive complex formation the series of stability constants can be evaluated. If the stabilities of the successive complexes are widely different, the $\Delta E_{1/2}$ versus $\log [L]$ curve consists of several segments, and the corresponding stability constants can be analogously obtained. In the case of overlapping equilibria the method developed by DeFord and Hume [320] can be applied. By considering the complex formation equilibria, the electrode reaction and the diffusion processes, it can be derived that

$$\text{antilog} \left(\frac{0.4343 \, zF}{RT} \Delta E_{1/2} + \log \frac{I_s}{I_c} \right) = \sum_0^N \beta_i [L]^i \qquad (5.114)$$

where I_s and I_c are the diffusion constants of the non-complexed and the complexed species. Equation 5.114 can be solved by the Leden method or by using computers [321, 322].

Relatively few metal ions can be reduced reversibly on the dropping mercury electrode, therefore the application of Eqs. 5.113 and 5.114 is fairly limited. There are two possibilities for the application of shift in half-wave potential to polarographic study of complex equilibria in the case of an irreversibly reduced central ion. First, the application of suitable competition reactions [323]; if to the solution of the irreversibly reduced metal ion and the ligand another metal ion is added which also forms a complex with the same ligand and which can be reversibly reduced at a more positive potential, the equilibrium constants can be simply evaluated. Secondly, the approach developed by Tamamushi and Tanaka [324] can be applied. In the case of an irreversible electrode process the following equation is valid

$$E_{d.e.} = (E_{1/2})_s - \frac{RT}{zF} \ln \frac{i}{i_d - i} \tag{5.115}$$

and the shift of the half-wave potential can be described by

$$\Delta E_{1/2} = (E_{1/2})_s - (E_{1/2})_c = \frac{2.303\ RT}{\alpha zF} \log \beta_n + \frac{2.303\ RT}{\alpha zF}\ n \ \log\ [L] \tag{5.116}$$

where α is the fraction of the total applied potential that favours the forward reaction; its deviation from 0.5 is the measure of the irreversibility and can be determined from the following relationship

$$E_{3/4} - E_{1/4} = \frac{2\ RT}{\alpha zF} \ln\ 3. \tag{5.117}$$

5.9.2 MEASUREMENT OF THE CURRENT

It is frequently assumed that the diffusion coefficients of the free metal ion and of the complex are the same. If this assumption is valid the limiting diffusion current does not depend on the concentration of the ligand. If the experiments indicate the contrary, the stability constant can be calculated from the change of the limiting diffusion current with the concentration of the ligand. The mean diffusion coefficient $(\bar{D})$ can be given [325] as

$$\bar{D} = \frac{D_0 + D_n \beta_n [L]^n}{1 + \beta_n [L]^n}. \tag{5.118}$$

The diffusion coefficient can be calculated from the diffusion current according to the well-known Ilkovič equation

$$i_d = 706\ z\ D^{1/2}\ v^{2/3}\ t^{1/6}\ c \tag{5.119}$$

where v is the rate of flow of mercury from the capillary, t is the time from the beginning the drop, c is the concentration of the depolarizer; in the case in question $c = T_{Me}$.

This method can be used irrespective of the irreversibility of the electrode reaction.

5.9.3 CHRONOPOTENTIOMETRY

In the case of chronopotentiometry the potential of the electrode is measured at constant current as a function of time. Tanaka and Yamada [326] derived the following relationship between the experimentally measurable chronopotentiometric transition time and the ligand concentration

$$\frac{[\text{Me}]}{T_{\text{Me}}} \frac{\tau}{\tau_0} = \frac{1 + \beta_1 \xi [\text{L}]}{1 + \beta_1 [\text{L}]} \tag{5.120}$$

where τ_0 is the transition time measured in the absence of the ligand and τ is the ratio of the diffusion coefficients of the complex and the free metal ion. The left side of Eq. 5.120 is experimentally measurable and the two constants (β_1 and ξ) can be calculated. This method was applied for the determination of the stability constants of several outer-sphere complexes.

5.10 Miscellaneous methods

Besides the methods already treated, there are many others which can be used to get quantitative or semi-quantitative information on the extent of complex formation. These are discussed briefly because of their more or less limited applicability or because they are less easily available than the other methods.

5.10.1 CONDUCTIVITY MEASUREMENTS

If the complex is an anion the complex formation evidently results in a decrease in the number of the charged particles and therefore a decrease in the conductivity. The application of this method is limited to the quantitative study of the formation of the complex $\text{MeL}^{(m-1)+}$. If more than one complex is formed the exact calculation of the stability constants is not possible. An inherent disadvantage of the method is that no neutral salt can be added to achieve a constant ionic medium because in this case the change of the conductivity due to the complex formation would be insignificant. A brilliant account is given of the theory and application of the conductivity method, by Davies [327].

Recently a number of high-frequency conductivity studies of complex formation have been made [328–330]. These, however, can only be used to learn the metal ion to ligand ratio in the case of formation of stable complexes.

5.10.2 NUCLEAR MAGNETIC RESONANCE STUDIES

The nuclear magnetic resonance spectrum is determined by the electronic environment of a nucleus due to the magnetic screening by the electron cloud. The changes in the electronic environment due to the formation of complexes are the measure of the stabilities of the complex species. The NMR method has been frequently applied to both inner-sphere and outer-sphere complexes. An excellent review by Hinton and Amis [331] treats all results published up to mid-1966.

The main importance of NMR studies is that they can give information as to which particular donor atom of a multidentate ligand is involved in the complex formation. Kula et al. [332] and Sudmeier and Reilley [333] could determine from NMR and pH measurement the successive protonation sites of EDTA and related ligands. Comparative NMR and pH studies clarified some structural and bonding characteristics of metal complexes of aminopolycarboxylic acids [334–336].

5.10.3 GAS CHROMATOGRAPHY

Gas chromatography offers a convenient possibility [337] for the determination of stability constants of complexes involving volatile ligands. Gil-Av and Herling [338] studied the equilibria between 16 different olefins and silver(I) ions. Cvetanović et al. [339] determined the equilibrium constants of a number of silver(I)–olefin complexes at different temperatures and studied the effect of deuteration of olefin on the stability. Falconer and Cvetanović [340] determined the stability of iodine–olefin complexes.

5.10.4 DENSITY OF SOLUTIONS

The density of solutions is determined by the interactions between the solutes and the solvent and by the interactions of different solutes. In aqueous solutions a contraction usually occurs, that is, the volume of the solution is less than the sum of volumes of the solvent and of the dissolved compound. The contraction is the consequence of the interaction of the water molecules with the cations and anions. Evidently if complex formation occurs the interaction of the complexes with the water will arise to a smaller extent and the contraction will be much smaller [341–342]. Čelada[343] has recently made semiquantitative conclusions on the complexes formed in solutions of different electrolytes.

5.10.5 MEASUREMENT OF SURFACE TENSION AND VISCOSITY

There are a great many studies [344–345] where the surface tension, the viscosity or some other properties of a solution containing the central ion were measured as a function of ligand concentration. The curves obtained had several minima and maxima and the authors concluded that in solution the composition of the complexes corresponds to the metal ion to ligand ratio at these particular points. However, as Heintz and Hume [346] pointed out, these methods are without any theoretical foundation. There is no reason to use such procedures in any kind of equilibrium studies.

REFERENCES

1. SOUCHAY, P., *Research 4*, 11 (1951).
2. KENTTÄMAA, J., *Acta Chem. Scand. 12*, 1323 (1958).
3. ROSSOTTI, F. J. C. and ROSSOTTI, H. S., *J. Phys. Chem. 63*, 1041 (1954).
4. TOBIAS, S. R., *J. Inorg. Nucl. Chem. 19*, 348 (1961).
5. BJERRUM, J., *Kgl. Dansk Vid. Selsk. Mat.-Fys. Medd. 11*, No. 5 (1931).
6. SCAIFE, D. B. and TYRRELL, H. J. V., *J. Chem. Soc.* 386 (1958).
7. SCAIFE, D. B. and TYRRELL, H. J. V., *J. Chem. Soc.* 392 (1958).
8. GOLDRING, L. S., HAWES, R. C., HARE, G. O., BECKMANN, A. O. and STICKNEY, M. E., *Anal. Chem. 25*, 869 (1953).
9. PRUE, J. E., *J. Chem. Soc.* 7534 (1965).
10. WOLDBYE, F. and BAGGER, S., *Proceedings of the Analytical Chemical Conference*, Budapest, 1966, Vol. 3, p. 139.
11. OSTROMYSSLENSKY, J., *Ber. 44*, 268 (1911).
12. DENISON, R. B., *Trans. Faraday Soc. 8*, 20 (1912).
13. JOB, P., *Ann. Chim. Phys. 9*, 113 (1928).
14. KLAUSEN, K. S. and LANGMYHR, F. J., *Anal. Chim. Acta 28*, 335 (1963).
15. JONES, M. M. and INNES, K. K., *J. Phys. Chem. 62*, 1005 (1958).
16. SOMMER, L. and HLNIČKOVÁ, M., *Bull. Soc. Chim. France*, 36 (1959).
17. WOLDBYE, F., *Acta Chem. Scand. 9*, 299 (1955).
18. VOSBURGH, W. C. and COOPER, G. R., *J. Am. Chem. Soc. 63*, 427 (1941).
19. KATZIN, L. J. and GEBERT, E., *J. Am. Chem. Soc. 72*, 5455 (1950).
20. BABKO, A. K., *Zh. Obshch. Khim. 16*, 1551 (1946).
20a. NEWMAN, L. and HUME, D. N., *J. Am. Chem. Soc. 79*, 4576 (1957).
21. YOE, J. A. and JONES, A. L., *Ind. Eng. Chem., Anal. Ed. 16*, 11 (1944).
22. MEYER, A. S. and AYRES, G. H., *J. Am. Chem. Soc. 79*, 49 (1957).
23. HARVEY, A. E. and MANNING, D. L., *J. Am. Chem. Soc. 72*, 4488 (1950).
24. BENT, H. E. and FRENCH, C. L., *J. Am. Chem. Soc. 63*, 1568 (1941).
25. KINGERY, W. D. and HUME, D. N., *J. Am. Chem. Soc. 71*, 2393 (1949).
26. ASMUS, E., *Z. Anal. Chem. 178*, 104 (1960).
27. MUSHRAN, S. P., SANYAL, P. and PANDEY, J. D., *J. Indian Chem. Soc. 43*, 273 (1966).
28. CILENTO, G. and SANIOTO, D. L., *Z. physik. Chem. Leipzig, 223*, 333 (1963).
29. ROSENSTEIN, L., *J. Am. Chem. Soc. 34*, 1117 (1912).
30. STEARNS, R. S. and WHELAND, G. W., *J. Am. Chem. Soc. 69*, 2025 (1947).
31. McCONNELL, H. and DAVIDSON, N., *J. Am. Chem. Soc. 72*, 3164 (1950).
32. McCULLOUGH, R. L., JONES, L. H. and PENNEMAN, R. A., *J. Inorg. Nucl. Chem. 13*, 286 (1960).
33. BENESI, H. A. and HILDEBRAND, J. H., *J. Am. Chem. Soc. 71*, 2703 (1949).
34. NASH, C. P., *J. Phys. Chem. 64*, 950 (1960).
35. SCOTT, R. L., *Rec. Trav. Chim. 75*, 787 (1956).
36. KEEFEN, R. M. and ANDREWS, L. J., *J. Am. Chem. Soc. 74*, 1891 (1952).
37. ROSE, N. J. and DRAGO, R. S., *J. Am. Chem. Soc. 81*, 6138 (1959).
38. KETELAAR, J. A. A., VAN DE STOLPE, C., GOUDSMIT, A. and DZUBES, W., *Rec. Trav. Chim. 71*, 1104 (1952).
39. PERSON, W. B., *J. Am. Chem. Soc. 87*, 167 (1965).
40. CARTER, S., MURRELL, J. N. and ROSCH, E. J., *J. Chem. Soc.* 2048 (1965).
41. NAKAGURA, S., *J. Am. Chem. Soc. 76*, 3070 (1954).
42. EDMONDS, S. N. and BIRNBAUM, N., *J. Am. Chem. Soc. 63*, 1471 (1941).
43. IRVING, H., ROSSOTTI, H. S. and HARRIS, G., *Analyst 80*, 83 (1955).
44. THAMER, B. J. and VOIGT, A. F., *J. Phys. Chem. 56*, 225 (1952).
45. ANG, K. P., *J. Phys. Chem. 62*, 1110 (1958).
46. THAMER, B. J., *J. Phys. Chem. 59*, 450 (1955).
47. NEWMAN, L. and HUME, D. N., *J. Am. Chem. Soc. 79*, 4571 (1957).
48. NEWMAN, L. and HUME, D. N., *J. Am. Chem. Soc. 79*, 4581 (1957).
49. JANSSEN, M. J., *Rec. Trav. Chim. 75*, 1397 (1956).
50. YATSIMIRSKII, K. B., *Zh. Neorgan. Khim. 1*, 2306 (1956).
51. MATTOO, B. N., *Z. Physik. Chem. Frankfurt 13*, 316 (1957).

52. KOMAR, N. P., *Dokl. Akad. Nauk SSSR, 72*, 565 (1950).
53. FRONAEUS, S., *Acta Chem. Scand. 5*, 139 (1951).
54. BJERRUM, J., *Kgl. Dansk. Vid. Selsk. Mat.-Fys. Medd. 21*, No. 4 (1941).
55. JEZOWSKA–TRZEBIATOWSKA, B., PAJDOWSKI, L. and STAROSTA, J., *Roczniki Chem. 35*, 445 (1961).
56. IRVING, H. and MELLOR, D. P., *J. Chem. Soc.* 3957 (1955).
57. HUGHES, V. L. and MARTELL, A. E., *J. Phys. Chem. 57*, 694 (1953).
58. MUSGRAVE, T. R. and KELLER, R. N., *Inorg. Chem. 4*, 1793 (1965).
59. JANSSEN, M. J., *J. Inorg. Nucl. Chem. 8*, 340 (1958).
60. KLEINER, K. E., *Zh. Obshch. Khim. 20*, 1747 (1950).
61. RŮŽIČKA, J. and STARÝ, J., *Talanta 14*, 909 (1967).
62. PENNEMAN, R. A. and JONES, L. H., *J. Chem. Phys. 24*, 293 (1956).
63. PENNEMAN, R. A. and JONES, L. H., *J. Inorg. Nucl. Chem. 20*, 19 (1961).
64. COLEMAN, J. S., PETERSEN, H., JR. and PENNEMAN, R. A., *Inorg. Chem. 4*, 135 (1965).
65. COONEY, R. P. L. and HALL, J. R., *J. Inorg. Nucl. Chem. 28*, 1675 (1966).
66. FRONAEUS, S. and LARSSON, R., *Acta Chem. Scand. 16*, 1433 (1962).
67. FRONAEUS, S. and LARSSON, R., *Proc. 8th Intern. Conf. Coord. Chem.* p. 383, Springer, Vienna, 1964.
68. TOBIAS, R. S., *J. Chem. Ed. 44*, 2, 70 (1967).
69. RAO, I. R., *Proc. Roy. Soc. 127A*, 279 (1930).
70. REDLICH, O., *Chem. Rev. 39*, 333 (1946).
71. DELWAULLE, M. L. and FRANCOIS, F., *J. Chim. Phys. 46*, 80 (1949).
72. DELWAULLE, M. L., FRANÇOIS, F., DELHAYE-BUISSET, M. B. and DELHAYE, M., *J. Phys. Radium 15*, 206 (1954).
73. DELWAULLE, M. L., *Compt. Rend. 224*, 389 (1947).
74. DELWAULLE, M. L., FRANCOIS, F. and WIEMANN, J., *Compt. Rend. 208*, 1818 (1939).
75. DELWAULLE, M. L., *Bull. Soc. Chim. France* 1294 (1955).
76. LONG, L. H. and DOLLIMORE, D., *J. Chem. Soc.* 4457 (1954).
77. GOUBEAU, J., RICHTER, D. E. and BECHER, H. J., *Z. anorg. allgem. Chem. 278*, 12 (1955).
78. ROLFE, J. A., SHAPPARD, D. E. and WOODWARD, L. A., *Trans. Faraday Soc. 50*, 1275 (1954).
79. YELLIN, W. and PLANE, R. A., *J. Am. Chem. Soc. 83*, 2448 (1961).
80. MORRIS, D. F. D., SHORT, E. L. and WATERS, D. N., *J. Inorg. Nucl. Chem. 25*, 375 (1963).
81. KECHI, Z. and MARKOWSKI, J., *Roczniki Chem. 36*, 345 (1965).
82. SPIRO, T. G., *Inorg. Chem. 4*, 731 (1965).
83. KELLER, O. L. and CHELTHAM-STRODE, A., *Inorg. Chem. 5*, 367 (1966).
84. FARRER, H. N., McGREDY, M. M. and TOBIAS, R. S., *J. Am. Chem. Soc. 87*, 5019 (1965).
85. NIXON, J. and PLANE, R. A., *J. Am. Chem. Soc. 84*, 4445 (1962).
86. HESTER, R. E., PLANE, R. A. and WALRAFEN, J. E., *J. Chem. Phys. 38*, 249 (1963).
87. HESTER, R. E. and PLANE, R A., *J. Chem. Phys. 40*, 411 (1964).
88. WOODWARD, L. A., *Proc. 8th Intern. Conf. Coord. Chem.* p. 15, Springer, Vienna, 1964.
89. MARONI, V. A. and SPIRO, T. G., *J. Am. Chem. Soc. 89*, 45 (1967).
90. MARONI, V. A. and SPIRO, T. G., *J. Am. Chem. Soc. 88*, 1410 (1967).
91. MILLER, J. T. and IRISH, D. E., *Can. J. Chem. 45*, 147 (1967).
92. KRISHNAN, K. and PLANE, R. A., *Inorg. Chem. 5*, 852 (1966).
92a JONES, M. M., JONES, E. A., HARMON, D. F. and SEMMES, R. T., *J. Am. Chem. Soc. 83*, 2038 (1961).
93. WOLDBYE, F., *Record of Chem. Progress 24*, 197 (1963).
94. WOLDBYE, F., *Technique of Inorganic Chemistry* (Edited by H. B. Jonassen and A. Weissberger), Vol. 4, Intersci. New York, 1964.
95. KATZIN, L. J. and GULYÁS, E., *J. Phys. Chem. 64*, 1739 (1960).
96. BECK, M. T., CSISZÁR, B. and SZARVAS, P., *Nature 188*, 846 (1960).
97. FREI, V., *Collection Czech. Chem. Commun. 27*, 2450 (1962).
98. NEBBIA, G., *Annali di Chimica 48*, 695 (1958).

99. NEBBIA, G. and PIZZOLI, E. M., *Bol. Sci. Far. Chim. Ind. 19*, 28 (1961).
100. NEBBIA, G., PIZZOLI, E. M. and FARILLI, M., *Bol. Sci. Far. Chim. Ind. 19*, 109 (1961).
101. CSISZÁR, B., HALMOS, M. and BECK, M. T., *Naturwiss. 48*, 571 (1961).
102. KATZIN, L. J. and GULYÁS, E., *J. Phys. Chem. 64*, 1347 (1960).
103. SZARVAS, P. and TIBÉLY, S., *Magy. Kém. Folyóirat 57*, 212 (1951).
104. CSISZÁR, B., unpublished results.
105. FREI, V. and SOLCOVA, A., *Collection Czech. Chem. Commun. 32*, 1815 (1967).
106. FREI, V. and SOLCOVA, A., *Collection Czech. Chem. Commun. 30*, 961 (1965).
107. GALLOIS, F., *J. Chim. Phys. 35*, 212, 249 (1938).
108. GALLOIS, F. and MOUNIER, J., *Compt. Rend. 223*, 790 (1946).
109. PLSKO, E. and OBERT, T., *Chem. Zvesti, 16*, 169 (1962).
110. LINDE, H. W., ROGERS, K. B. and HUME, D. N., *Anal. Chem. 25*, 404 (1953).
111. JORDAN, J. and ALLEMAN, T. G., *Anal. Chem. 29*, 9 (1957).
112. SCHLYTER, K. and SILLÉN, L. G., *Acta Chem. Scand. 13*, 385 (1959).
113. SCHLYTER, K., *Kungl. Tekniska Högskolans Handlingar* No. 132 (1959).
114. MEAD, T. E., *J. Phys. Chem. 66*, 2149 (1962).
115. JOHANSSON, S., *Arkiv Kemi 24*, 189 (1965).
116. CHRISTENSEN, J. J., IZATT, R. M. and HANSEN, L. D., *Rev. Sci. Instr. 36*, 779 (1965).
117. BJÖRKMAN, M. and SILLÉN, L. G., *Kungl. Tekniska Högskolans Handlingar* No. 199 (1963).
118. ARNEK, R., *Arkiv Kemi, 24*, 531 (1965).
118a TYRRELL, H. J. V. and BEEZER, A. E., *Thermometric Titrimetry*, Chapman and Hall, London, 1968.
119. BECKER, F., BARTHEL, J., SCHMAHL, N. G. and LÜSCHOW, H. M., *Z. Physik. Chem. Frankfurt 37*, 52 (1963).
120. BOLLES, T. F. and DRAGO, R. S., *J. Am. Chem. Soc. 87*, 5015 (1965).
121. BRENNER, A., *J. Electrochem. Soc. 112*, 611 (1965).
122. CHRISTENSEN, J. J., IZATT, R. M., HANSEN, L. D. and PARTRIDGE, J. A., *J. Phys. Chem. 70*, 2003 (1966).
123. WEISSBERGER, A. (Editor), *Technique of Organic Chemistry*, Vol. VIII, Part II, 2nd Ed. Interscience, New York, 1963.
124. EIGEN, M. and TAMM, K., *Z. Elektrochem. 66*, 107 (1962).
125. BECK, M. T. and GÖRÖG, S., *Proc. Symp. Chem. Coord. Comp. Agra*, 1959, Part II, p. 195.
126. POSTMUS, C. and KING, E. L., *J. Phys. Chem. 59*, 1208 (1955).
127. POSTMUS, C. and KING, E. L., *J. Phys. Chem. 59*, 1216 (1955).
128. LEE, T. S., KOLTHOFF, I. M. and LEUSSING, D. L., *J. Am. Chem. Soc. 70*, 3596 (1948).
129. BRANDT, W. W. and GELLSTROM, D. K., *J. Am. Chem. Soc. 74*, 3532 (1952).
130. BAXENDALE, I. H. and GEORGE, P., *Trans. Faraday Soc. 46*, 736 (1950).
131. ATKINSON, G. and KOR, S. L., *J. Phys. Chem. 71*, 673 (1967).
132. HURWITZ, P. and KUSTIN, K., *J. Phys. Chem. 71*, 324 (1907).
133. PETRUCCI, S., *J. Phys. Chem. 71*, 1174 (1967).
134. BEHR, B. and WENDT, H., *Z. Elektrochem. 66*, 223 (1962).
135. WENDT, H. and STREHLOW, H., *Z. Elektrochem. 66*, 228 (1962).
136. BECK, M. T. and GÉHER, J., unpublished results.
137. YATSIMIRSKII, K. B., *Zh. Analit. Khim. 10*, 344 (1955).
138. PUNGOR, E., *Ann. Univ. Sci. Budapest Rolando Eötvös Nominatae Sect. Chim. 2*, 69 (1960).
139. SYKES, K. W., *J. Chem. Soc.* 125 (1952).
140. SHERILL, M. S., *Z. Phys. Chem. 43*, 765 (1903).
141. KOELICHEN, K., *Z. Phys. Chem. 33*, 129 (1900).
142. AUSTERWEIL, G., *Magy. Kém. Folyóirat 14*, 26, 41, 58 (1908).
143. BELL, R. P. and PRUE, J. E., *J. Chem. Soc.* 362 (1949).
144. BELL, R. P. and WAIND, G. M., *J. Chem. Soc.* 1979 (1950).
145. BELL, R. P. and PANCKHUST, M. H., *J. Chem. Soc.* 2836 (1956).
146. KOLTUN, W. L., DEXTER, R. N., CLARK, R. E. and GURD, F. R. N., *J. Am. Chem. Soc. 80*, 4188 (1958).

147. KOLTUN, W. L., CLARK, R. E., DEXTER, R. N., KATSOYANNIS, P. G. and GURD, F. R. N., *J. Am. Chem. Soc. 81*, 295 (1959).
148. KING, E. L., *Catalysis*, Vol. 2, (Edited by P. H. Emmett) Reinhold, New York, 1955, p. 337.
149. HUDIS, J. and WAHL, A. C., *J. Am. Chem. Soc. 75*, 4153 (1953).
150. DUKE, F. R. and PARCHER, F. Z., *J. Am. Chem. Soc. 78*, 1540 (1956).
151. BRUBAKER, C. H. and MICHEL, J. P., *J. Inorg. Nucl. Chem. 4*, 55 (1957).
152. COHEN, D., SULLIVAN, J. C. and HINDMAN, J. C., *J. Am. Chem. Soc. 77*, 4964 (1955).
153. HUFFMAN, R. F. and DAVIDSON, N., *J. Am. Chem. Soc. 78*, 4836 (1956).
154. LACHMAN, S. J. and TOMPKINS, F. C., *Trans. Faraday Soc. 40*, 130 (1944).
155. YATSIMIRSKII, K. B. and ALEKSEEVA, I. I., *Zh. Neorgan. Khim. 1*, 952 (1956).
156. JONES, P., TOBE, M. L. and WYNNE-JONES, W. F. K., *Trans. Faraday Soc. 55*, 91 (1959).
157. YATSIMIRSKII, K. B. and KARACHEVA, G. A., *Zh. Neorgan. Khim. 3*, 352 (1956).
158. BECK, M. T., GÖRÖG, S. and KISS, Z., *Acta Chim. Acad. Sci. Hung. 42*, 321 (1964).
159. BECK, M. T. and GIMESI, I., *Magy. Kém. Folyóirat 69*, 552 (1963).
160. PETERS, E. and HALPERN, J., *Can. J. Chem. 34*, 554 (1956).
161. TAUBE, H., *J. Am. Chem. Soc. 70*, 3928 (1948).
162. LISTER, M. W., *Can. J. Chem. 31*, 638 (1953).
163. BECK, M. and GÖRÖG, S., *Magy. Kém. Folyóirat 65*, 55 (1959).
164. JONES, M. M., *Biochim. Biophys. Acta 39*, 385 (1960).
165. MALMSTRÖM, B. G., *Thesis*, Uppsala, 1956.
166. BECK, M. T. and DÓZSA, L., *Inorg. Chim. Acta 1*, 134 (1967).
167. DÓZSA, L., DURHAM, D. A. and BECK, M. T., *J. Inorg. Chem.* in press.
168. REINHARDT, H. and RYDBERG, J. in ref. 169. p. 612.
169. DYRSSEN, D., LILJENZIN, J. O. and RYDBERG, J. (editors), *Solvent Extraction Chemistry*, North-Holland, Amsterdam, 1967.
170. HENDIXSON, W. S., *Z. Anorg. Allgem. Chem. 13*, 73 (1897).
171. HÖK, B., *Svensk. Kem. Tidskr. 65*, 182 (1953).
172. DYRSSEN, D. and JOHANSSON, D., *Acta Chem. Scand. 9*, 763 (1955).
173. DAWSON, H. M. and McCRAE, J., *J. Chem. Soc. 79*, 1069 (1900).
174. SEWAND, R. P., *J. Am. Chem. Soc. 76*, 4850 (1954).
175. LEUSSING, D. L. and HANSEN, R. C., *J. Am. Chem. Soc. 79*, 4270 (1957).
176. WINSTEIN, S. and LUCAS, H. J., *J. Am. Chem. Soc. 60*, 836 (1938).
177. TRUEBLOOD, K. N. and LUCAS, H. J., *J. Am. Chem. Soc. 74*, 1338 (1952).
178. DYRSSEN, D. and LIEN, D., unpublished work, quoted by Dyrssen, D. in *Nucl. Sci. Eng. 16*, 448 (1963).
179. MORSE, H., *Z. Physik. Chem. Leipzig, 41*, 709 (1902).
180. RYDBERG, J., *Svensk. Kem. Tidskr. 67*, 499 (1955).
181. RYDBERG, J., *Acta Chem. Scand. 4*, 1503 (1950).
182. RYDBERG, J., *Arkiv Kemi 8*, 101 (1955).
183. RYDBERG, J., *Arkiv Kemi 8*, 113 (1955).
184. DYRSSEN, D., *Svensk. Kem. Tidskr. 67*, 213 (1952).
185. DYRSSEN, D., *Acta Chem. Scand. 8*, 1394 (1953).
186. DYRSSEN, D., *Rec. Trav. Chim. 75*, 753 (1956).
187. IRVING, H. and BELL, C. F., *J. Chem. Soc.* 1216 (1952).
188. IRVING, H. and ROSSOTTI, F. J. C., *J. Chem. Soc.* 1927, 1946 (1955).
189. MARCUS, Y., *Chem. Rev. 63*, 139 (1963).
189a FREISER, H., *Anal. Chem. 38*, 131R (1966).
189b PEPPARD, D. F., *Adv. Inorg. Chem. Radiochem.* (Ed., H. J. Emeleus and A. G. Sharpe), Vol. 9, Academic Press, New York, 1966, p. 1.
190. BLAKE, C. A., BAES, D. F., BROWN, K. B., COLEMAN, C. F. and WHITE, J. C., *Proc. 2nd Intern. Conf. Peaceful Uses Atomic Energy*, 15/P/1550 (1958).
191. IRVING, H. and EDGINGTON, D. N., *J. Inorg. Nucl. Chem. 15*, 158 (1960); *20*, 314 (1961); *20*, 321 (1961); *21*, 169 (1961).
192. IRVING, H., in ref. 169, p. 91.
193. HEALY, T. V., *J. Inorg. Nucl. Chem. 19*, 314, 328 (1961).
194. HEALY, T. V., in ref. 169, 119.
195. NEWMAN, L., *J. Inorg. Nucl. Chem. 25*, 304 (1963).

196. WANG, S. M., PARK, D. Y. and LI, N. C., in ref. 169, p. 111.
197. TAUBE, M. and SIEKIERSKI, S., *Nukleonika 6*, 489 (1961).
198. AL-NIAIMI, N. S., *Thesis*, Oxford (1964), quoted in [192].
199. IRVING, H., *Proc. Symp. Coord. Chem. Tihany, Hungary*, 1965, Akadémiai Kiadó, Budapest, 1966, p. 219.
200. CONNICK, R. E. and MCVEY, W. H., *J. Am. Chem. Soc. 71*, 3182 (1949).
201. SULLIVAN, J. C. and HINDMAN, J. C., *J. Am. Chem. Soc. 76*, 5931 (1954).
202. ROSSOTTI, F. J. C. and ROSSOTTI, H. S., *Acta Chem. Scand. 10*, 779 (1953).
203. ANTIKAINEN, P. J. and DYRSSEN, D., *Acta Chem. Scand. 14*, 86 (1960).
204. SEKINE, T., *Acta Chem. Scand. 19*, 1526 (1965).
205. STARÝ, J., in ref. 169, p. 630.
206. BODLÄNDER, G., *Z. Physik. Chem. Leipzig, 9*, 730 (1892).
207. REYNOLDS, C. A. and ARGERSINGER, W. J., *J. Phys. Chem. 56*, 417 (1952).
208. BECK, M. T., *Acta Chim. Acad. Sci. Hung. 4*, 227 (1954).
209. BECK, M. T. and GÖRÖG, S., *Magy. Kém. Folyóirat 65*, 413 (1959).
210. OLSON, D. C. and MARGERUM, D. W., *J. Am. Chem. Soc. 82*, 5602 (1960).
211. SANO, K., *Biochem. Z. 168*, 14 (1926).
212. IRVING, H., EWART, J. A. D. and WILSON, J. T., *J. Chem. Soc.* 2672 (1949).
213. LEDEN, I., *Svensk. Kem. Tidskr. 64*, 249 (1952).
214. BERNE, I. and LEDEN, I., *Svensk. Kem. Tidskr. 65*, 88 (1953).
215. LEDEN, I., *Proc. Symp. Coord. Chem. Copenhagen* 1953, p. 77.
216. AHRLAND, S. and GRENTHE, I., *Acta Chem. Scand. 11*, 1111 (1957).
217. LIESER, K. H., *Z. Anorg. Allgem. Chem. 292*, 97 (1957).
218. KRATOCHVIL, I., TEŽAK, B. and VOUK, V. B., *Arhiv Kemiju, 26*, 191 (1954).
219. HAIGHT, G. P. JR., *Acta Chem. Scand. 16*, 209 (1962).
220. HAIGHT, G. P. JR., ZOLTEVICZ, J. and EVANS, W., *Acta Chem. Scand. 16*, 311 (1962).
221. HAIGHT, G. P. JR., SPRINGER, C. H. and HEILMANN, O. J., *Inorg. Chem. 3*, 195 (1964).
222. PREER, J. R. and HAIGHT, G. P., *Inorg. Chem. 5*, 656 (1966).
223. KING, E. L., *J. Am. Chem. Soc. 71*, 319 (1949).
224. SAMUELSON, O., *Ion Exchangers in Analytical Chemistry*, Wiley, New York, 1953.
225. HELFFERICH, F., *Ionenaustauscher*, Bd. 1, Verlag Chemie, Weinberg, 1959.
226. FRONAEUS, S., *Svensk. Kem. Tidskr. 65*, 1 (1953).
227. MARCUS, Y., in MARINSKY, J. A. (Ed.) *Ion Exchange*, Marcel Dekker, New York, 1967, p. 101.
228. SCHUBERT, J. and RICHTER, J. W., *J. Am. Chem. Soc. 70*, 4259 (1948).
229. SCHUBERT, J., LIND, E. L., WESTFALL, W. N., PFLEGER, R. and LI, N. C., *J. Am. Chem. Soc. 80*, 4799 (1958).
230. FRONAEUS, S., *Acta Chem. Scand. 5*, 859 (1951).
231. FRONAEUS, S., *Proc. Symp. Coord. Chem. Copenhagen*, 1953, p. 61.
232. WHITEKER, R. A. and DAVIDSON, N., *J. Am. Chem. Soc. 75*, 3081 (1953).
233. CARLESON, B. G. F. and IRVING, H., *J. Chem. Soc.* 4390 (1954).
234. LOMAN, H. and VAN DALEN, E., *J. Inorg. Nucl. Chem. 28*, 2027 (1966); *29*, 699 (1967).
235. LEDEN, I., *Svensk. Kem. Tidskr. 64*, 145 (1952).
236. FRONAEUS, S., *Svensk. Kem. Tidskr. 65*, 1 (1953).
237. MARCUS, Y. and CORYELL, L. D., *Bull. Res. Commit. Israel, A8*, 1 (1959).
238. KRAUS, K. A. and NELSON, F., in *The Structure of Electrolytic Solutions*, (Ed. W. J. Hamer), Wiley, New York, 1959, p. 340.
239. KRAUS, K. A. and NELSON, F., *Proc. Intern. Conf. Peaceful Uses Atomic Energy, 7*, 113 (1955).
420. JONES, L. H. and PENNEMAN, R. A., *Chem. Phys. 22*, 965 (1954).
241. SONESSON, A., *Acta Chem. Scand. 15*, 1 (1961).
242. FRONAEUS, S., LUNDQUIST, I. and SONESSON, A., *Acta Chem. Scand. 16*, 196g (1962).
243. STOKES, R. H. and WALTON, H. F., *J. Am. Chem. Soc. 76*, 3327 (1954).
244. COCKERELL, L. and WALTON, H. F., *J. Phys. Chem. 66*, 75 (1962).
245. SURYARAMAN, M. G. and WALTON, H. F., *J. Phys. Chem. 66*, 78 (1962).
246. HELFFERICH, F., *J. Am. Chem. Soc. 84*, 3237 (1962).
247. WALLACE, R. M., *J. Phys. Chem. 71*, 1271 (1967).

248. FORSLING, W., HIETANEN, S. and SILLÉN, L. G., *Acta Chem. Scand. 6*, 901 (1952).
249. FLENGAS, S. N., *Trans. Faraday Soc. 51*, 62 (1955).
250. LEDEN, I., *Z. Physik. Chem. Leipzig, A 188*, 160 (1941).
251. BETHGE, P. O., JONEVALL-WASTÖÖ, I. and SILLÉN, L. G., *Acta Chem. Scand. 2*, 828 (1948).
252. QVARFORT, I. and SILLÉN, L. G., *Acta Chem. Scand. 3*, 505 (1949).
253. NILSSON, R. O., *Arkiv Kemi 12*, 337 (1958).
254. NÁRAY-SZABÓ, I. and SZABÓ, Z., *Z. Physik. Chem. Leipzig A 166*, 228 (1933).
255. GEORGE, R. and BJERRUM, J., *Acta Chem. Scand.* in press.
256. BIEDERMANN, G., *Arkiv. Kemi 9*, 277 (1956).
257. GRANÉR, F. and SILLÉN, L. G., *Acta Chen. Scand. 1*, 631 (1947).
258. OLIN, A., *Acta Chem. Scand. 11*, 1445 (1957).
259. AHRLAND, S. and GRENTHE, I., *Acta Chem. Scand. 11*, 1111 (1957).
260. BJERRUM, J. and WRIGHT, G. A., *Acta Chem. Scand. 16*, 159 (1962).
261. BUDEVSKY, O. and PLATIKANOVA, E., *Talanta, 14*, 901 (1967).
262. LAMB, A. B. and LARSSON, A. T., *J. Am. Chem. Soc. 42*, 2024 (1920).
263. DE, A. K., GHOSH, N. N. and RAY, P., *J. Indian Chem. Soc. 27*, 493 (1950).
264. BROSSET, C. and ORRING, J., *Svensk. Kem. Tidskr. 55*, 101 (1943).
265. SCHMID, R. W. and REILLEY, C. N., *J. Am. Chem. Soc. 678*, 5513 (1951).
266. PUNGOR, E. and HAVAS, J., *Acta Chim. Acad. Sci. Hung. 50*, 78 (1966).
267. PALATY, V., *Can. J. Chem. 41*, 18 (1963).
268. RECHNITZ, G. A. and BRAUNER, J., *Talanta, 11*, 617 (1964).
269. RECHNITZ, G. A. and ZAMOCHNIK, S. B., *Talanta, 11*, 1061 (1964).
270. *Orion Research Inc. Bulletin.* No. 92-20 B.
271. NAKAYAMA, F. S. and RASNICK, B. A., *Anal. Chem. 39*, 1022 (1967).
272. LUTHER, R., *Z. Physik. Chem. Leipzig, 27*, 364 (1898).
273. BERECZ, E. and SZITA, L. to be published.
274. LEDEN, I., *Thesis*, Lund, 1943.
275. FRONAEUS, S., *Acta Chem. Scand. 4*, 72 (1950).
276. AHRLAND, S., CHATT, J., DAVIES, N. R. and WILLIAMS, A. A., *J. Chem. Soc.* 264 (1958).
277. GRYDER, J. W., *Proc. Nat. Acad. Sci. U. S. 46*, 952 (1960).
278. IVES, D. J. G. and JANZ, S. J., *Reference Electrodes*, Academic Press, New York, 1961.
279. ANDEREGG, G., *Helv. Chim. Acta 42*, 344 (1959).
280. PUNGOR, E., *Anal. Chem. 39*, 29A (1967).
281. BURGER, E. and PINTÉR, B., *Hung. Sci. Instr.* No. 8, 11 (1966).
282. FRANT, M. S. and ROSS, J. W., *Science 154*, 1553 (1966).
283. LINGANE, J. J., *Anal. Chem. 39*, 881 (1967).
284. SHULMAN, V. M., KRAMARYOVA, T. V., LARIONOV, S. V. and DUBINSKIJ, V. I., *Proc. 10th Intern. Conf. Coord. Chem.* (edited by K. Yamasaki) Chemical Society of Japan, Tokyo, 1967, p. 63.
285. BRITTON, H. T. S., *Hydrogen Ions*, Chapman and Hall, London, 1955.
286. BATES, R. G., *Electrometric pH Determinations*, Chapman and Hall, London, 1954.
287. BIEDERMANN, G., *Acta Chem. Scand. 10*, 1340 (1956).
288. BERECZKI-BIEDERMANN, K., *Arkiv Kemi 9*, 175 (1956).
289. VAN UITERT, L. G. and HAAS, C. G., *J. Am. Chem. Soc. 75*, 451 (1953).
290. VAN UITERT, L. G. and FERNELIUS, W. C., *J. Am. Chem. Soc. 76*, 5887 (1954).
291. SCHWARZENBACH, G. and ACKERMANN, H., *Helv. Chim. Acta 32*, 1543 (1949).
292. BJERRUM, J., *Thesis*, Copenhagen, 1941.
293. SPIKE, C. G. and PARRY, R. W., *J. Am. Chem. Soc. 75*, 2726 (1953).
294. GELLES, E. and HAY, R. M., *J. Chem. Soc.* 3673 (1958).
295. MC INTYRE, G. H., BLOCK, B. P. and FERNELIUS, W. C., *J. Am. Chem. Soc. 81*, 529 (1959).
296. NÄSÄNEN, R., KOSKINEN, M., SALONEN, R. and KIIRSKI, A., *Suomen Kemistilehti B 38*, 81 (1965).
297. MALEY, L. E. and MELLOR, D. P., *J. Austr. Sci. Res. A2*, 579 (1949).
298. ALBERT, A. and SERJEANT, E. P., *Biochem. J., 76*, 621 (1960).
299. DATTA, S. P. and GRZYBOWSKI, A. K., *J. Am. Chem. Soc. A* 1058 (1966).

300. BOCHKOVA, V. M. and PESHKOVA, V. M., *Zh. Neorgan. Khim. 3*, 1132 (1958).
301. DATTA, S. P. and RABIN, B. R., *Trans. Faraday Soc. 52*, 1117 (1956).
302. MALEY, L. E. and MELLOR, D. P., *J. Australian Sci. Res. A2*, 92 (1949).
303. GRENTHE, I. and FERNELIUS, W. C., *J. Am. Chem. Soc. 82*, 6258 (1960).
304. JONASSEN, H. B., BERTRAND, J. A., GROVES, F. R., and STEAMS, R. I., *J. Am. Chem. Soc. 79*, 4279 (1957).
305. SCHWARZENBACH, G. and BIEDERMANN, W., *Helv. Chim. Acta 31*, 331 (1948).
306. IRVING, H. M. N. H. and MILES, M. G., *J. Chem. Soc. A* 727 (1966).
307. IRVING, H. M. N. H. and MILES, M. G., *J. Chem. Soc. A* 1268 (1966).
308. SCHURMANS, H., THUN, H. and VERBECK, F., *J. Inorg. Nucl. Chem. 29*, 1759 (1967).
309. IRVING, H. and ROSSOTTI, H. S., *J. Chem. Soc.* 2910 (1954).
310. MCBRYDE, W. A. E., *Can. J. Chem. 45*, 2093 (1967).
311. SCHWARZENBACH, G., GUT, R. and ANDEREGG, G., *Helv. Chim. Acta 37*, 937 (1954).
312. THOMPSON, L. C. and KUNDRA, S. K., *J. Inorg. Nucl. Chem. 28*, 2945 (1966).
313. DURHAM, E. J. and RYSKIEWICH, D. P., *J. Am. Chem. Soc. 80*, 4812 (1959).
314. BOHIGIAN, T. A., JR., and MARTELL, A. E., *Inorg. Chem. 4*, 1264 (1965).
315. IRVING, H., *Advances in Polarography* (Edited by I. S. Langmuir), Vol. 1, Pergamon, Oxford, 1960, p. 42.
316. VLČEK, A. A., *Progress in Inorganic Chemistry* (Edited by F. A. Cotton), Vol. 5, Interscience, New York, 1963, p. 211.
317. CROW, D. R. and WESTWOOD, J. V., *Quart. Rev. 19*, 57 (1965).
318. HEYROVSKÝ, J. and ILKOVIČ, D., *Collection Czech. Chem. Commun. 7*, 198 (1935).
319. LINGANE, J. J., *Chem. Rev. 29*, 1 (1941).
320. DEFORD, D. D. and HUME, D. N., *J. Am. Chem. Soc. 73*, 5321 (1951).
321. MCMASTERS, D. L. and SCHAAP, W. B., *Proc. Indiana Acad. Sci. 67*, 111, 117 (1958).
322. MOMOKI, K., SATO, H. and OGEWA, H., *Anal. Chem. 39*, 1072 (1967).
323. RINGBOM, A. and ERIKSON, L., *Acta Chem. Scand. 87*, 1105 (1953).
324. TAMAMUSHI, R. and TANAKA, N., *Bull. Chem. Soc. Japan, 22*, 227 (1949).
325. KORYTA, J., *Progress in Polarography* (Ed. P. Zuman), Vol. 1, Interscience, New York, 1962, p. 291.
326. TANAKA, N. and YAMADA, A., *Z. Anal. Chem. 224*, 117 (1967).
327. DAVIES, C. W., *Ion Association*, Butterworths, London, 1962.
328. HARA, R. and WEST, P., *Anal. Chim. Acta 11*, 264 (1954).
329. PUNGOR, E. and ZAPP, N. E., *Acta Chim. Acad. Sci. Hung. 25*, 133 (1960).
330. RYABCHIKOV, D. I. and ZARINSKY, V. A., *Talanta, 14*, 133 (1967).
331. HINTON, J. F. and AMIS, E. S., *Chem. Rev. 67*, 367 (1967).
332. KULA, R. J., SAWYER, D. T., CHAN, S. I. and FINLEY, C. M., *J. Am. Chem. Soc. 85*, 2930 (1963).
333. SUDMEIER, J. L. and REILLEY, C. N., *Anal. Chem. 36*, 1698 (1964).
334. KULA, R. J., *Anal. Chem. 38*, 1581 (1966).
335. AOCHI, Y. O. and SAWYER, D. T., *Inorg. Chem. 5*, 2085 (1966).
336. KULA, R. J., *Anal. Chem. 39*, 1171 (1967).
337. NORMAN, R. O. C., *Proc. Chem. Soc.* 151 (1958).
338. GIL-AV, E. and HERLING, J., *J. Phys. Chem. 66*, 1208 (1962).
339. CVETANOVIČ, R. J., DUNCAN, F. J., FALCONER, W. E. and IRWIN, R. S., *J. Am. Chem. Soc. 87*, 1827 (1965).
340. FALCONER, W. E. and CVETANOVIČ, R. J., *J. Chromatog. 27*, 20 (1967).
341. LENGYEL, S., *Ann. Univ. Sci. Budapest, Rolando Eötvös Nominatae, Ser. Chim. 1*, 96 (1959).
342. BECK, M. T., *Thesis*, Szeged, 1957.
343. JEDINAKOVA, V. and ČELADA, J., *Sci. Papers Inst. Chem. Techn. Prague 137*, 79 (1966).
344. KAZI, H. J. and DESAI, K., *J. Indian Chem. Soc. 31*, 163, 165, 329, 331, 769 (1964).
345. NAYAR, M. R. and NAYAR, K. D., *I. Indian Chem. Soc. 29*, 250 (1952).
346. HEINTZ, H. A. and HUME, D. N., *J. Phys. Chem. 61*, 462 (1957).

Chapter 6

THE PROTONATION OF COMPLEXES

Ligands are bases in the sense of the definitions by Brønsted and Lewis. Therefore in general a competition must be considered between protons and metal ions for the ligand. A partial protonation of the ligand does not rule out the possibility of complex formation. On the contrary, it may result in the formation of protonated species. It is helpful to distinguish between the protonation of complexes of unidentate and multidentate ligands.

6.1 Protonation of complexes of unidentate ligands

If the ligand has only one lone pair of electrons, the protonation results in the rupture of the metal-ligand bond. For example

$$Cu(H_2O)_5(NH_3)^{2+} + H_3O^+ \rightleftharpoons Cu(H_2O)_6^{2+} + NH_4^+ .$$

If, however, the unidentate ligand has more than one lone pair of electrons, the protonated ligand may remain in the co-ordination sphere, e.g.

$$Cu(H_2O)_5(OH)^+ + H_3O^+ \rightleftharpoons Cu(H_2O)_6^{2+} + H_2O .$$

It is probable that in concentrated acid solutions the undissociated acid molecule is co-ordinated to the metal ion. The complex acid $HFeCl_4$ is probably a protonated complex $FeCl_3HCl$. The fact that Ag_2SO_4 is fairly soluble in sulphuric acid may be the consequence of the formation of silver ions solvated by sulphuric acid molecules.

6.2 Protonation of complexes of multidentate ligands

There are many possibilities for the protonation of chelate complexes. The protonated donor group may remain bonded to the central ion or the protonation may result in the rupture of the metal ion–donor atom bond and to the successive decomposition of the metal chelate. The protonation scheme depends on the stoichiometry of the complex, the stability of the chelate, and the protonation scheme of the free ligand itself.

6.2.1 THE PROTONATION SCHEME OF FREE LIGANDS

The macroscopic dissociation constants of polybasic acids are, in fact, composite constants because protons bound to different donor groups may dissociate simultaneously [1]. For example the ionization of cysteine can be represented by the following scheme:

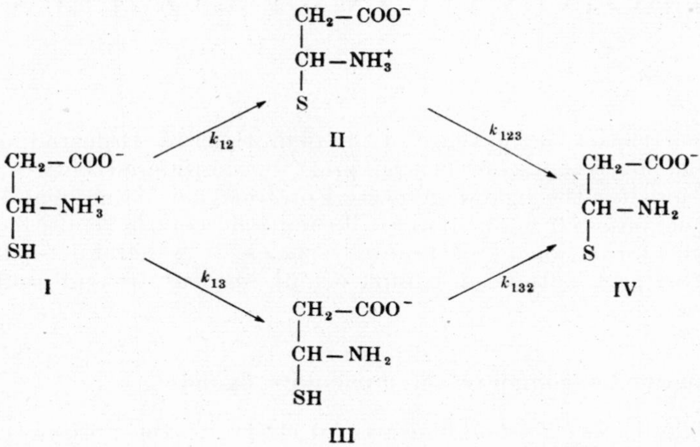

$$\tag{6.1}$$

The macroscopically determined macroconstants are related to the 'microconstants' by the expression

$$K_{a2} = \frac{[\text{II} + \text{III}][\text{H}^+]}{[\text{I}]} = k_{12} + k_{13} \tag{6.2}$$

and

$$K_{a2} \cdot K_{a3} = \frac{[\text{IV}][\text{H}^+]^2}{[\text{I}]} . \tag{6.3}$$

Evidently the ratio of the concentrations of different isomeric species is independent of the pH.

Wrathall, Izatt and Christensen [2] calculated the values of the four microconstants from a comparative calorimetric and potentiometric study of cysteine, mercaptoacetic acid and S-methyl-L-cysteine. The study of these latter compounds gave information on the ionization tendency of the SH and NH_3^+ groups in cysteine.

Similarly, for citric acid we have

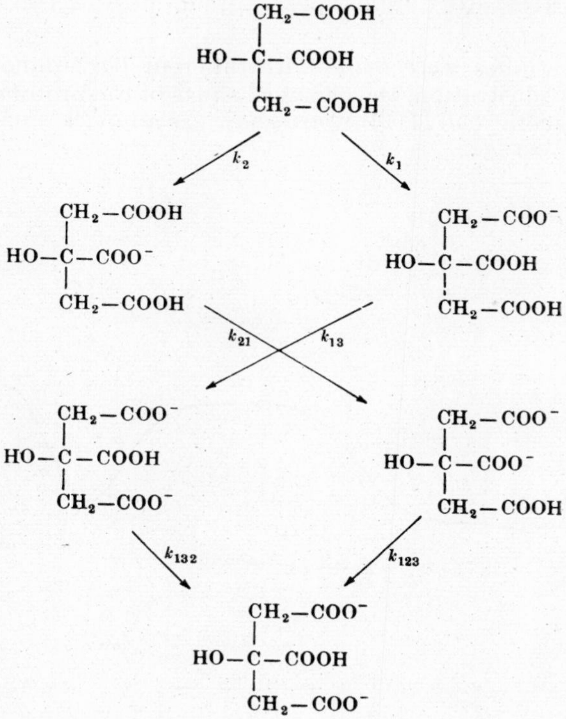

$$
\begin{array}{c}
\text{CH}_2-\text{COOH} \\
| \\
\text{HO}-\text{C}-\text{COOH} \\
| \\
\text{CH}_2-\text{COOH}
\end{array}
$$

k_2 k_1

$$
\begin{array}{c}
\text{CH}_2-\text{COOH} \\
| \\
\text{HO}-\text{C}-\text{COO}^- \\
| \\
\text{CH}_2-\text{COOH}
\end{array}
\qquad\qquad
\begin{array}{c}
\text{CH}_2-\text{COO}^- \\
| \\
\text{HO}-\text{C}-\text{COOH} \\
| \\
\text{CH}_2-\text{COOH}
\end{array}
$$

k_{21} k_{13}

$$
\begin{array}{c}
\text{CH}_2-\text{COO}^- \\
| \\
\text{HO}-\text{C}-\text{COOH} \\
| \\
\text{CH}_2-\text{COO}^-
\end{array}
\qquad\qquad
\begin{array}{c}
\text{CH}_2-\text{COO}^- \\
| \\
\text{HO}-\text{C}-\text{COO}^- \\
| \\
\text{CH}_2-\text{COOH}
\end{array}
$$

k_{132} k_{123}

$$
\begin{array}{c}
\text{CH}_2-\text{COO}^- \\
| \\
\text{HO}-\text{C}-\text{COOH} \\
| \\
\text{CH}_2-\text{COO}^-
\end{array}
$$

(6.4)

There are the following relationships between the macroconstants and the microconstants:

$$K_1^{\text{H}} = 2k_1 + k_2 \tag{6.5}$$

$$K_1^{\text{H}} K_2^{\text{H}} = k_1 k_{13} + 2 k_2 k_{21} = 2 k_1 k_{12} + k_1 k_{13} \tag{6.6}$$

$$K_1^{\text{H}} K_2^{\text{H}} K_3^{\text{H}} = k_1 k_{12} k_{123} = k_1 k_{13} k_{132} = k_2 k_{21} k_{123} \tag{6.7}$$

The microconstants were evaluated from the NMR study [3] of the different methyl esters of citric acid. The NMR method is ideally suited for this purpose because the pH-dependence of the chemical shift of the resonance signals of the different covalently bound protons of the molecule indicates the site of protonation. The experiments showed that the ionizations of the terminal carboxylic acid groups are predominant. Martin [4] pointed out that this conclusion is wrong and is the consequence of the consideration of erroneous macroconstants. The titrimetric study of the different methyl esters of citric acid indicated that the symmetrical singly ionized species predominates. The values of the microconstants are as

10*

follows: $pk_1 = 3.85$, $pk_2 = 3.35$, $pk_{12} = 4.60$, $pk_{21} = 5.10$, $pk_{123} = 5.85$, $pk_{132} = 6.05$.

Many NMR studies were made with different polyaminopolycarboxylic acids [5–7]. For illustration, the chemical shifts of the protons of diethylene-triaminepenta-acetic acid [DTPA] are shown as a function of pH in Fig. 6.1

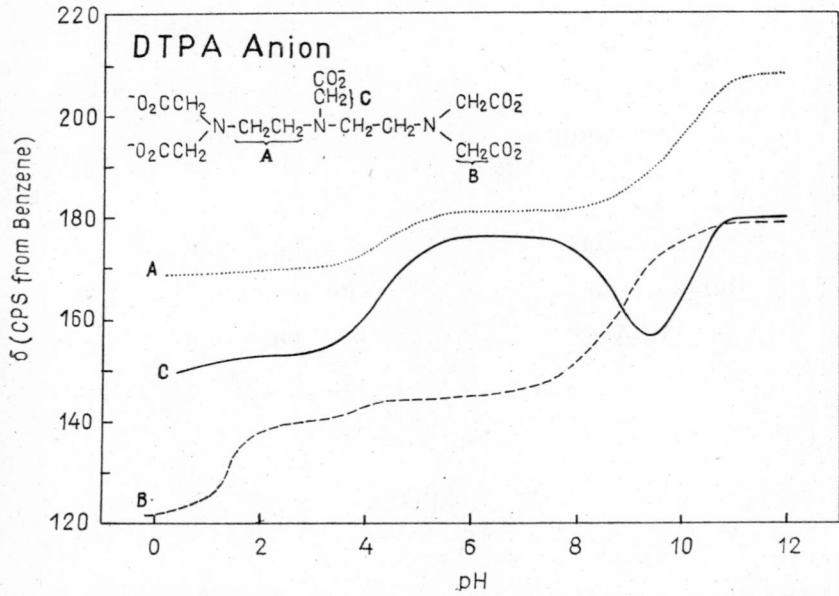

Fig. 6.1 Chemical shifts as a function of pH for the protons of diethylenetriamine-penta-acetate, 54.4 Mc. (Reproduced with permission from *Inorg. Chem.* **3**, 458 (1964))

[6], where the chemical shifts of the various methylene groups (A, B and C) of the DTPA molecule

$$^-OOC-CH_2 \diagdown \atop ^-OOC-CH_2 \diagup N-CH_2-CH_2-N-CH_2-CH_2-N \diagup CH_2-COO^- \atop \diagdown CH_2-COO^-$$

with the central branch COO^- — CH_2 — attached to the central N.

are plotted

It can be concluded that the first proton is taken up by the DTPA anion at the central nitrogen atom. The uptake of the second proton results in a migration of the already bound proton, leading to the formation of a symmetrical species. The third proton is bound to the central nitrogen atom, and further protonation follows the greatest possible symmetry.

6.2.2 PROTONATION OF 1:1 METAL CHELATES

In this case the relationships are relatively simple. This type of complex is formed if the number of donor groups of the chelating agent is great enough to saturate or nearly saturate the co-ordination sphere of the central ion. The protonation occurs in z successive steps:

$$MeL \rightarrow MeHL \rightarrow \ldots MeH_{z-1}L \rightarrow Me + H_zL \qquad (6.8)$$

The essence of the potentiometric method of Schwarzenbach [8–10] is that the experiments are made in the presence of a large excess of the metal ion ($T_{Me} : T_L \geq 10$), hence

$$[Me] = T_{Me}. \qquad (6.9)$$

Considering the following equilibria

$$Me + L \rightleftharpoons MeL; \qquad K_{H,0} = \frac{[MeL]}{[Me][L]} \qquad (6.10)$$

$$Me + HA \rightleftharpoons MeHL; \qquad K_{H,1} = \frac{[MeHL]}{[Me][HL]} \qquad (6.11)$$

$$H_2L \rightleftharpoons H^+ + HA; \qquad K_1^H = \frac{[H^+][HL]}{[H_2L]} \qquad (6.12)$$

$$HL \rightleftharpoons H^+ + L; \qquad K_2^H = \frac{[H^+][L]}{[HL]} \qquad (6.13)$$

and the following mass balances

$$T_L = [H_2L] + [HL] + [L] + [MeHL] + [MeL] \qquad (6.14)$$

$$T_{Me} = [Me] + [MeL] + [MeHL] \qquad (6.15)$$

$$(2 - a)T_L = [H^+] - [OH^-] + 2[H_2L] + [HL] + [MeHL] \qquad (6.16)$$

where a is the number of equivalents of base added, it follows that

$$\left(2 - a - \frac{[H^+] - [OH^-]}{T_L}\right)(1 + K_{H,0}T_{Me}) +$$

$$+ \left(1 - a + \frac{[H^+] - [OH^-]}{T_L}\right)(1 + K_{H,1}T_{Me})[H^+] +$$

$$+ \left(-a - \frac{[H^+] - [OH^-]}{T_L}\right)K_{a1}K_{a2}[H^+]^2 = 0. \qquad (6.17)$$

Equation 6.17 can be solved numerically. The same method can be applied where the number of the protons taken up by the chelate exceeds three, but in this case the application of computers is helpful. An inherent limitation of this and the similar titration method of Chaberek, Courtney and Martell [11] is that at low pH values they cannot be applied to follow the protonation. Another method [12–13] can be used with fairly acidic solutions, assuming that the concentration of the free metal ion or the sum of the concentrations of the complexes can be determined. The sum of the concentrations of the different protonated complexes is

$$\sum_0^{z-1} [\mathrm{MeH_iL}] = [\mathrm{MeL}] + [\mathrm{MeHL}] + \ldots + [\mathrm{MeH_{z-1}L}] \qquad (6.18)$$

where z is the number of the proton acceptor groups of the multidentate ligand L. Let us define the following function

$$\varkappa(\mathrm{H^+}) = \frac{\displaystyle\sum_0^{z-1} [\mathrm{MeH_iL}]}{[\mathrm{Me}][\mathrm{L}]} = \frac{[\mathrm{MeL}]}{[\mathrm{Me}][\mathrm{L}]} + \frac{[\mathrm{MeHL}]}{[\mathrm{Me}][\mathrm{L}]} + \ldots + \frac{[\mathrm{MeH_{z-1}L}]}{[\mathrm{Me}][\mathrm{L}]}. \qquad (6.19)$$

The first term on the right-hand side of Eq. 6.19 is the stability constant of the non-protonated complex, and the further terms only differ from the constants of the protonated complexes

$$K_{\mathrm{H},n} = \frac{[\mathrm{MeH_nL}]}{[\mathrm{Me}][\mathrm{H_nL}]} \qquad (6.20)$$

in that in the denominator $[\mathrm{L}^{l-}]$ is written instead of $[\mathrm{H_nL}^{n-l}]$. Expressing the concentration of $\mathrm{MeH_nL}^{n+m-y}$ by Eq. 6.20 and substituting into Eq. 6.19 we get

$$\varkappa(\mathrm{H^+}) = \sum_0^{z-1} K_{\mathrm{H},i} \frac{[\mathrm{H_iL}]}{[\mathrm{L}]} = K_{\mathrm{H},0} + K_{\mathrm{H},1} \frac{[\mathrm{HL}]}{[\mathrm{L}]} + \ldots + K_{\mathrm{H},z-1} \frac{[\mathrm{H_{z-1}L}]}{[\mathrm{L}]}. \qquad (6.21)$$

Considering that

$$\frac{[\mathrm{H_nL}]}{[\mathrm{L}]} = \frac{[\mathrm{H^+}]^n}{K_{z+1} \cdot K_z \cdot K_{z-1} \ldots K_{z-\mathrm{N}+1}} \qquad (6.22)$$

$\varkappa(\mathrm{H^+})$ can be written as follows

$$\varkappa(\mathrm{H^+}) = \sum_0^{z-1} K_{\mathrm{H},i} \frac{[\mathrm{H^+}]^i}{K_z \cdot K_{z-1} \cdot \ldots K_{z-i+1}} =$$

$$= K_{\mathrm{H},0} + K_{\mathrm{H},1} \frac{[\mathrm{H^+}]}{K_z} + \ldots + K_{\mathrm{H},z} \frac{[\mathrm{H^+}]}{K_z \cdot K_{z-1} \ldots K_2}. \qquad (6.23)$$

Three kinds of behaviour can be distinguished. In the first, the uptake of one proton results in the splitting off of the ligand, therefore the formation of a protonated complex in detectable amount is excluded. Plotting $\varkappa(\mathrm{H^+})$ as a function of the hydrogen ion concentration gives a straight line parallel to the abscissa.

$$\varkappa(\mathrm{H^+}) = K_{\mathrm{H},0}.$$

If besides the non-protonated complex only a monoprotonated complex is formed in appreciable concentration, the uptake of the second proton leads to a total decomposition. Plotting $\varkappa(H^+)$ as a function of the hydrogen ion concentration gives a straight line with intercept $K_{H,0}$, and $K_{H,1}$ can be directly calculated from the slope.

If more than one protonated complex exists a curve of higher order is obtained by plotting $\varkappa(H^+)$ versus $[H^+]$. The stability constants of different complexes can be obtained from an analysis of this curve, most conveniently by means of computers.

Approximate values of the stability constants of the protonated Fe(III)-EDTA and Fe(III)DCTA complexes are given in Table 6.1.

TABLE 6.1

Stability constants of the protonated Fe(III)EDTA
and Fe(III)DCTA complexes*

$K_{H,i}$	Fe(III)EDTA	Fe(III)DCTA
$K_{H,0}$	1.68×10^{24}	3.0×10^{27}
$K_{H,1}$	1.80×10^{15}	1.0×10^{17}
$K_{H,2}$	5.26×10^{8}	9.0×10^{10}
$K_{H,3}$	—	8.7×10^{7}

* EDTA = ethylenediaminetetra-acetate
DCTA = 1,2-diaminocyclohexanetetra-acetate

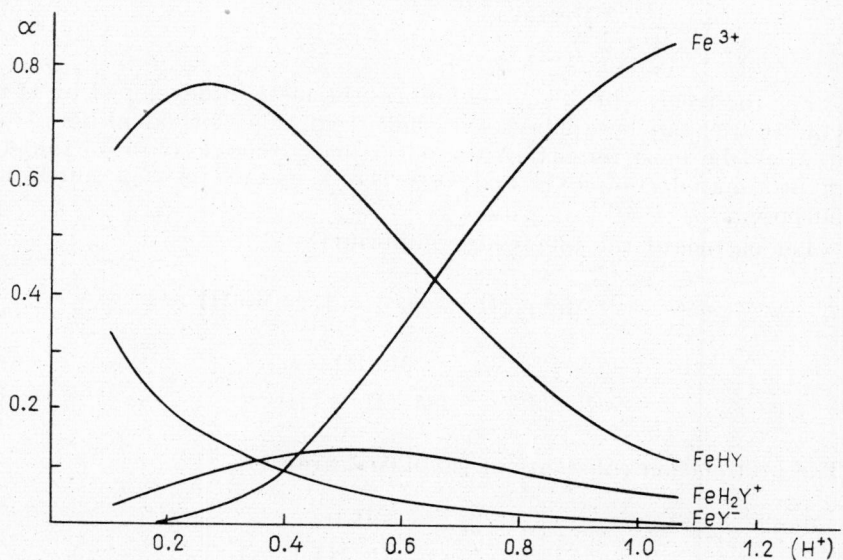

Fig. 6.2 Distribution of complexes as a function of hydrogen ion concentration at $T_{Fe} = 2.102 \times 10^{-3}$ M; $T_{EDTA} = 3 \times 10^{-3}$ M

Knowing the stability constants, the acidic dissociation constants and the total concentrations it is possible and instructive to calculate the distribution of the different complexes as a function of pH. Figure 6.2 presents such a diagram for the Fe(III)EDTA system.

6.2.3 PROTONATION OF METAL CHELATES IN GENERAL

If several ligand molecules are bound in the molecule of the metal chelate the protonation scheme is more complicated. The following chelate types are the most frequent. For co-ordination number four, MeX$_2$ (X is a bidentate ligand); for co-ordination number six, MeX$_3$ and MeY$_2$ (Y is a terdentate ligand). The overall protonation schemes are as follows

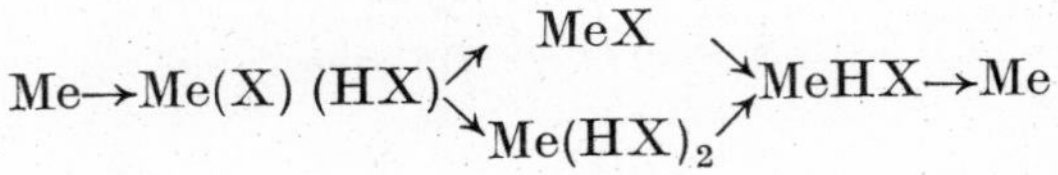

$$\text{Me} \rightarrow \text{Me(X) (HX)} \underset{\searrow\ \text{Me(HX)}_2}{\overset{\nearrow\ \text{MeX}}{}} \overset{\searrow}{\underset{\nearrow}{}} \text{MeHX} \rightarrow \text{Me} \tag{6.24}$$

$$\text{Me} \rightarrow \text{Me(HX)} \qquad \text{Me} \longrightarrow \text{Me(H)} \longrightarrow \text{Me(H)}_2 \qquad \text{MeH} \rightarrow \text{Me} \tag{6.25}$$
$$\text{Me(HX)}_2 \rightarrow \text{Me(HX)}_3 \qquad \text{Me}$$

$$\text{Me} \rightarrow \text{MeY(HY)} \overset{\text{MeX(H}_2\text{Y)} \longrightarrow \text{MeY} \longrightarrow \text{MeHY}}{\underset{\text{Me(HY)}_2}{\text{Me(HY)(H}_2\text{Y)} \rightarrow \text{Me(H}_2\text{Y)}_2 \rightarrow \text{Me(H}_2\text{Y)} \rightarrow \text{Me}}} \tag{6.26}$$

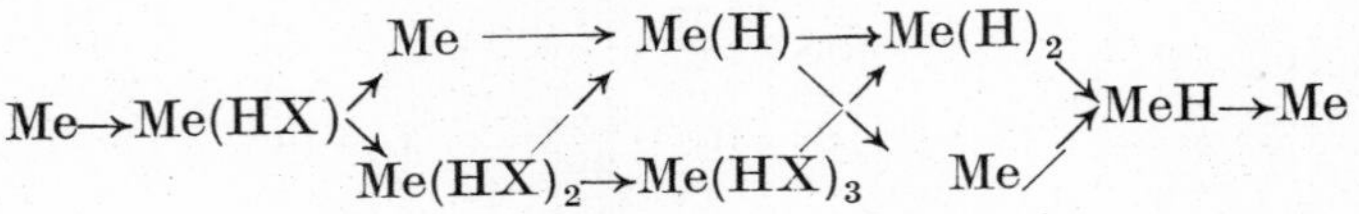

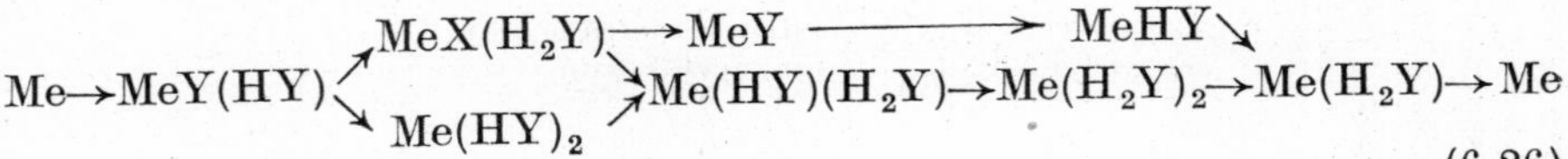

For the treatment of such equilibria the method developed by Österberg [14, 16–18] can be applied. We shall treat this method in detail because it would be most important to collect many reliable data on the stability of protonated complexes and Österberg's method is well suited for this purpose.

Let us regard the following equilibrium

$$\text{Me} + q\text{H}^+ + (q + r)\,\text{L} = \text{Me(HL)}_q\text{L}_r \tag{6.27}$$

$$\beta_{q,r} = \frac{\text{Me(HL)}_q\,\text{L}_r}{[\text{M}][\text{H}^+]^q[\text{L}]^{q+r}}\ . \tag{6.28}$$

The protonation constants of the ligand are

$$\beta_i^{\text{H}} = \frac{[\text{H}_i\text{L}]}{[\text{H}^+]^i[\text{L}]} \tag{6.29}$$

and the total concentrations

$$T_{Me} = Me\,(1\, + \varSigma\,\beta_{qr}\,[H]^q[L]^{q+r}). \tag{6.30}$$

$$T_L = [L]\,\varSigma\,(q + r)\,\beta_r[Me]\,[H^+]^q[L]^{q+r} + \varSigma\,i\,\beta_i^H\,[H^+]^i[L] \tag{6.31}$$

$$T_H = 2T_L - [NaOH] = [H^+] - [OH^-] + \varSigma q\,\beta_{qr}\,[Me]\,[H^+]^q\,[L]^{q+r}\, +$$
$$+ \varSigma\,i\,\beta_i^H\,[H^+]\,[L]. \tag{6.32}$$

The average co-ordination number

$$\bar{n} = \frac{T_L - [L] - \sum \beta_i^H\,[H^+]\,[L]}{T_{Me}} = \frac{\sum (q+r)\,\beta_q.[H^+]^q[L]^{q+r}}{1 + \sum \beta_{qr}[H^+]^q[L]^{q+r}}. \tag{6.33}$$

It follows from Hedström's theory [15] that

$$- \log\,[L] = -\Bigg[\int_{pH_0}^{pH}\Bigg(\frac{\partial\,T_H}{\partial\,T_L}\Bigg)_{[H^+]}\,d\,pH\Bigg]_{T_L} - \log\,[L]_0. \tag{6.34}$$

The value of $\log\,[L]_{\!\shortmid}/[L]$ can be obtained by graphical integration. The procedure is illustrated by Österberg's data for the copper(II)–O-phosphoryl-ethanolamine system. First, pH is plotted as a function of T_H at different T_L (Fig. 6.3) and from these curves a T_H–T_L diagram is constructed

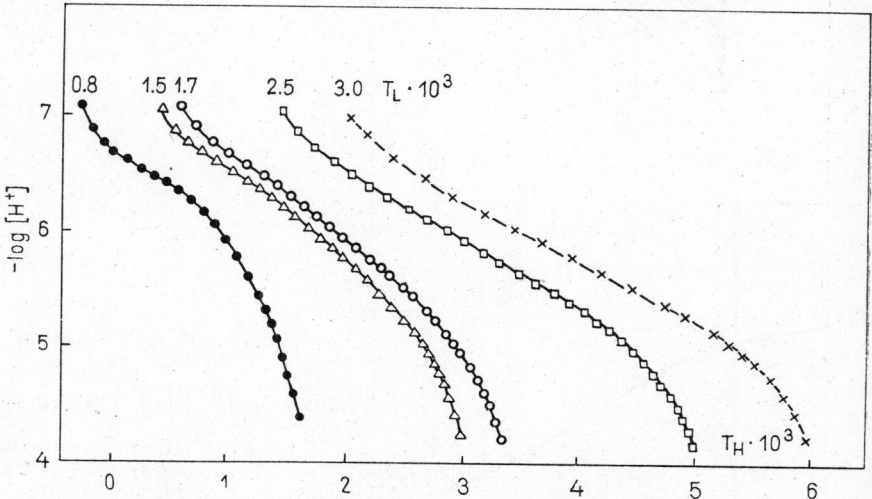

Fig. 6.3 pH as a function of T_H at different concentrations (T_L) of O-phosphoryl-ethanolamine. (Reproduced with permission from *Acta Chem. Scand.* 14, 471 (1960))

(Fig. 6.4). From these the values of $\left(\dfrac{\partial T_H}{\partial T_L}\right)_{[H^+]}$ as a function of pH are directly obtained. It appears from Fig. 6.4a that in the copper(II)–O-phosphorylethanolamine system $pH_0 = 4.3$. Log $[L]_0/[L]$ is given by the area under the curve of Fig. 6.4a between pH_0 and the corresponding pH. $[L]_0$ can

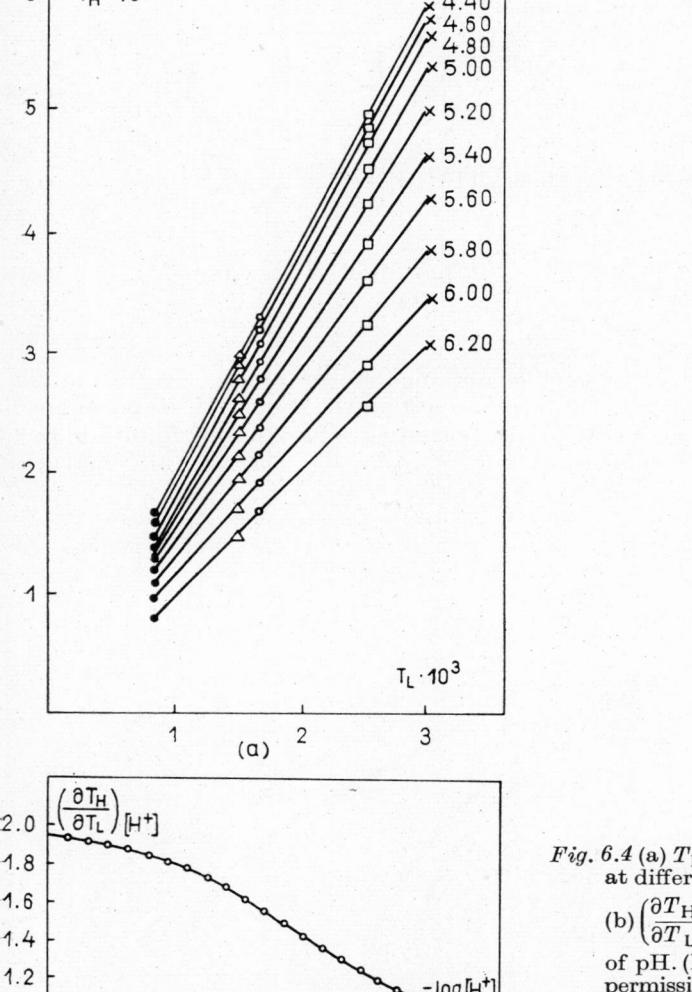

Fig. 6.4 (a) T_H as a function of T_L at different pH values;

(b) $\left(\dfrac{\partial T_H}{\partial T_L}\right)$ [4$^+$] as a function

of pH. (Reproduced with permission from *Acta Chem. Scand.* **14**, 471 (1960))

be calculated from Eq. 6.34:

$$[L]_0 = \frac{T_L}{1 + \beta_1^H[H^+] + \beta_2^H[H^+]^2} \; . \tag{6.35}$$

Since the value of $\bar{n}$ in this case even at the highest T_L and pH values does not exceed unity, it is reasonable to assume that the value of $(q + r)$ is not greater than 2. Therefore, it follows from Eq. 6.33 that

$$\frac{\bar{n}}{(1-\bar{n})[L]} \varphi = \beta_{10}[H^+] + \beta_{01} + \frac{(2-\bar{n})[L]}{1-\bar{n}} (\beta_{20}[H^+]^2 + \beta_{11}[H^+] + \beta_{02}) \tag{6.36}$$

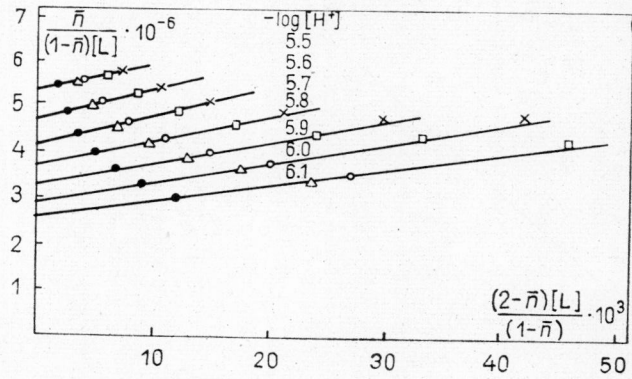

Fig. 6.5 $\bar{n}/(1-\bar{n})$ [L] as a function of $(2-\bar{n})$ [L]/$1-\bar{n}$. (Reproduced with permission from *Acta Chem. Scand.* **14**, 471 (1960))

where $\varphi = 1 + \beta_1^H [H^+]^{-1}$. By plotting $\bar{n}/(1 - \bar{n})[L]$ as a function of $(2 - \bar{n})[L]/(1 - \bar{n})$ a straight line is obtained (Fig. 6.5), the intercept and the slope of which are

$$I\varphi = \beta_{1,1}[H^+] + \beta_{0,1} \tag{6.37}$$

and

$$S\varphi = \beta_{2,0}[H^+]^2 + \beta_{1,2}[H^+] + \beta_{0,2}. \tag{6.38}$$

The solution of Eqs. 6.37 and 6.38 provides the unknown constants. The simplest way to solve these equations is to plot I and S as functions of $[H^+]$. (Fig. 6.6.) The fact that in both cases straight lines are obtained shows that the quadratic term in Eq. 6.32 is negligible; that is, the $Cu(HL)_2$ complex is not present in appreciable quantity. Österberg improved this

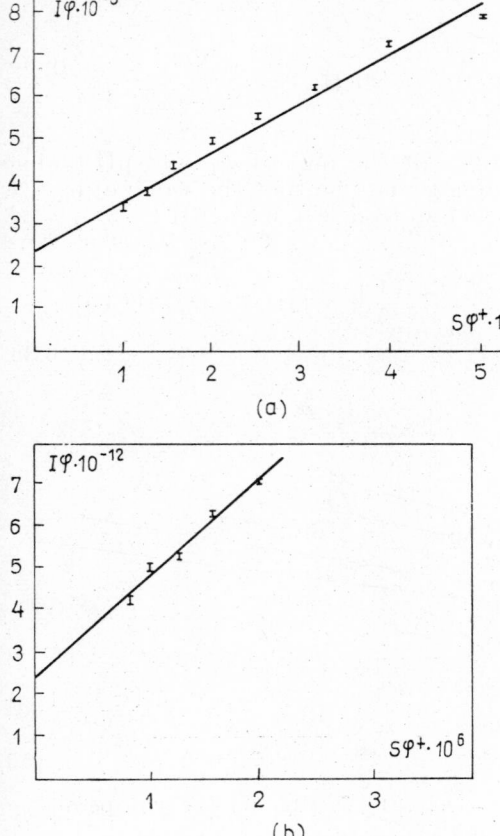

Fig. 6.6 (a) $I\varphi$ as a function of hydrogen ion concentration; (b) $S\varphi$ as a function of hydrogen ion concentration. (Reproduced with permission from *Acta Chem. Scand.* **14**, 471 (1960))

method by using computers. This latter method is suitable for the evaluation of constants of much more complicated systems, including the formation of polynuclear species. His data are summarized in Table 6.2.

TABLE 6.2

Ligand	Ion	Medium	Species (log β_{qr})
O-Phosphoryl-ethanolamine	H^+	0.15–(K)Cl	HL^- (10.13), H_2L (15.69)
	Mg^{2+}	0.15–(KCl)	MgL (1.70), $MgHL^+$ (11.37)
	Ca^{2+}	0.15–(KCl)	CaL (1.56), $CaHL^+$ (11.25)
	Mn^{2+}	0.15–(KCl)	MnL (2.55), $MnHL^+$ (11.86)
	Cu^{2+}	0.15–(K)Cl	CuL (6.39), $CuHL^+$ (12.07), (CuL_2^{2-} (12.4), $CuHL_2^-$ (18))

TABLE 6.2 *(continued)*

Ligand	Ion	Medium	Species (log β_{qr})
O-Phosphorylserine	H^+	0.15–K(Cl)	HL^{2-} (9.71), H_2L^- (15.36), H_3L (17.44)
	Mg^{2+}	0.15–(KCl)	MgL^- (2.39), MgHL (11.3)
	Ca^{2+}	0.15–(KCl)	CaL^- (2.15), CaHL (11.1)
	Mn^{2+}	0.15–(KCl)	MnL^- (3.8), MnHL (11.6)
	Cu^{2+}	0.15–(KCl)	CuL^- (9.65), CuHL (11.55), CuL_2^{4-} (15.5)
	Fe^{3+}	0.15–K(Cl)	FeL (14.0), ($FeHL^+$ (16.3), FeL_2^{3-} (21.4))
O-Phosphoryl-D,L-serylglycine	H^+	0.15–(K)Cl	HL^{2-} (8.01), H_2L^- (13.42), H_3L (16.55)
	H^+	0.15–K(NO_3)	HL^{2-} (8.02), H_2L^- (13.46), H_3L (16.66)
	Mg^{2+}	0.15–(KCl)	MgL^- (1.94), MgHL (9.41), Mg_2L^+ (3.2)
	Ca^{2+}	0.15–(KCl)	CaL^- (1.85), CaHL (9.31)
	Mn^{2+}	0.15–(KCl)	MnL^- (2.63), MnHL (9.90), Mn_2L^+ (4.2)
	Cu^{2+}	0.15–K(NO_3)	CuL^- (7.88), CuHL (11.58), CuHL (11.58), $CuHL_2^{3-}$ (18.0), (CuL_2^{4-} (11.6))
O-Phosphoryl-L-seryl-L-glutamic acid	H^+	0.15–(K)Cl	HL^{3-} (8.25), H_2L^{2-} (13.94), H_3L^- (18.33), H_4L (21.35)
	H^+	0.15–K(NO_3)	HL^{3-} (8.19), H_2L^{2-} (13.88), H_3L^- (18.28), H_4L (21.33)
	Mg^{2+}	0.15–(KCl)	MgL^{2-} (2.09), $MgHL^-$ (9.88), MgH_2L (14.9), Mg_2L (3.9), Mg_2HL^+ (11.4)
	Ca^{2+}	0.15–(KCl)	CaL^{2-} (2.14), $CaHL^-$ (9.89), CaH_2L (15.0), Ca_2L (3.6), $CaHL^+$ (10.9)
	Mn^{2+}	0.15–(KCl)	MnL^{2-} (2.98), $MnHL^-$ (10.49), MnH_2L (15.4), Mn_2L (4.9), Mn_2HL^+ (11.8)
	Cu^{2+}	0.15–K(NO_3)	CuL^{2-} (8.34), $CuHL^-$ (13.04), CuH_2L (16.65), CuL_2^6 (12.36), $CuH_2L_2^{4-}$ (24.3)
O-Phosphoryl-L-seryl-L-lysine	H^+	0.15–(K)Cl	HL^- (7.58), H_2L (12.92), H_3L^+ (15.90)
	Mg^{2+}	0.15–(KCl)	MgL (1.63), $MgHL^+$ (8.81)
	Ca^{2+}	0.15–(KCl)	CaL (1.53), $CaHL^+$ (8.72)
	Mn^{2+}	0.15–(KCl)	MnL (2.33), $MnHL^+$ (9.36)
Glycyl-O-phosphoryl-D,L-serine	H^+	0.15–(K)Cl	HL^{2-} (8.42), H_2L^- (14.45), H_3L (17.37)
	Mg^{2+}	0.15–(KCl)	MgL^- (1.86), MgHL (10.06), Mg_2L^+ (3.2)
	Ca^{2+}	0.15–(KCl)	CaL^- (1.77), CaHL (9.90)
Glycyl-O-phosphoryl-D,L-serylglycine	H^+	0.15–(K)Cl	HL^{2-} (8.22), H_2L^- (13.99), H_3L (17.28)
	Mg^{2+}	0.15–(KCl)	MgL^- (1.79), MgHL (9.70)
	Ca^{2+}	0.15–(KCl)	CaL^- (1.81), CaHL (9.67)
	Mn^{2+}	0.15–(KCl)	MnHL (10.36)

6.3 Protonation schemes of metal chelates

The stability constant ($\beta_{H,n} = [MeH_ML][Me]^{-1}[H_nL]^{-1}$) and the protonation constant ($\beta_{H,n}^H = [MeH_nL][MeL]^{-1}[H^+]^{-n}$) of the MeH_nL complex, the protonation constant of the ligand ($\beta_n^H = [H_nL][L]^{-1}[H^+]^{-n}$) and the stability constant of the metal chelate are in the simple relationship

$$\beta_{H,n} \cdot \beta_n^H = \beta_{H,n}^H \cdot \beta_{H,0}. \tag{6.39}$$

It must be borne in mind, however, that there may be several isomeric ligands of the composition H_nL, and the corresponding β_n^H must be considered. Sometimes it is ambiguous which acidic dissociation or protonation constant has to be used. For example the ligand 4-(2-pyridylazo)-resorcinol (PAR) dissociates according to the following scheme

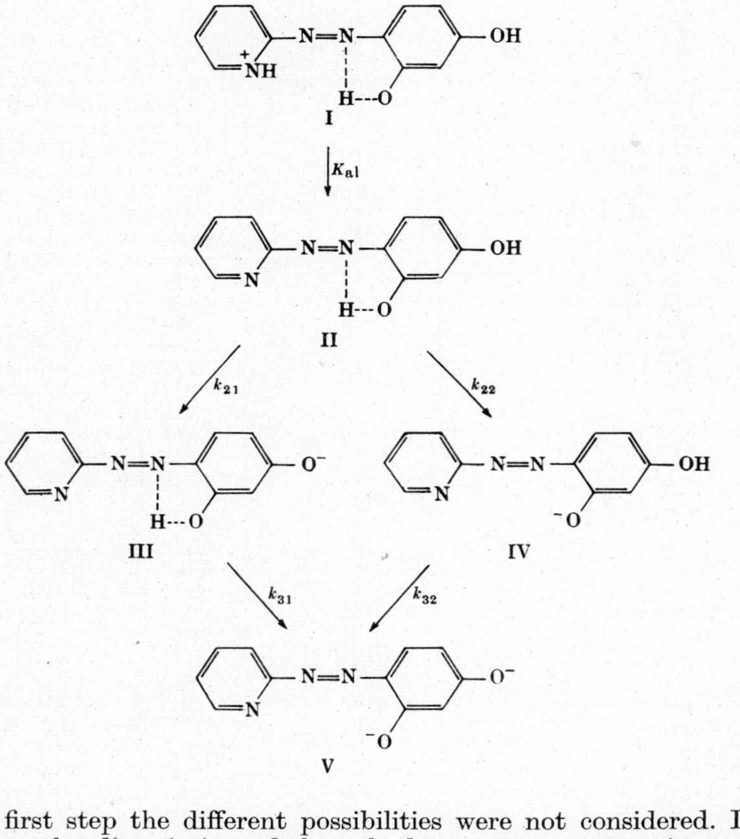

In the first step the different possibilities were not considered. In the second step the dissociation of the p-hydroxy group proton is preferred because of the internal hydrogen bridge involving the o-hydroxy group.

Therefore the experimentally determined K_{a2} value is approximately equal to k_{21}. However, PAR forms complexes with metal ions in the form of the isomeric species IV, and the corresponding k_{22} is not available. It has generally been considered that $k_{21} = K_{a3}$ (the potentiometrically determined K_{a3} is practically equal to k_{32}) e.g. [19, 20]. As Chalmers [21] has pointed out, however, this view is not necessarily correct, and a better approximation would be to take k_{22} as equal to K_{a2}, since (i) the contribution of the internal hydrogen bridge to the increase in pK for the o-hydroxy proton would be expected to be small compared with the inductive effect of the dissociated p-hydroxy group, and (ii) there seems to be no reason why, in the absence of the hydrogen-bonding effect, the o-hydroxy group should not be at least as acidic as the p-hydroxy group.

All the equilibrium constants which can be assigned to the protonation schemes given by Eqs. 6.8, 6.24–6.26 are, in fact, overall constants. That is, the protonation may occur at different sites leading to the formation of isomeric protonated metal complexes. It is fairly difficult and — at least at present — in most cases not even possible to decide which isomer is predominant in the solution, and there is no system for which the 'microconstants' are available. The factors which have to be considered in choosing the predominant forms of the protonated metal chelate species are illustrated by the example of the protonation of cysteine complexes. The donor atoms being different, the decomposition of even the 1:1 chelate is very complicated. The following scheme, where cysteine is represented by $\widehat{ONS}$ shows all the possibilities.

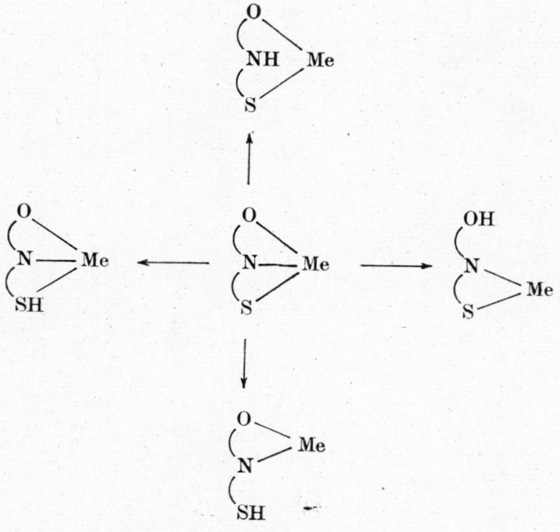

$$(6.40)$$

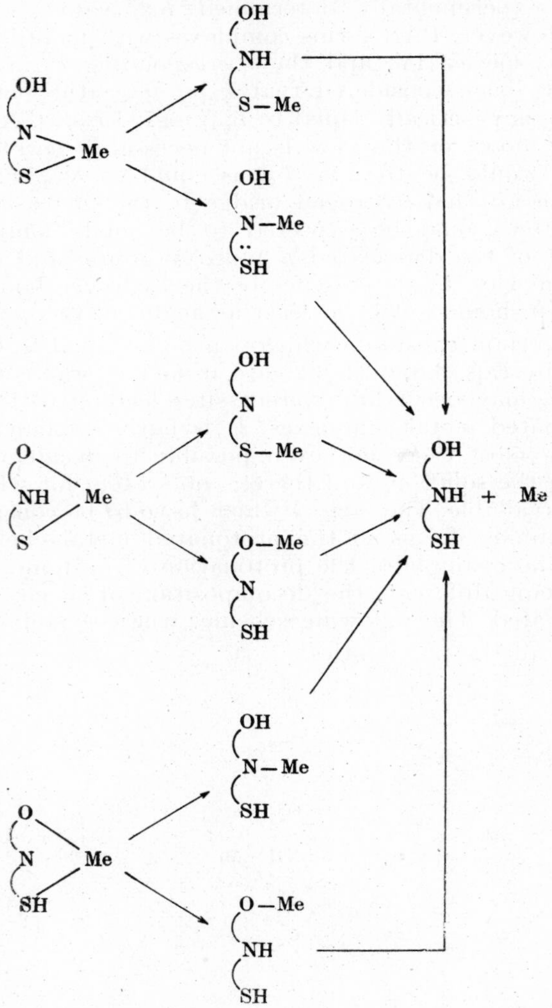

$$
\begin{array}{c}
\text{OH} \\
\text{N} \\
\text{S}
\end{array}\!\!\text{Me}
\quad
\begin{array}{c}
\text{OH} \\
\text{NH} \\
\text{S}-\text{Me}
\end{array}
\qquad
\begin{array}{c}
\text{OH} \\
\text{N}-\text{Me} \\
\text{SH}
\end{array}
$$

$$
\begin{array}{c}
\text{O} \\
\text{NH} \\
\text{S}
\end{array}\!\!\text{Me}
\quad
\begin{array}{c}
\text{OH} \\
\text{N} \\
\text{S}-\text{Me}
\end{array}
\qquad
\begin{array}{c}
\text{OH} \\
\text{O}-\text{Me} \\
\text{N} \\
\text{SH}
\end{array}
\qquad
\begin{array}{c}
\text{OH} \\
\text{NH} \\
\text{SH}
\end{array}
+ \quad \text{Me}
\qquad (6.41)
$$

$$
\begin{array}{c}
\text{O} \\
\text{N} \\
\text{SH}
\end{array}\!\!\text{Me}
\quad
\begin{array}{c}
\text{OH} \\
\text{N}-\text{Me} \\
\text{SH}
\end{array}
\qquad
\begin{array}{c}
\text{O}-\text{Me} \\
\text{NH} \\
\text{SH}
\end{array}
$$

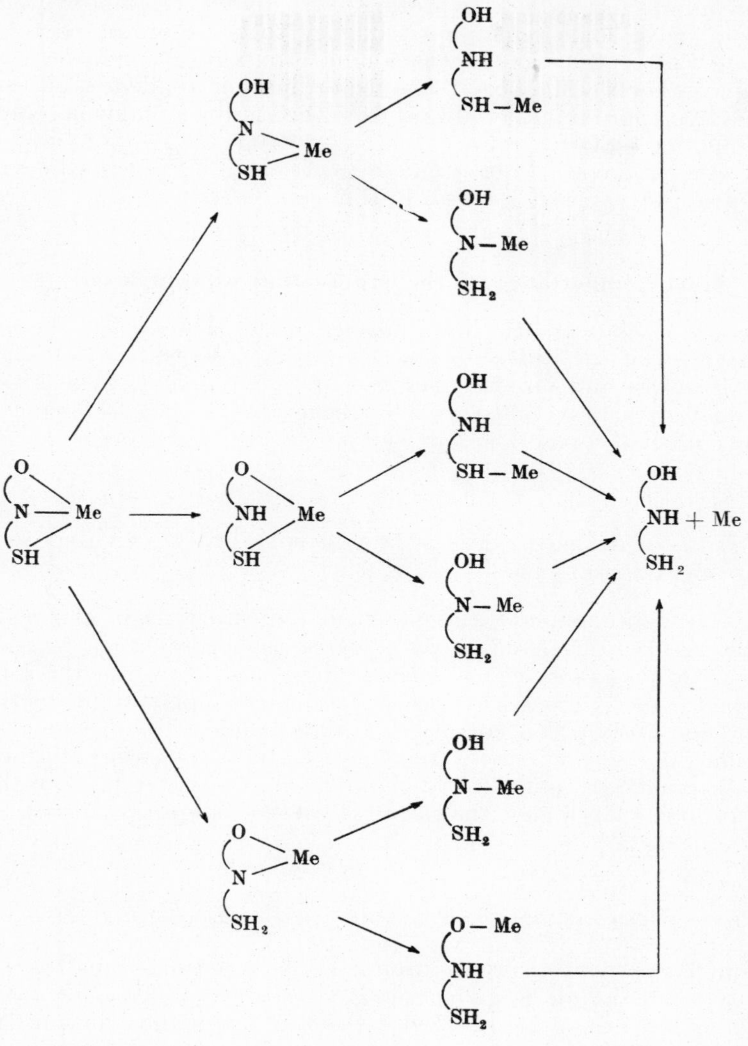

$$
\begin{array}{c}
\text{OH} \\
\overset{\displaystyle(}{\text{NH}} \\
\text{SH}-\text{Me}
\end{array}
$$

$$
\begin{array}{c}
\text{OH} \\
\overset{\displaystyle(}{\text{N}} \\
\text{SH}
\end{array}\!\!\Big\rangle\text{Me}
$$

$$
\begin{array}{c}
\text{OH} \\
\overset{\displaystyle(}{\text{N}}-\text{Me} \\
\text{SH}_2
\end{array}
$$

$$
\begin{array}{c}
\text{O} \\
\overset{\displaystyle(}{\text{N}}-\text{Me} \\
\text{SH}
\end{array}
$$

$$
\begin{array}{c}
\text{O} \\
\overset{\displaystyle(}{\text{NH}}\;\;\text{Me} \\
\text{SH}
\end{array}
$$

$$
\begin{array}{c}
\text{OH} \\
\overset{\displaystyle(}{\text{NH}} \\
\text{SH}-\text{Me}
\end{array}
$$

$$
\begin{array}{c}
\text{OH} \\
\overset{\displaystyle(}{\text{N}}-\text{Me} \\
\text{SH}_2
\end{array}
$$

$$
\begin{array}{c}
\text{OH} \\
\overset{\displaystyle(}{\text{NH}}+\text{Me} \\
\text{SH}_2
\end{array} \qquad (6.42)
$$

$$
\begin{array}{c}
\text{O} \\
\overset{\displaystyle(}{\text{N}}\!\!\Big\rangle\text{Me} \\
\text{SH}_2
\end{array}
$$

$$
\begin{array}{c}
\text{OH} \\
\overset{\displaystyle(}{\text{N}}-\text{Me} \\
\text{SH}_2
\end{array}
$$

$$
\begin{array}{c}
\text{O}-\text{Me} \\
\overset{\displaystyle(}{\text{NH}} \\
\text{SH}_2
\end{array}
$$

It follows from the microconstants of the dissociation of cysteine represented by Eq. 6.1 that at the protonation of IV the amine group is slightly preferred in comparison with the sulphide group. In the protonation of a cysteine complex it must be considered, however, that the protonation of the co-ordinated amine leaves a seven-membered ring, while the protonation of the terminal sulphide group leaves a five-membered ring. Further there is a great difference between the affinities of the amine and the sulphide groups for different metal ions (see Chapter 11). Finally, it must be

11

borne in mind that even the protonated sulphide thioalkoxy (i.e. thiol) group may have some remaining tendency to be co-ordinated to a metal ion, while the protonation of the amine group results necessarily in the splitting of the $N-Me$ bond. The co-ordination tendency of the $-SH$ group may be approximately judged from the study of analogous compounds containing the $-SCH_3$ group.

NMR studies have also been made to determine the co-ordination sites at different pH values for metal-EDTA complexes [22–23].

6.4 The kinetic importance of the protonation of complexes

There are obvious kinetic consequences of the protonation of complexes. The reactivity of the different metal complexes depends on the nature of the donor atoms and on the structure of the complexes. The kinetic role of protonation will be illustrated by several examples referring to substitution and redox reactions, as well as the reactions of the co-ordinated ligands.

6.4.1 THE ROLE OF PROTONATION IN THE FORMATION AND DECOMPOSITION OF METAL CHELATES

There are many kinetic studies which indicate that in the mechanism of acidic decomposition of metal chelates the protonated species play a role [24–27]. These species are intermediates and their kinetic role is well established even in cases where their thermodynamic stability is too small for their equilibrium concentrations to be considerable.

It seems to be very likely that protonated species are formed when metal ions react with multidentate ligands. However, the rates of the chelation steps are much higher than that of the first step, the formation of the first bond.

6.4.2 THE EFFECT OF PROTONATION IN REDOX REACTIONS

Although the effect of protonation in redox reactions must be very general, only one example is given here [28]. For thermodynamic reasons the Co(II) aquo-ion cannot be oxidized with Cr(VI) in acidic solution. However, in the presence of suitable ligands, e.g. aminopolycarboxylic acids, the potential of the Co(III)–Co(II) system decreases so greatly that the oxidation may take place. According to kinetic measurements the rate of oxidation shows a maximum as a function of pH, the pH of the rate maximum being shifted to lower values in the order EDTA < DCTA < DTPA. These and some other findings can be explained as follows. Complex formation is a necessary condition for oxidation. However, if all the six co-ordination sites of the cobalt(II) are occupied by the donor atoms of the aminopolycarboxylic acids the complex formation prevents the electron-transfer from Co(II) to the oxidant, that is, the co-ordinated aminopolycarboxylic acid plays the role of an 'insulator'. Therefore a further necessary condition of the redox process is that the donor groups of the ligand molecule be partially removed from the co-ordination sphere. This is achieved by the protonation of the complex making possible the direct or water-bridged interaction of the reactants (Fig. 6.7).

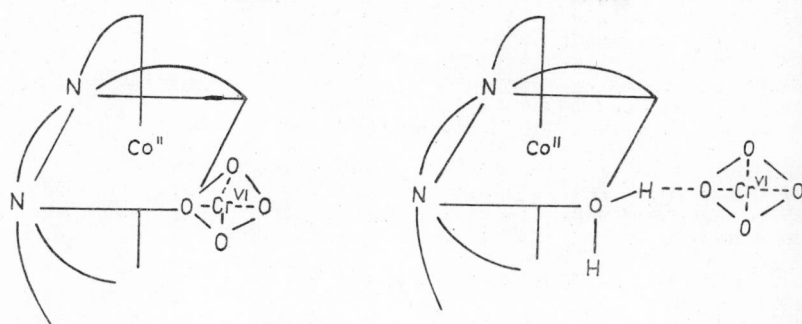

Fig. 6.7 Probable structure of the activated binuclear complex

6.4.3 THE EFFECT OF PROTONATION ON LIGAND REACTIVITY

Co-ordination always modifies and sometimes dramatically effects the reactivity of the co-ordinated ligands. The oxidation by permanganate of EDTA bound to different metal ions reveals the role of protonation [29]. The oxidation of the Cr(III)EDTA complex is faster than that of the Bi(III)-EDTA complex. In the Cr(III) complex EDTA behaves as a quinque-dentate ligand. The pH dependence of the rate of oxidation of the Cr(III)-EDTA complex shows that the protonated species is less reactive than the non-protonated complex containing free carboxylate groups. This and similar systems deserve a more thorough study.

REFERENCES

1. BJERRUM, N., *Z. Physik. Chem. Leipzig 106*, 219 (1923), for an English translation see N. BJERRUM, *Selected Papers*, Einar Munksgaard, Copenhagen, 1949, p. 198.
2. WRATHALL, D. O., IZATT, R. M. and CHRISTENSEN, J. J., *J. Am. Chem. Soc. 86*, 4779 (1964).
3. LOEWENSTEIN, A. and ROBERTS, J. D., *J. Am. Chem. Soc. 82*, 2705 (1960).
4. MARTIN, R. B., *J. Phys. Chem. 65*, 2053 (1961).
5. CHAPMAN, D., LLOYD, D. R. and PRINCE, R. H., *J. Chem. Soc.* 3645 (1963).
6. KULA, R. J. and SAWYER, D. T., *Inorg. Chem. 3*, 458 (1964).
7. SUDMEIER, J. L. and REILLEY, C. N., *Anal. Chem. 36*, 1698, 1707 (1964).
8. SCHWARZENBACH, G., WILLI, A. and BACH, R. O., *Helv. Chim. Acta 30*, 1303 (1947).
9. SCHWARZENBACH, G., *Helv. Chim. Acta 33*, 947 (1950).
10. SCHWARZENBACH, G. and ACKERMANN, A., *Helv. Chim. Acta 31*, 1029 (1948).

11. CHABEREK, S., COURTNEY, R. C. and MARTELL, A. E., *J. Am. Chem. Soc.* 75, 2185 (1956).
12. BECK, M. T. and GÖRÖG, S., *Acta Chim. Acad. Sci. Hung.* 22, 159 (1960).
13. BECK, M. T. and GÖRÖG, S., *J. Inorg. Nucl. Chem.* 12, 353 (1960).
14. ÖSTERBERG, R., *Acta Chem. Scand.* 14, 471 (1960).
15. HEDSTRÖM, B., *Acta Chem. Scand.* 9, 613 (1955).
16. ÖSTERBERG, R., *Acta Chem. Scand.* 19, 1445 (1965).
17. ÖSTERBERG, R., *Arkiv Kemi* 25, 177 (1966).
18. ÖSTERBERG, R., *Thesis*, Göteborg, 1966.
19. HLINIČKOVÁ, M. and SOMMER, L., *Collection Czech. Chem. Commun.* 26, 278 (1961).
20. STANLEY, R. W. and CHENEY, G. E., *Talanta* 13, 1619 (1966).
21. CHALMERS, R. A., *Talanta* 14, 527 (1967).
22. KULA, R. J., SAWYER, D. T., CHAN, S. I. and FINLEY, C. M., *J. Am. Chem. Soc.* 85, 2930 (1963).
23. AOCHI, Y. O. and SAWYER, D. T., *Inorg. Chem.* 5, 2085 (1966).
24. GRAZIANO, F. D. and HARRIS, G. M., *J. Phys. Chem.* 63, 330 (1959).
25. KRISHNAMURTY, K. V. and HARRIS, G. M., *J. Phys. Chem.* 64, 346 (1960).
26. SCHLÄFER, H. L., KLING, O., MÄHLER, L. and OPITZ, H. P., *Z. Physik. Chem.* Frankfurt 24, 307 (1960).
27. WELCH, G. L. and HAMM, R. E., *Inorg. Chem.* 2, 295 (1963).
28. BECK, M. T., SERES, I. and BÁRDI, I., *Acta Chim. Acad. Sci. Hung.* 41, 231 (1964).
29. BECK, M. T. and KLING, O., *Acta Chem. Scand.* 15, 453 (1961).

Chapter 7

FORMATION AND STABILITY OF AQUO-COMPLEXES

As has already been stressed, the representation of successive complex formation by the reactions

$$Me + L \rightleftharpoons MeL$$
$$\vdots$$
$$MeL_{N-1} + L \rightleftharpoons MeL_N$$

(7.1)

is a simplification. In solutions metal ions always exist as complexes, their co-ordination sphere being saturated with suitable ligands. If the counter-ion is not a good donor and other complex-forming ligands are absent, the solvent molecules themselves are co-ordinated: solvo-complexes are formed. Complex formation therefore means the stepwise displacement of the co-ordinated solvent molecules by the ligand

$$MeS_N + L \rightleftharpoons MeS_{N-1}L + S$$
$$\vdots$$
$$MeSL_{N-1} + L \rightleftharpoons MeL_N + S.$$

(7.2)

The stability constant of the ith complex therefore is

$$K_i^* = \frac{[MeS_{N-i}L_i][S]}{[MeS_{N-i+1}L_{i-1}][L]}$$

(7.3)

instead of the usual expression

$$K_i = \frac{[MeL_i]}{[MeL_{i-1}][L]}.$$

(7.4)

Since in dilute solutions the concentration of the solvent is practically constant, the constants defined by Eq. 7.4 can be used to describe the distribution of complexes and to characterize their thermodynamic stabil-

165

ity. The comparison of Eqs. 7.3 and 7.4 shows the relationship between K_i and K_i^*

$$K_i[S] = K_i^* \qquad (7.5)$$

and similarly

$$\beta_i[S]^i = \beta_i^*. \qquad (7.6)$$

7.1 Hydration of ions

There is a fair amount of experimental evidence that ions are hydrated in aqueous solution. Some methods can even give quantitative information on the extent of hydration. Hydration numbers for different ions have been evaluated from the transport of solvent during ionic migration in an electrostatic field [1]; from the effect of electrolytes on the solubility of non-electrolytes [2]; from the activity of the water in a given electrolyte solution [3]; from the density of salt solutions [4, 5]; from the amount of water extracted together with a given salt into an organic solvent [6, 7]; etc. Frequently there are large differences between the hydration numbers determined by different methods. The basic source of these differences is that the different methods give information on different types of hydration. The solution of the problem cannot be approached without considering the structure of water. This problem, however, is beyond the scope of this book, so only a brief account is given here. The reader may consult the excellent review by Wicke [8] and the original papers cited therein. The structure of a single water molecule is well known and explains the strong interaction of water and metal ions. Of the two lone pairs of electrons only one is used when water is co-ordinated to a metal ion. The views on the structure of liquid water and that of aqueous solutions are very controversial. It is agreed that owing to hydrogen bridges there are associated water molecules present, but the authors disagree on the number of the associated molecules and the structure of these species. According to Eucken [9], besides single water molecules, dimers, tetramers and octamers are the main constituents. In most of the recent models the existence of large clusters consisting of 25–100 water molecules is assumed, and sometimes even the existence of monomeric water molecules is denied. According to the infrared studies by Yamatera, Fitzpatrick and Gordon [10] three kinds of water molecules can be distinguished depending on whether the molecule uses none, one or both of the available electron pairs for making hydrogen bonds with the adjacent water molecule(s).

It is evident that a fraction of the water molecules in the environment cf a metal ion is directly bonded by use of one of the available electron pairs of the oxygen atom of the water molecule. This fraction can be regarded as co-ordinated water molecules. The number of directly co-ordinated water molecules can be determined from isotopic exchange experiments [11], from NMR studies [12] and from density measurements [4, 5]. As expected, the number of co-ordinated water molecules is equal to the characteristic co-ordination number of the ion. To the directly-attached water molecules further water molecules may be bonded through hydrogen bridges. The energy of these hydrogen bonds is evidently a function of the distance from the centre of the metal ion. The different methods for the determination of hydration numbers differentiate between molecules held at different strengths.

7.2 Stability constants of aquo-complexes

There is no sense in speaking of the successive formation of aquo-complexes in solution, the co-ordination sphere of the metal ions being always saturated. The stability constant of the aquo-complex of maximum co-ordination number can be calculated from data for the gas phase by means of thermodynamic cycles. Such calculations were made by Yatsimirskii [13]. His results are given in Table 7.1.

TABLE 7.1

Thermodynamic constants for the reaction

$$Me^{2+}_{(g)} + 6\ H_2O_{(g)} = Me(H_2O)_{6(g)}$$

Cations	$[Me(H_2O)_6]^{2+}$			
	$-\Delta H$ kcal·mole^{-1}	$-\Delta S$ cal·mole^{-1}·deg^{-1}	$-\Delta G$ kcal·mole^{-1}	$\log \beta_6$
Ca^{2+}	291	138	250	184
Cr^{2+}	...	138	...	...
Mn^{2+}	308	138	267	196
Fe^{2+}	...	138	...	...
Co^{2+}	365	138	324	238
Ni^{2+}	373	138	332	244
Cu^{2+}	374	138	333	244
Zn^{2+}	359	138	318	233

In mixed solvents the concentration of the water can be adjusted to a value at which the co-ordination sites are only partially occupied by water molecules. However, if the counter-ion cannot be co-ordinated by the central ion, the remaining co-ordination sites are occupied by the molecules of the other solvents and increase of the water concentration results in the gradual displacement of the co-ordinated solvent molecules. In mixed solvents therefore it is possible to determine the equilibrium constants of displacement reactions of this kind. Such systems were studied first by Bjerrum and Jørgensen [14]. They determined the absorbance of methanolic solutions of the salts of a number of transition metal ions as a function of water concentration. From their results they evaluated the equilibrium constants, defined as

$$K_i^* = \frac{[MeS_{N-i}(H_2O)_i]\,[S]}{[MeS_{N-i+1}(H_2O)_{i-1}]\,[H_2O]} \tag{7.7}$$

by the method of corresponding solutions. Later these investigations were extended by Jørgensen [15]. Katzin and Gebert [16] have criticized these

TABLE 7.2

Stability constants defined by Eq. 7.7

Me	S	$\log K_1^*$	$\log K_2^*$	$\log K_3^*$	$\log K_4^*$	$\log K_5^*$	$\log K_6^*$	Method	Remarks	Ref.
Cu^{2+}	nitromethane	4.12	3.23	2.41	2.29	1.79	1.77	polarography	0.1 M (C $(C_2H_5)_4NClO_4$	[18]
Cu^{2+}	nitromethane	4.21	3.16	2.58	2.12	1.79	—	spectrophotometry	$K_6^* > K_5^*$	[18]
Cu^{2+}	ethanol	0.98	0.70	0.35	0.28	0.07	0.04	spectrophotometry	From data of Mine and Libus [19]	[20]
Cu^{2+}	ethanol	—	—	—	—	1.41	−0.26	spectrophotometry		[20]
Cu^{2+}	acetone	2.88	2.63	2.13	1.88	—		polarography	0.1 M $LiClO_4$	[21]
Cu^{2+}	acetone	2.88	2.38	1.93	1.78	—		polarography	0.1 M $(C_2H_5)_4NClO_4$	[21]
Cu^{2+}	acetone	—	—	—	—	2.51	−0.57	spectrophotometry	—	[20]
Co^{2+}	ethanol	—	—	—	—	2.22	−0.46	spectrophotometry	—	[22]
Co^{2+}	acetone	—	—	—	—	1.82	−0.56	spectrophotometry	—	[22]
Ni^{2+}	ethanol	—	—	—	—	2.31	0.24	spectrophotometry	—	[22]
Ni^{2+}	acetone	—	—	—	—	1.68	−0.79	spectrophotometry	—	[22]

studies and attributed the changes in the absorption of light to the inter-action between the metal ion and the counter-ion, which is influenced by the water content of the solvent. They pointed out that the spectra of $Co(NO_3)_2$ in ethanolic and in aqueous solution are considerably different, but the addition of 10% of water to the ethanolic solution results in the spectrum of the hexa-aquocobalt(II) ion. Jørgensen [17] concluded that the spectral behaviour of the salts of transition metal ions in mixed solvents cannot be adequately described by simple equilibrium considerations. He suggested the term *Katzin effect* for the phenomenon in which anions are unusually strongly bound to certain metal ions if the solvent does not contain at least minor quantities of privileged solvating ligands, such as water and ammonia. It is very probable that for the peculiar phenomena found in mixed solvents, the specific interactions of the different solvents and their mutual effect on each other's activities is responsible.

The stability constants determined by Bjerrum and Jørgensen [14] and by Jørgensen [15] are so-called average constants. Later the constants referring to the stepwise displacement of the organic solvent by water were polarographically and spectrophotometrically determined. A survey of these constants is given in Table 7.2. It must be mentioned that the constants were recalculated from the published data according to Eq. 7.7, taking into account the concentrations of the organic solvents.

Recently an extrapolation method was suggested for the determination of the stability constants of aquo-complexes. Swinarski [23] plotted the successive stability constants as a function of the number of ligands taken up and stated that for many systems the values extrapolated to zero for the same central ion are independent of the nature of the ligand. This extrapolated value is attributed to the stability constant of the correspond-ing aquo-complex. A good agreement was found between these values and those determined in solvent mixtures. It appears that this procedure has no sound basis. The systems considered by Swinarski did not contain any other solvent but water, so the comparison of these numerical values with the constants referring to solvent mixtures is not correct. The weakest point of this extrapolation method is that the stability con-stants have physical meaning only for co-ordination numbers from one to the maximum value, and the value of the zero-th stability constant is unity by definition.

As has already been mentioned, in dilute or in moderately concentrated solutions, stepwise complex formation can be adequately described by the stability constant defined by Eq. 7.4. If, however, the water concentra-tions in different systems are substantially different, the stability con-stants cannot be directly compared. The concentration of the water is greatly altered if neutral salts are present in high concentration. This change in the water concentration results in the decrease of the mole fraction of the water and — to a much greater extent — in the decrease of the amount of free water, owing to the hydration of the neutral salt. This appears from Table 7.3, the data of which refer to sodium perchlorate solutions, assuming the values 0, 1, 3 and 10 for the total hydration number of the salt [24].

TABLE 7.3

The molarity of water in sodium perchlorate solutions

M_{NaClO_4}	M_{H_2O}	$M_{H_2O} - M_{NaClO_4}$	$M_{H_2O} - 3M_{NaClO_4}$	$M_{H_2O} - 10M_{NaClO_4}$
0	55.51	55.51	55.51	55.51
0.082	55.23	55.15	54.98	54.41
1.008	52.74	51.68	49.57	42.16
2.952	47.51	44.51	38.52	17.55
9.396	30.06	20.74	2.094	—

The data show that when the concentration of the salt exceeds 1 M the water concentration is appreciably smaller, even disregarding the decrease due to hydration. In concentrated sodium perchlorate solutions, if hydration by only three molecules of water per pair of ions is assumed there is practically no 'free' water in the system. This change of the water concentration must be considered when stability constants determined at widely different ionic strengths are compared. Systematic investigations, even for simple complex systems, to trace the effect of salt concentration on stability constants, are lacking and research in this direction would be very important.

Finally, it must be stressed that every calculation using the concentration of water has to be regarded with some reservation. In the calculations it would be necessary to use the concentration of monomeric water, the existence of which is denied by some of the theories. Evidently the relative fraction of this species will increase with decreasing water concentration because of the decreasing probability of association. Consequently the effect of the increasing neutral salt concentration on the formation of complexes is even more complicated than that taken into account previously. The concentration of the free water decreases owing to the hydration of the neutral salt, but the distribution of the different polymeric water molecules within the 'free water' is modified, and, because of the decreasing concentration, the monomeric species becomes more dominant. The situation is the same as in solvent mixtures. At high concentration of an organic solvent, that is at low water concentration, the monomeric water species will be predominant [25], but in such solutions the interaction between the organic solvent and the water molecules will greatly affect the activity of the water. Conversely, the interactions between the organic solvent molecules themselves decrease with increasing concentration of water, but the interactions between the molecules of water and the organic solvent will more strongly influence the activity of the organic solvent. These complicated interactions may be responsible for the interesting phenomena observed by Katzin and Jørgensen.

REFERENCES

1. BUCHBÖCK, G., Z. Physik. Chem. Leipzig 55, 563 (1906).
2. KOSAKEWICH, P. P. and ISMAILOW, N. A., Z. Physik. Chem. Leipzig 150A, 308 (1930).
3. BJERRUM, N., Z. Anorg. Allgem. Chem. 109, 275 (1920).
4. BERNAL, B. J. and FOWLER, R. H., J. Chem. Phys. 1, 515 (1933).
5. CELADA, J., Sci. Papers Inst. Chem. Technol. Pardubice 3, 15 (1959).
6. KATZIN, L. I. and SULLIVAN, J. C., J. Phys. Colloid Chem. 55, 346 (1951).
7. YATES, P. C., LARAN, R., WILLIAMS, R. E. and MOORE, T. E., J. Am. Chem. Soc. 75, 2212 (1953).
8. WICKE, E., Angew. Chem. Intern. Ed. Engl. 5, 106 (1966).
9. EUCKEN, A., Z. Electrochem. 52, 255 (1948).
10. YAMATERA, H., FITZPATRICK, B. and GORDON, G., J. Mol. Spectr. 14, 268 (1964).
11. HUNT, J. P. and TAUBE, H., J. Chem. Phys. 19, 602 (1951).
12. JACKSON, J. A., LEMONS, J. F. and TAUBE, H., J. Chem. Phys. 32, 553 (1960).
13. YATSIMIRSKII, K. B., Advances in the Chemistry of the Coordination Compounds (edited by S. Kirschner), MacMillan, New York, 1961, p. 96.
14. BJERRUM, J. and JØRGENSEN, C. K., Acta Chem. Scand. 7, 951 (1953).
15. JØRGENSEN, C. K., Acta Chem. Scand. 8, 175 (1954).
16. KATZIN, L. I. and GEBERT, E., Nature, 175, 425 (1955).
17. JØRGENSEN, C. K., Inorganic Complexes, Academic Press, New York, 1963, p. 100.
18. LARSON, R. C. and IWAMOTO, R. T., Inorg. Chem. 1, 316 (1962).
19. MINC, S. and LIBUS, W., Roczniki Chem. 29, 1073 (1955).
20. FRIEDMAN, M. J. and PLANE, R. A., Inorg. Chem. 2, 11 (1963).
21. NELSON, I. V. and IWAMOTO, R. T., Inorg. Chem. 3, 661 (1964).
22. PASTERNACK, R. F. and PLANE, R. A., Inorg. Chem. 4, 1171 (1965).
23. SWINARSKI, A., Theory and Structure of Complex Compounds (edited by B. J. Trzebiatowska), Pergamon Press, London, 1964, p. 487.
24. BECK, M. T., Kémiai Közl. 26, 353 (1966).
25. JOHNSON, J. R., CHRISTIAN, S. D. and AFFSPRUNG, H. E., J. Chem. Soc. (A) 77 (1966).

Chapter 8

MIXED LIGAND COMPLEXES

The co-ordination sphere of a central ion is homogeneous if all donor atoms are identical, whether the ligands are unidentate or multidentate. For example

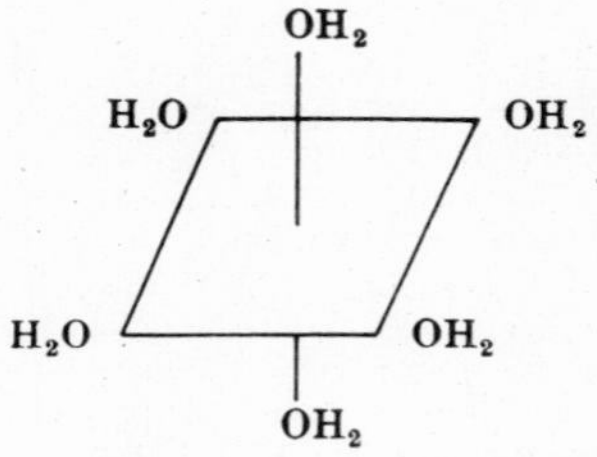

I

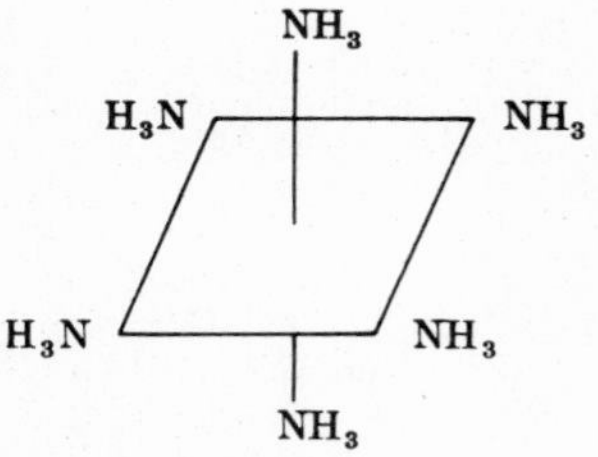

II

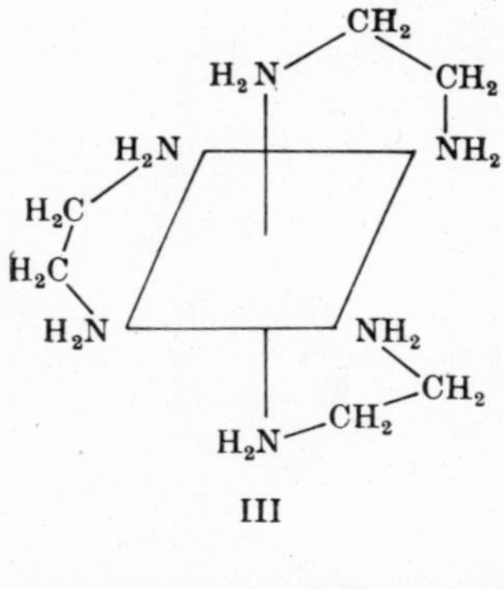

III

If the donor atoms are different the co-ordination sphere is heterogeneous. There is a broad spectrum of difference between donor atoms. In the extreme case the atomic numbers of the donor atoms are different: N, O, S etc.; in other cases the reason is more subtle: there may be a difference in the oxidation state of the donor atom (for example NH_3 or NO_2^-, NCS^-) or in the adjacent or more remote environment of the donor atom [e.g. NH_3,

$NH_2CH_2CH_2NH_2$, $CH_3NHCH_2CH_2NH_2$, $CH_3CH(NH_2)CH_2NH_2$]. If the different donor atoms belong to the same ligand molecule the complex may be termed a *mixed donor complex* (e.g. IV); if the different donor atoms belong to separate ligand molecules the complex may be termed a *mixed ligand complex* (e.g. V and VI). Mixed ligand complexes are very common among the inert complexes. Their synthesis, kinetic behaviour

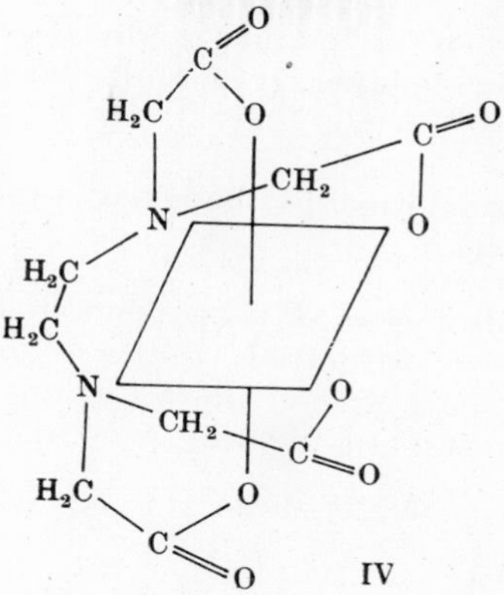

IV

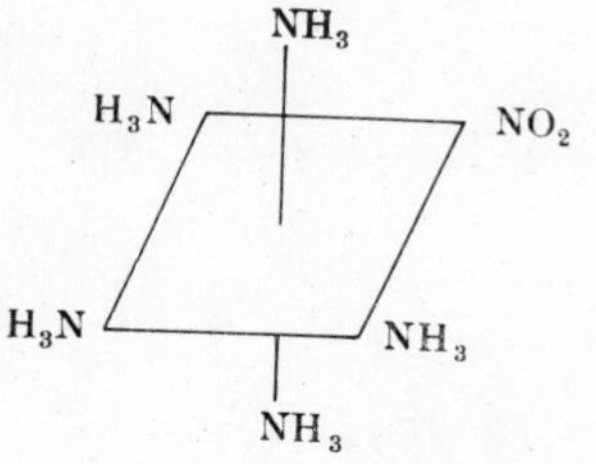

V

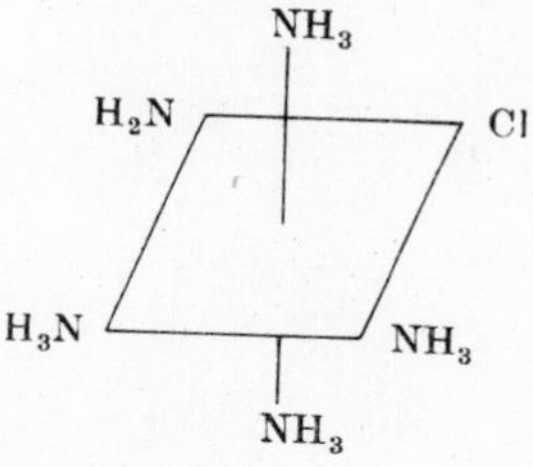

VI

and isomerism are treated in many old and recent papers and monographs [1–5]. In this Chapter only the equilibrium problems associated with the formation of mixed ligand complexes are dealt with. The range of the mixed ligand complexes may be reduced if co-ordinated water or other solvent molecules are not regarded as foreign ligands. In some cases this treatment is justified.

It is evident that the number of mixed ligand complexes is extremely high and the possibility of their formation must always be taken into consideration if more than two kinds of ligands are present. Among the inert complexes there are many containing even more than two different ligands. The number of possible variations of different co-ordinatively saturated complexes of unidentate ligands can be given [5] as

$$\binom{N+x-1}{x-1} = \frac{(N+x-1)(N+x-2)\ldots(N+1)}{(x-1)!} \tag{8.1}$$

where N is the co-ordination number and x the number of different ligands participating in formation of the complex. For $N = 6$ and $x = 3$ the number of possible species is 28 (3 parent complexes, 15 complexes containing two different ligands and 10 complexes containing three). At present only the equilibria of mixed ligand complexes involving two different ligands can be quantitatively treated.

It is helpful to distinguish three types of mixed ligand complex depending on whether the ligands involved are (i) unidentate–unidentate; (ii) unidentate–multidentate; (iii) multidentate–multidentate.

8.1 Some relationships between equilibrium constants of mixed ligand complexes

The simplest case of mixed ligand complex formation is when only two ligands are co-ordinated to the central ion. If the different ligands are symbolised as A and B the following equilibria and equilibrium constants must be considered: for the parent complexes:

$$\text{Me} + \text{A} \rightleftharpoons \text{MeA}; \qquad K_{10} = [\text{MeA}]\,[\text{Me}]^{-1}\,[\text{A}]^{-1} \tag{8.2}$$

$$\text{MeA} + \text{A} \rightleftharpoons \text{MeA}_2; \quad K_{20} = [\text{MeA}_2]\,[\text{MeA}]^{-1}\,[\text{A}]^{-1} \tag{8.3}$$

$$K_{10} \cdot K_{20} = \beta_{20} \tag{8.4}$$

$$\text{Me} + \text{B} \rightleftharpoons \text{MeB}; \qquad K_{01} = [\text{MeB}]\,[\text{Me}]^{-1}\,[\text{B}]^{-1} \tag{8.5}$$

$$\text{MeB} + \text{B} \rightleftharpoons \text{MeB}_2; \quad K_{02} = [\text{MeB}_2]\,[\text{MeB}]^{-1}\,[\text{B}]^{-1} \tag{8.6}$$

$$K_{01} \cdot K_{02} = \beta_{02} \tag{8.7}$$

For the mixed ligand complexes:

$$\text{MeAB} \rightleftharpoons \text{MeA} + \text{B}; \qquad K_{\text{A}} = [\text{MeAB}]\,[\text{MeA}]^{-1}\,[\text{B}]^{-1} \tag{8.8}$$

$$\text{MeAB} \rightleftharpoons \text{MeB} + \text{A}; \qquad K_{\text{B}} = [\text{MeAB}]\,[\text{MeB}]^{-1}\,[\text{B}]^{-1} \tag{8.9}$$

$$\text{MeAB} \rightleftharpoons \text{Me} + \text{A} + \text{B}; \quad \beta_{11} = [\text{MeAB}]\,[\text{Me}]^{-1}\,[\text{A}]^{-1}\,[\text{B}]^{-1} \tag{8.10}$$

$$2\,\text{MeAB} \rightleftharpoons \text{MeA}_2 + \text{MeB}_2; \quad K_r = [\text{MeAB}]^2\,[\text{MeA}_2]^{-1}\,[\text{MeB}_2]^{-1}. \tag{8.11}$$

The relationships between the equilibrium constants of the parent and of the mixed ligands complexes are:

$$K_r = \frac{\beta_{11}^2}{\beta_{20} \cdot \beta_{02}} \tag{8.12}$$

or

$$\beta_{11} = \sqrt{K_r \cdot \beta_{20} \cdot \beta_{02}} \tag{8.13}$$

K_{A} and K_{B} are given by the following ratios

$$K_{\text{A}} = \frac{\beta_{11}}{K_{10}} \tag{8.14}$$

and

$$K_{\mathrm{B}} = \frac{\beta_{11}}{K_{01}} . \tag{8.15}$$

The comparison of the equilibrium constants K_{A} and K_{02} shows the effect of the displacement of ligand B with ligand A on the strength of the $\mathrm{Me-B}$ bond. Analogous information can be obtained from the comparison of K_{B} and K_{20}.

Similar relationships can be simply obtained for more complicated systems.

8.2 Study of reactions of the type $i\mathrm{MeA}_h + j\mathrm{MeB}_h \rightleftharpoons h\mathrm{MeA}_i\mathrm{B}_j$

If the stability constant of the MeA_h and MeB_h parent complexes (where h is less than N, the maximum co-ordination number) is markedly greater than that of all the other successive complexes, in the solution of these complexes the concentrations of the dissociation products, including those of the free ligands, as well as the concentrations of the complexes of higher co-ordination number, are negligibly small. Therefore the reactions of these parent complexes can be fully characterized by the following general reaction equation:

$$i\,\mathrm{MeA}_h + j\,\mathrm{MeB}_h \rightleftharpoons h\,\mathrm{MeA}_i\,\mathrm{B}_j. \tag{8.16}$$

The number of mixed ligand complexes is then equal to $h-1$. To calculate the $h-1$ equilibrium constants the ratio of the concentrations of any two complex species must be determined as a function of the ratio of the initial total concentrations of the two parent complexes [6]. So far only a fairly limited number of such systems has been studied. The parent complexes of co-ordination number h are more stable than the mixed ligand complexes in the case of some characteristic class (b) metal ions (see Chapter 11) such as Hg^{2+}, Bi^{3+}, Tl^{3+}.

8.2.1 $\mathrm{MeA}_2 + \mathrm{MeB}_2$ SYSTEM

The quantitative treatment is evidently the simplest if $h = 2$. The most common and thoroughly studied systems are the pairs of the different halide and pseudohalide salts of mercury(II) [6]. For the determination of the single K_r equilibrium constant, spectrophotometric [7, 8] and liquid–liquid distribution [9] methods have been developed. The following simple balances are valid in such systems

$$T_{\mathrm{A}} = 2\mathrm{MeA}_2 + \mathrm{MeAB} \tag{8.17}$$

$$T_{\mathrm{B}} = 2\mathrm{MeB}_2 + \mathrm{MeAB} \tag{8.18}$$

$$T_{\mathrm{Me}} = \frac{T_{\mathrm{A}} + T_{\mathrm{B}}}{2} . \tag{8.19}$$

T_A and T_B are the original total concentrations of the MeA_2 and MeB_2 parent complexes. Their ratio is denoted by R:

$$R = T_A/T_B. \tag{8.20}$$

The partial mole fraction of the mixed ligand complex is given by

$$\alpha_{11} = \frac{[MeAB]}{T_{Me}} = \frac{[MeAB]}{[MeA_2] + [MeAB] + [MeB_2]}. \tag{8.21}$$

It follows from Eqs. 8.11 and 8.17–8.21 that

$$K_r = \frac{4\,\alpha_{11}^2}{(1 - \alpha_{11})^2 - \left(\frac{1 - R}{1 + R}\right)^2} \tag{8.22}$$

and

$$\alpha_{11} = \frac{1 \pm \sqrt{1 - \frac{4R}{(1 + R)^2}(1 - 4\,K_r^{-1})}}{1 - 4\,K_r^{-1}}. \tag{8.23}$$

K_r can be calculated if α_{11} is known at any values of R; α_{11} can be determined spectrophotometrically. The absorbance A of the three-component system in a 1-cm cell is given by

$$A = \varepsilon_{20}[MeA_2] + \varepsilon_{11}[MeAB] + \varepsilon_{02}[MeB_2] \tag{8.24}$$

where ε_{20}, ε_{11} and ε_{02} are the molar absorptivities of the corresponding complexes. Introducing the symbols

$$\Delta\varepsilon = \varepsilon_{11} - \frac{\varepsilon_{20} + \varepsilon_{02}}{2} \tag{8.25}$$

and

$$\Delta A = A - (\varepsilon_{20}T_A + \varepsilon_{02}T_B) \tag{8.26}$$

it follows that

$$\alpha_{11} = \frac{\Delta A}{\Delta\varepsilon \cdot T_M}. \tag{8.27}$$

The prerequisite for spectrophotometric determination of the stability constants of these mixed ligand complexes and even of the detection of their formation is to find a wavelength where $\Delta\varepsilon \neq 0$. Evidently it is helpful to choose a wavelength where $\Delta\varepsilon$ is maximal. The molar absorptivities of the parent complexes can be obtained directly, and that of the mixed ligand complex from experiments at the extreme values of R, namely, if $T_A \gg T_B$ it can be expected that the total amount of T_B is present as MeAB and vice versa. That is, at extreme values of R two-component systems are in effect formed and ε_{11} becomes available. Spiro and Hume [6] have evaluated ε_{11} from continuous variation experiments. T_M was kept constant while R was varied. At the extreme values of R we can write

$$A - \varepsilon_{20} T_{\mathrm{B}} = (2\varepsilon_{11} - \varepsilon_{02})T_{\mathrm{A}} \quad \text{for large } R \qquad (8.28)$$

and

$$A - \varepsilon_{02} T_{\mathrm{A}} = (2\varepsilon_{11} - \varepsilon_{20})T_{\mathrm{B}} \quad \text{for small } R. \qquad (8.29)$$

That is, by plotting $(A - \varepsilon_{20} T_{\mathrm{B}})$ versus T_{A} and $(A - \varepsilon_{02} T_{\mathrm{A}})$ versus T_{B}, straight lines are obtained and from the intercepts ε_{11} can be calculated. Once ε_{11} is known, the partial mole fraction of the mixed ligand complex can be obtained from Eq. 8.27 at the intermediary R values, and then K_r can be calculated from Eq. 8.25. Spiro and Hume made their experiments with mixed halide complexes of mercury(II). Figures 8.1 and 8.2 refer to the Hg(II)–Br$^-$–I$^-$ system.

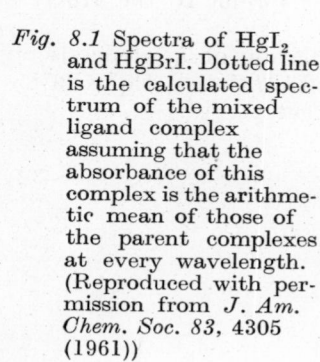

Fig. 8.1 Spectra of HgI$_2$ and HgBrI. Dotted line is the calculated spectrum of the mixed ligand complex assuming that the absorbance of this complex is the arithmetic mean of those of the parent complexes at every wavelength. (Reproduced with permission from *J. Am. Chem. Soc. 83*, 4305 (1961))

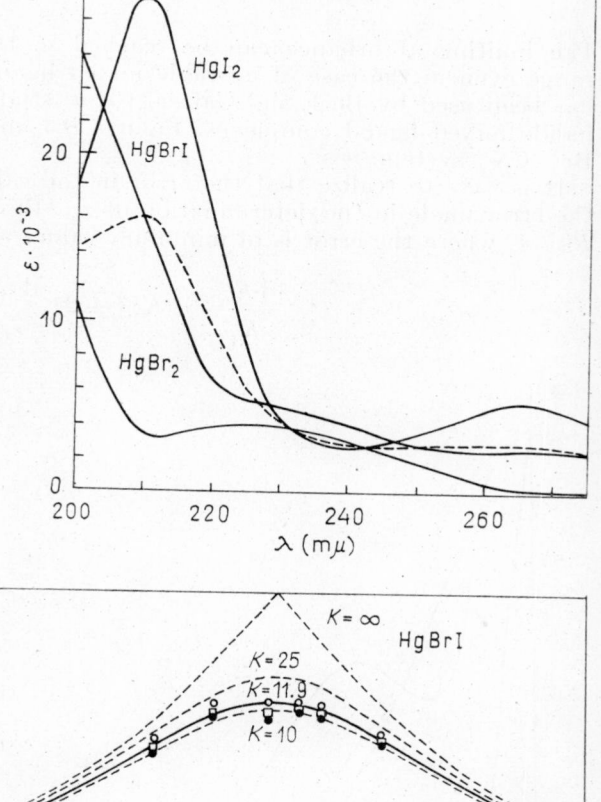

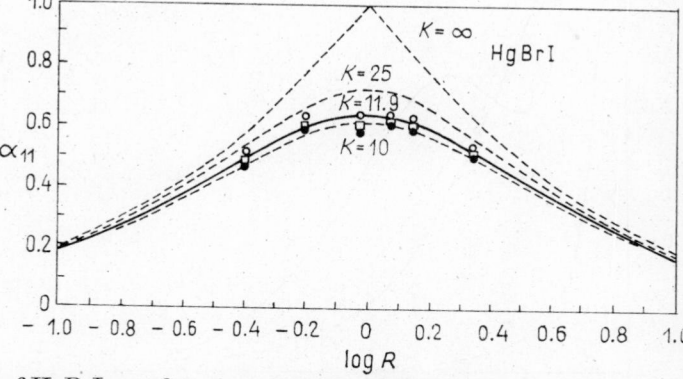

Fig. 8.2 Mole fraction of HgBrI as a function of log R; ● 219 nm; □ 220 nm; ○ 221 nm. (Reproduced with permission from *J. Am. Chem. Soc. 83*, 4305 (1961))

Another possibility for the determination of ε_{11} is to measure the absorbance at constant T_A as a function of T_B and vice versa. When T_B is constant and $T_A \to \infty$, the absorbance approaches a limiting value A_∞. From this limiting value ε_{11} can be simply calculated:

$$\varepsilon_{11} = \frac{A_\infty}{[2\,T_B} + \frac{\varepsilon_{20}}{2}. \tag{8.30}$$

If T_A is constant and T_B increases, ε_{11} can be analogously obtained:

$$\varepsilon_{11} = \frac{A_\infty}{2\,T_A} + \frac{\varepsilon_{02}}{2}. \tag{8.31}$$

The limiting absorbance can be reached in the available concentration range even in the case of unstable mixed ligand complexes. This method has been used by Beck and Gaizer [10] in studying mercury(II)–cyanide–halide mixed ligand complexes. Figures 8.3 and 8.4 refer to the Hg(II)–Br^-–CN^- system.

It is easy to realize that the error in the calculation of K_r depends on the error made in the determination of α_{11}. By differentiating Eq. 8.23 for $R = 1$, where the error is of minimum value, we get

$$\frac{d\,K_r}{K_r} = (K_r^{1/2} + 2)\frac{da_{11}}{\alpha_{11}}. \tag{8.32}$$

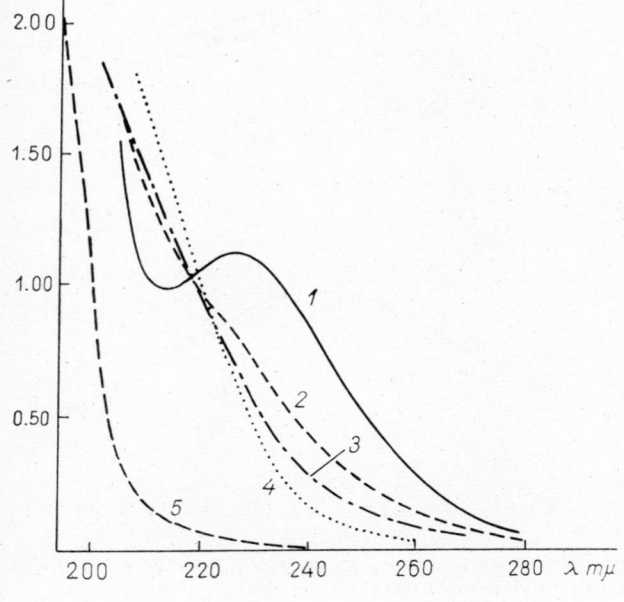

Fig. 8.3 Spectra of $HgBr_2$, $Hg(CN)_2$ and their mixtures.
Curve 1 — $T_{HgBr_2} = 3\times10^{-4}$ M;
Curve 2 — $T_{HgBr_2} = 3\times10^{-4}$ M.
$T_{Hg(CN)_2} = 8\times10^{-4}$ M;
Curve 3 — $T_{HgBr_2} = 3.10^{-4}$ M, $T_{Hg(CN)_2} = 2\times10^{-3}$ M;
Curve 4 — $T_{HgBr_2} = 3\times10^{-4}$ M, $T_{Hg(CN)_2}$ 2×10^{-2} M;
Curve 5 — $T_{Hg(CN)_2} = 2\times10^{-2}$ M

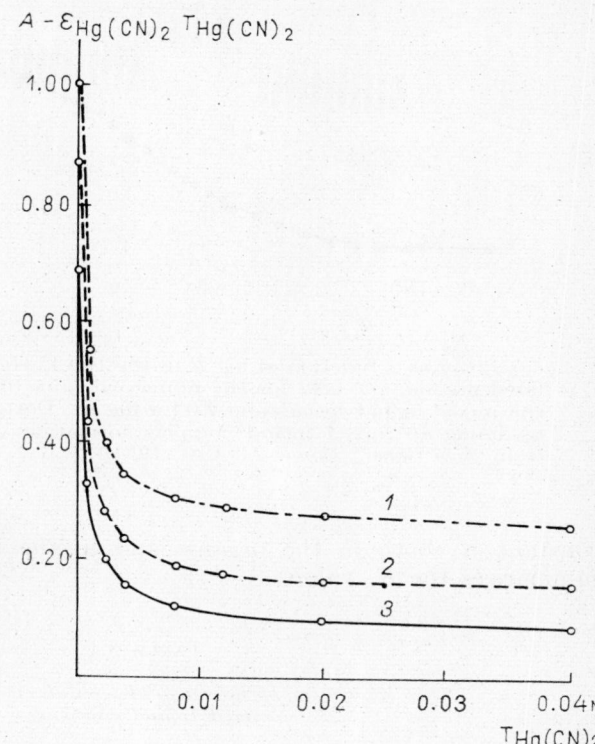

Fig. 8.4 Change of corrected absorbance as a function of $T_{Hg(CN)_2}$ in the system $HgBr_2 - Hg(CN)_2 . T_{HgBr_2} = 3 \times 10^{-4}$ M;
Curve *1* — 235 nm;
Curve *2* — 240 nm;
Curve *3* — 245 nm

This means that the error in α_{11} is multiplied by $(K_r^{1/2} + 2)$ in calculation of K_r. On the other hand the error in the measurement of α_{11} is determined by the absolute value of $\Delta\varepsilon$ and the ratio of ε_{11} and the arithmetic mean of the molar absorptivities of the parent complexes:

$$\frac{d\Delta\varepsilon}{\Delta\varepsilon} = \frac{\dfrac{d\varepsilon_{11}}{\varepsilon_{11}} + \dfrac{\varepsilon_{20} + \varepsilon_{02}}{2\,\varepsilon_{11}} \left(\dfrac{d(\varepsilon_{20} + \varepsilon_{02})}{\varepsilon_{20} + \varepsilon_{02}} \right)}{1 - \dfrac{\varepsilon_{20} + \varepsilon_{02}}{2\,\varepsilon_{11}}}. \tag{8.33}$$

Marcus [9] has developed a liquid–liquid distribution method for the study of such systems. The mathematical treatment is very similar to that of the spectrophotometric methods, but the experimental uncertainty seems to be higher. This appears from Fig. 8.5 which shows how small the difference is between the two theoretical curves for $K_r = 0$ and $K_r = \infty$. The liquid–liquid distribution method gives information on the equilibrium

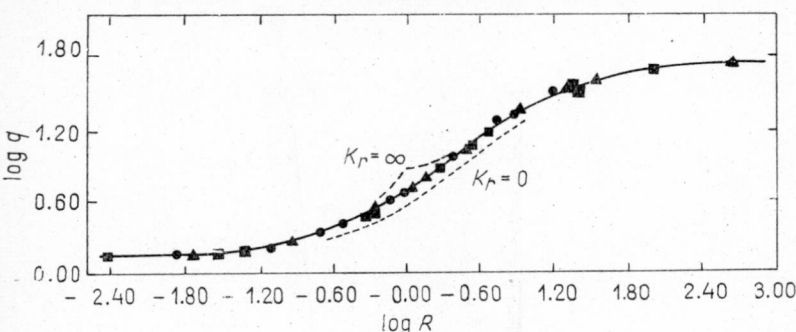

Fig. 8.5 Log q as a function of log R in the $HgCl_2$–$HgBr_2$ system. Full line is calculated assuming -0.42 for the numerical value of the distribution coefficient of the mixed ligand species and 2.0 for log K_r. Dotted line is the calculated curve assuming no mixed ligand complex formation. (Reproduced with permission from *Acta Chem. Scand.* **11**, 610 (1957))

constant K_r both in the organic and in the aqueous phase. Table 8.1 summarizes the K_r values.

TABLE 8.1

K_r values of mercury(II)–halide–pseudohalide mixed ligand complexes

Complex	Method	Medium	log K_r	Ref.
HgClBr	sp	$I = 10^{-3}$	1.14 ± 0.11	[7]
HgClBr	dist	$0.5\ M$ NaClO$_4$	$2.0\ \pm 0.5$	[9]
HgClBr	dist	benzene	$2.0\ \pm 0.5$	[9]
HgClI	sp	$I = 10^{-3}$	1.35 ± 0.17	[7]
HgClI	dist	$0.5\ M$ NaClO$_4$	1.75 ± 0.20	[9]
HgClI	dist	benzene	1.51 ± 0.20	[9]
HgBrI	sp	$I = 10^{-3}$	1.07 ± 0.08	[7]
HgBrI	dist	$0.5\ M$ NaClO$_4$	1.10 ± 0.20	[9]
HgBrI	dist	benzene	0.76 ± 0.20	[9]
HgClCN	sp	$\rightarrow 0*$	0.93 ± 0.03	[8]
HgClCN	sp	dioxane	1.08 ± 0.06	[10]
HgClSCN	sp	$\rightarrow 0$	0.91 ± 0.06	[11]
HgBrCN	sp	$\rightarrow 0$	0.29 ± 0.02	[8]
HgBrCN	sp	dioxane	0.12 ± 0.02	[10]
HgBrSCN	sp	$\rightarrow 0$	0.10 ± 0.02	[11]
HgICN	sp	$\rightarrow 0$	-0.96 ± 0.03	[8]
HgICN	sp	dioxane	-0.70 ± 0.05	[10]
HgISCN	sp	$\rightarrow 0$	-0.96 ± 0.04	[11]
HgCNSCN	sp	$\rightarrow 0$	-0.18 ± 0.04	[11]

$* \rightarrow 0$ signifies extrapolation to infinite dilution.

8.2.2 $\mathrm{MeA_3 + MeB_3}$ SYSTEM [12]

In such systems two mixed ligand complexes may be formed and the following equilibria characterize the system:

$$\mathrm{MeA_3 + 2MeB_3 = 3MeAB_2} \tag{8.34}$$

$$\mathrm{2MeA_3 + MeB_3 = 3MeA_2B}. \tag{8.35}$$

The corresponding equilibrium constants are

$$K_{r12} = \frac{[\mathrm{MeAB_2}]^3}{[\mathrm{MeA_3}]\,[\mathrm{MeB_3}]^2}. \tag{8.36}$$

and

$$K_{r21} = \frac{[\mathrm{MeA_2B}]^3}{[\mathrm{MeA_3}]^2\,[\mathrm{MeB_3}]}. \tag{8.37}$$

In the system of four components the original concentrations of the parent complexes can be expressed as follows:

$$T_\mathrm{A} = [\mathrm{MeA_3}] + 2/3\,[\mathrm{MeA_2B}] + 1/3\,[\mathrm{MeAB_2}] \tag{8.38}$$

$$T_\mathrm{B} = [\mathrm{MeB_3}] + 2/3\,[\mathrm{MeAB_2}] + 1/3\,[\mathrm{MeA_2B}]. \tag{8.39}$$

The absorbance of the solutions (1-cm cells) is given by

$$A = \varepsilon_{30}\,[\mathrm{MeA_3}] + \varepsilon_{21}\,[\mathrm{MeA_2B}] + \varepsilon_{12}\,[\mathrm{MeAB_2}] + \varepsilon_{03}\,[\mathrm{MeB_3}]. \tag{8.40}$$

If T_A is kept constant and T_B is increased, the concentration of $\mathrm{MeA_3}$ gradually decreases, first $\mathrm{MeA_2B}$ and then $\mathrm{MeAB_2}$ being formed. If $T_\mathrm{B} \gg T_\mathrm{A}$, the system becomes essentially two-component, and then we can write

$$T_\mathrm{A} = 1/3\,[\mathrm{MeAB_2}] \tag{8.41}$$

$$T_\mathrm{B} = [\mathrm{MeB_3}] + 2/3[\mathrm{MeAB_2}]. \tag{8.42}$$

The absorbance of such solutions is

$$A = \varepsilon_{12}[\mathrm{MeAB_2}] + \varepsilon_{03}[\mathrm{MeB_3}]. \tag{8.43}$$

From Eqs. 8.41–8.43

$$A - \varepsilon_{03}\,T_\mathrm{B} = T_\mathrm{A}(3\varepsilon_{12} - 2\varepsilon_{03}). \tag{8.44}$$

If T_B is increased with T_A constant the condition for validity of Eqs. 8.41–8.44 is more and more closely approached. Thus plotting the left-hand side of Eq. 8.44 as a function of T_B gives a curve which approaches a limiting value from which ε_{12} can be calculated by means of the formula

$$\varepsilon_{12} = \frac{A_\alpha}{3T_\mathrm{A}} + \frac{2}{3}\,\varepsilon_{03}. \tag{8.45}$$

When the curve does not reach a limiting value in the applicable concentration range, $A\alpha$ can be extrapolated with more or less accuracy. The value of ε_{21} can be calculated in the same way. At constant T_B the value of T_A is increased and ε_{21} can be calculated:

$$\varepsilon_{21} = \frac{A_\alpha}{3T_B} + \frac{2}{3}\,\varepsilon_{30}. \tag{8.46}$$

For the calculation of the equilibrium constants we choose systems which consist essentially of three components. That is, the concentrations of the individual complexes can be calculated from the curved section immediately preceding the limiting value. For this part of the curve the material balance is expressed by the equations:

$$T_A = \frac{2}{3}\,[MeA_2B] + \frac{1}{3}\,[MeAB_2] \tag{2.47}$$

$$T_B = [MeB_3] + \frac{2}{3}\,[MeAB_2] + \frac{1}{3}\,[MeA_2B]. \tag{8.48}$$

The absorbance of the system is

$$A = \varepsilon_{21}[MeA_2B] + \varepsilon_{12}[MeAB_2] + \varepsilon_{03}[MeB_3]. \tag{8.49}$$

From Eqs. 8.47–8.49, the concentrations of the individual complexes can be given as

$$[MeAB_2] = \frac{A - \varepsilon_{03}T_B + T_A\,\dfrac{\varepsilon_{03} - 3\varepsilon_{21}}{2}}{\varepsilon_{12} - \dfrac{\varepsilon_{21} + \varepsilon_{03}}{2}} \tag{8.50}$$

$$[MeA_2B] = 3/2\,T_A - 1/2\,[MeAB_2] \tag{8.51}$$

$$[MeB_3] = T_B - 1/2\,T_A - 1/2\,[MeAB_2]. \tag{8.52}$$

Thus K_{12} can be explicitly calculated. To calculate K_{21} analogous measurements must be performed at constant T_B and increasing T_A. The corresponding expressions valid for the region immediately preceding the limiting value of the absorbance are as follows

$$[MeA_2B] = \frac{A - \varepsilon_{30}T_A + T_B\,\dfrac{\varepsilon_{30} - 3\varepsilon_{12}}{2}}{\varepsilon_{21} - \dfrac{\varepsilon_{12} + \varepsilon_{30}}{2}} \tag{8.53}$$

$$[MeAB_2] = 3/2\, T_B - 1/2\, [MeA_2B] \qquad (8.54)$$

$$[MeA_3] \;\; = T_A - 1/2\, T_B - 1/2\, [MeA_2B]. \qquad (8.55)$$

From these K_{21} can be directly obtained.

Such a type of complex equilibrium can be expected to occur in the case of complexes of some tervalent cations with univalent unidentate or univalent bidentate ligands. So far the $BiCl_3$–BiI_3 system has been studied in dioxan solution. The co-ordination numbers of the complexes formed are evidently not three, because solvent molecules are also co-ordinated. This, however, does not influence the evaluation of the equilibrium constants. Figure 8.6 gives evidence for the formation of mixed ligand complexes; Fig. 8.7 illustrates the reliability of the constants calculated. The constants at 25°C ($A = Cl^-$, $B = I^-$) are $K_{r12} = 0.09 \pm \pm 0.02$; $K_{r21} = 2.07 \pm 0.2$.

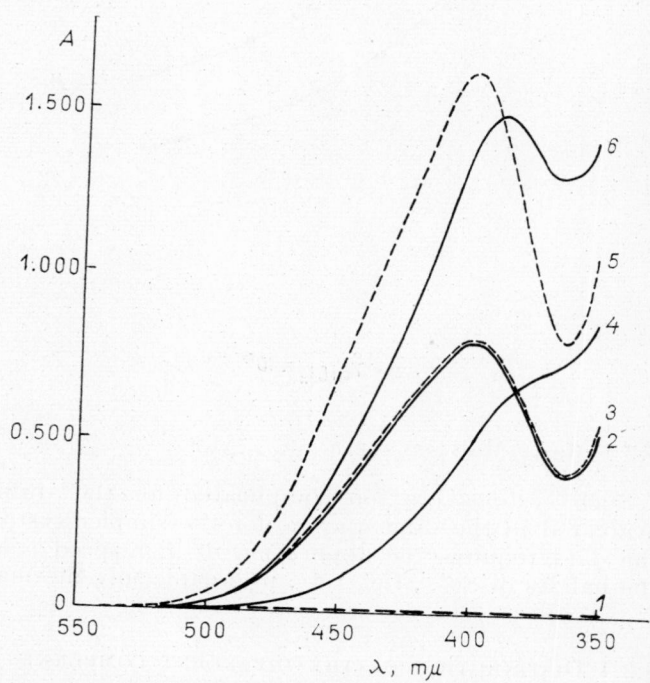

Fig. 8.6 Spectra of BiI_3, $BiCl_3$ and their mixtures in absolute dioxan.
Curve 1 — $T_{BiCl_3} = 1.2 \times 10^{-4}$ M; Curve 2 — $T_{BiI_3} = 1.2 \times 10^{-4}$ M;
Curve 3 — $T_{BiCl_{3s}} = 2.4 \times 10^{-4}$ M, $T_{BiI_3} = 1.2 \times 10^{-4}$ M, curve calculated assuming no interaction;
Curve 4 — The same solution as in Curve 3 measured curve;
Curve 5 — $T_{BiCl_3} = 1.2 \times 10^{-4}$ M, $T_{BiI_3} = 2.4 \times 10^{-4}$ M, curve calculated assuming no interaction;
Curve 6 — The same solution as in Curve 5 measured curve. (Reproduced with permission from J. Inorg. Nucl. Chem. 28, 503 (1966))

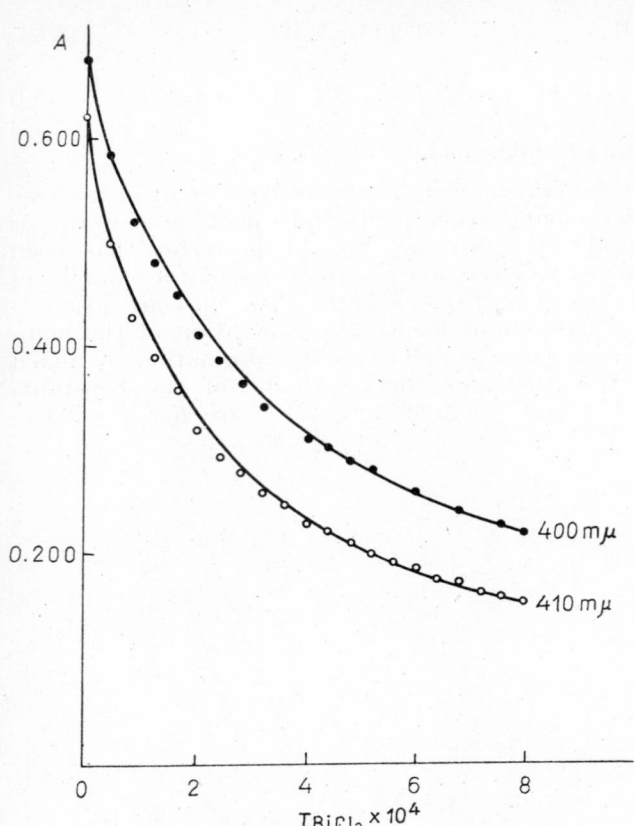

Fig. 8.7 The effect of $BiCl_3$ on the absorbance of BiI_3 in dioxan. $T_{BiI_3} = = 10^{-4}$ M. Dots are measured values, curves are calculated. (Reproduced with permission from *J. Inorg. Nucl. Chem. 28*, 503 (1966))

8.2.3 MeA$_4$–MeB$_4$ SYSTEM

Such systems are too complicated for the stability constants to be evaluated in the same way as for the simpler systems. The treatment of the data requires the application of high-speed computers. The pairs of the halides of Sn^{4+}, Ge^{4+} etc. in organic solvents may form such systems.

8.2.4 DISTRIBUTION OF THE DIFFERENT COMPLEXES

As all the equilibrium constants characterizing the systems in question are dimensionless quantities the distribution of the complexes is independent of the absolute value of the sum of the total concentrations $(T_A + T_B)$. The partial mole fractions of the different complex species are determined by the values of the equilibrium constants and by R. While the partial mole fractions of the parent complexes change monotonically with increasing or decreasing R, those of the mixed ligand species pass through a maximum, the position of which is determined by R. The position of the maximum mole fraction of the MeA_iB_j complex is at

$$R = \frac{i}{j}. \qquad (8.56)$$

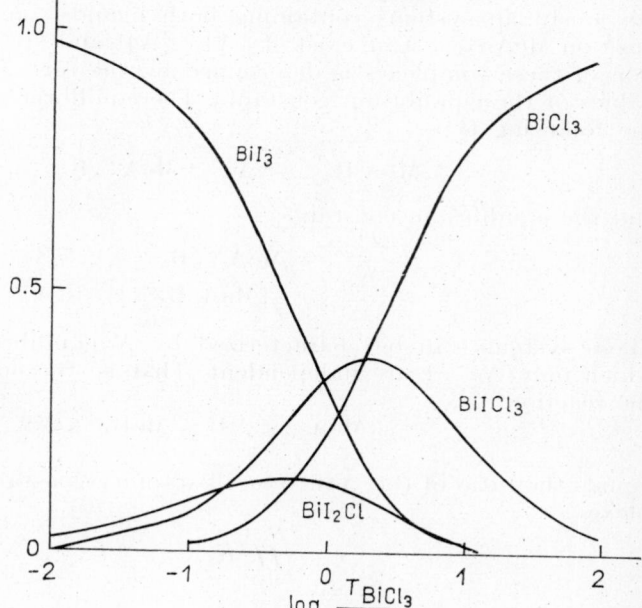

Fig. 8.8 Distribution of complexes as a function of log R in the BiI_3–$BiCl_3$ system. (Reproduced with permission from J. Inorg. Nucl. Chem. 28, 503 (1966))

The shape and the height of these curves are determined by the value of the equilibrium constants K_{rij}.

In the system MeA_i–MeB_j it is very simple to calculate the mole fractions of the three species if K_r and R are known. α_{11} is obtained by Eq. 8.23; α_{20} and α_{02} can be directly calculated:

$$\alpha_{20} = \frac{T_A}{T_A + T_B} - \frac{1}{2}\alpha_{11} \qquad (8.57)$$

$$\alpha_{02} = \frac{T_B}{T_A + T_B} - \frac{1}{2}\alpha_{11}. \qquad (8.58)$$

It is helpful to construct diagrams like that in Fig. 8.2.

The calculation of the mole fraction of the complex species in the MeA_3M–MeB_3 system is also very simple but it is time-consuming and laborious. It is recommended to use computers for this purpose. Figure 8.8 shows the distribution diagram for the $BiCl_3$–BiI_3 system.

8.3 Study of mixed ligand complexes of the type $MeA_i B_{N-i}$

If the concentration of the ligand A exceeds a certain limit ($[A]^*$) in the system consisting of central ion Me and ligand A, the solution contains only one complex ion MeA_N, or more precisely the concentrations of the other species are negligible. Similarly, in a system containing metal ion Me and ligand B, if $[B] > [B]^*$ the only complex ion present is MeB_N. Obviously, in systems containing both ligands, only complexes of composition MeA_iB_{N-i} can exist if $[A] > [A]^*$ and $[B] > [B]^*$. The distribution of these complexes is determined by the ratio of $[A]/[B]$ and by the values of the equilibrium constants. The equilibria in such systems are of the following type

$$MeA_iB_{N-i} + A \rightleftharpoons MeA_{i+1}B_{N-i-1} + B \qquad (8.59)$$

and the equilibrium constants

$$K_{ij} = \frac{[MeA_{i+1}B_{N-i-1}]}{[MeA_i B_{N-i}]} \cdot \frac{[B]}{[A]}. \qquad (8.60)$$

These systems can be characterized by N equilibrium constants among which only $N - 1$ are independent. That is, the equilibrium constant of the reaction

$$MeA_N + NB = MeB_N + NA \qquad (8.61)$$

equals the ratio of the Nth overall stability constants of the parent complexes:

$$\prod_0^{N-1} K_{i,j} = \frac{\beta_{N,0}}{\beta_{0,N}}. \qquad (8.62)$$

The equilibrium constants can be calculated if the concentration of any of the complexes of the type MeA_iN_{N-i} ($N < i < 0$) is known as a function of the ratio $[B]/[A]$. The measurement of the $[B]/[A]$ ratios belonging to the maximum mole fractions of the mixed ligand species gives a further possibility for the evaluation of the equilibrium constants.

8.3.1 Spectrophotometric determination of equilibrium constants

Spectrophotometry is well suited for the study of such systems. The method of calculation depends on the spectral characteristics of the system. The following examples will illustrate the application of spectrophotometry.

The treatment is particularly easy if only one mixed ligand complex is formed and the system exhibits two isosbestic points. It was found by De Witt and Watters [13] that Cu^{2+}–oxalate–ethylenediamine is such a system (Fig. 8.9). The appearance of the first isosbestic point (ε_1^*) at 600 nm means that at this wavelength the molar absorptivities of $Cu(en)_2^{2+}$ and $Cu(en)(ox)$ are the same: $\varepsilon_{20} = \varepsilon_{11} = \varepsilon_1^*$; the second isosbestic point ε_*^2 indicates that at 700 nm the molar absorptivities of $Cu(en)(ox)$ and $Cu(ox)_2^{2-}$ complexes are equal: $\varepsilon_{11} = \varepsilon_{02} = \varepsilon_2^*$. It is possible to calculate the partial

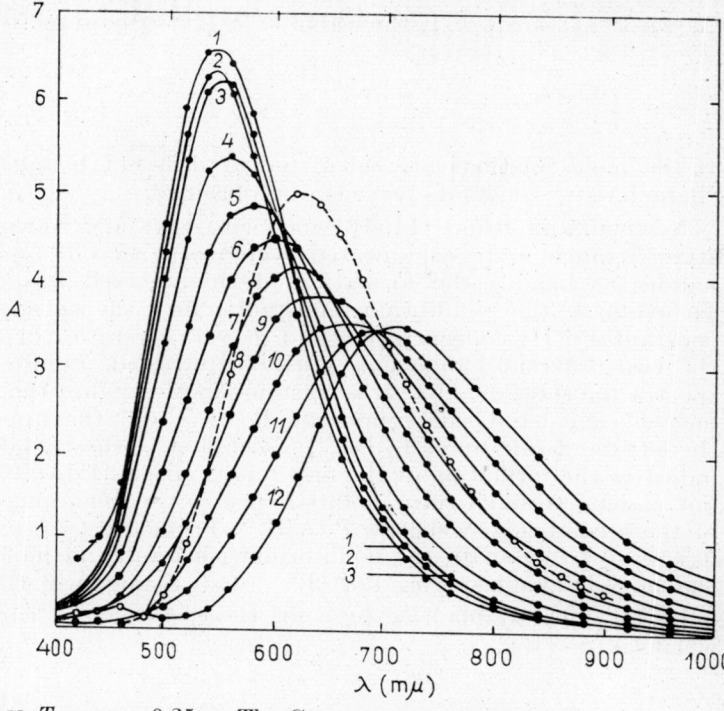

Fig. 8.9 Spectra of solutions containing copper(II), oxalate and ethylenediamine.

$T_{CuSO_4} = 10^{-2}$ M, $T_{oxalate} = 0.25$ M, The Cu : ox : en ratios are :
Curve *1*—1 : 250 : 3; Curve *2* — 1 : 0 : 3; Curve *3* — 1 : 250 : 2;
Curve *4* — 1 : 250 : 1.75; Curve *5* — 1 : 250 : 1.58; Curve *6* — 1 : 250 : 1.25;
Curve *7* — 1 : 250 : 1;
Curve *8* — calculated spectrum of the mixed ligand complex;
Curve *9* — 1 : 250 : 0.75; Curve *10* — 1 : 250 : 0.5;
Curve *11* — 1 : 250 : 0.25; Curve *12* — 1 : 250 : 0
(Reproduced with permission from *J. Am. Chem. Soc. 76*, 3810 (1954))

mole fraction of the Cu(en)$_2^{2+}$ complex from the first isosbestic point, and the partial mole fraction of the Cu(ox)$_2^{2-}$ complex from the second. If $\bar{\varepsilon}$ stands for the apparent molar absorptivity ($\bar{\varepsilon} = A/d \cdot T_{Cu}$, where d is the light-path length)

$$\alpha_{02} = \frac{\varepsilon_1^* - \bar{\varepsilon}_{600}}{\varepsilon_1^* - \varepsilon_{02}} \tag{8.63}$$

and

$$\alpha_{20} = \frac{\varepsilon_2^* - \bar{\varepsilon}_{700}}{\varepsilon_2^* - \varepsilon_{20}}. \tag{8.64}$$

With α_{20} and α_{02} known, the molar absorptivity of the mixed ligand complex at any wavelength can be simply obtained:

$$\varepsilon_{11} = \frac{\bar{\varepsilon} - \alpha_{20} \cdot \varepsilon_{20} - \alpha_{02} \cdot \varepsilon_{02}}{1 - \alpha_{20} - \alpha_{02}}. \tag{8.65}$$

As $\alpha_{20} + \alpha_{02} + \alpha_{11} = 1$, the value of K_{r11} can be directly calculated

$$K_{r11} = \frac{(1 - \alpha_{02} - \alpha_{20})^2}{\alpha_{20} \cdot \alpha_{02}}. \tag{8.66}$$

If the mole fractions are calculated at different ligand ratios, the other characteristic constants can also be obtained.

Newman and Hume [14, 15] and Srivastava and Newman [16, 17] have studied much more complicated systems. With Pd^{2+} as central ion and halides as ligands the formation of three mixed ligand species occurs. To evaluate the equilibrium constants, first the spectra of solutions of constant Pd(II) concentration and of varying ratios of concentrations of the two different halide ions must be recorded. Figure 8.10 shows such spectra for the Pd^{2+}–Br^-–Cl^- system. In this figure the uppermost curve on the right-hand side represents $PdBr_4^{2-}$ and the uppermost curve on the left-hand side refers to $PdCl_4^{2-}$. The first isosbestic point on either side indicates the formation of the first mixed species PdA_3B^{2-}. The first curve not passing through these points indicates the beginning of the formation of the second mixed species $PdA_2B_2^{2-}$. From the two-component systems it is easy to evaluate the equilibrium constant and molar absorptivity of the mixed ligand species PdA_3B^{2-} (and analogously those of $PdAB_3^{2-}$).

If $[B]/[A]$ is symbolised by r for these systems, it can be written that when $[A] \gg [B]$

$$\bar{\varepsilon} = \frac{\varepsilon_{40} + \varepsilon_{31} \cdot K_{31}r}{1 + K_{31}r} \tag{8.67}$$

and when $[A] \ll [B]$

$$\bar{\varepsilon} = \frac{\varepsilon_{04} + \varepsilon_{13} \cdot K_{04}^{-1} \cdot r^{-1}}{1 + K_{04}^{-1}r^{-1}}. \tag{8.68}$$

Srivastava and Newman evaluated the unknown constants by a curve-fitting method, but the application of the following linearized equations seems to be more convenient. We can write

$$\frac{1}{\varepsilon - \varepsilon_{40}} = \frac{1}{(\varepsilon_{31} - \varepsilon_{40})K_{31}r} + \frac{1}{\varepsilon_{31} - \varepsilon_{40}} \quad \text{(for } [A] \gg [B]) \tag{8.69}$$

and

$$\frac{1}{\bar{\varepsilon} - \varepsilon_{04}} = \frac{1}{(\varepsilon_{13} - \varepsilon_{04})K_{04}^{-1}r^{-1}} + \frac{1}{\varepsilon_{13} - \varepsilon_{04}} \quad \text{(for } [A] \ll [B]). \tag{8.70}$$

The evaluation of the unknown constants from Eqs. 8.69 and 8.70 is a simple procedure.

For the calculation of the molar absorptivities and stability constants of the two other mixed ligand complexes Srivastava and Newman suggested the following procedure. For certain r values the systems can be considered as consisting of three complex species (for $[A] > [B]$: MeA_4, MeA_3B,

Fig. 8.10 Absorption spectra of 1.5×10^{-4} M Pd(II) at ionic strength of 4.5. Each solution is 0.5 M in HClO$_4$, 2.0 M in LiClO$_4$ and 2.0 M in halide (LiCl + LiBr); logarithms of ratios of bromide to chloride in various curves are given as follows.

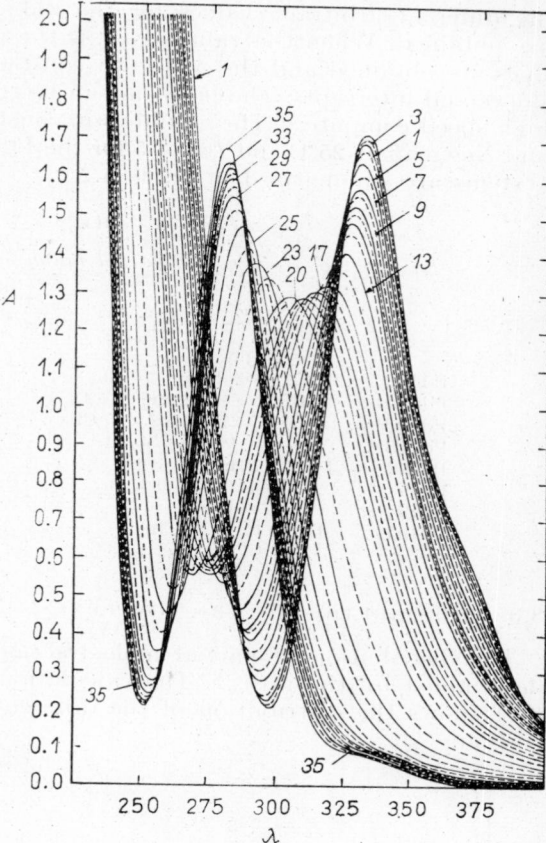

Curve 1: no chloride;
Curve 2: 1.398
Curve 3: 0.921
Curve 4: 0.699
Curve 5: 0.398
Curve 6: 0.222
Curve 7: 0.097
Curve 8: 0.000
Curve 9: −0.079
Curve 10: −0.146
Curve 11: −0.222
Curve 12: −0.301
Curve 13: −0.398
Curve 14: −0.523
Curve 15: −0.620
Curve 16: −0.699
Curve 17: −0.745
Curve 18: −0.796
Curve 19: −0.854
Curve 20: −0.921
Curve 21: −1.000
Curve 22: −1.222
Curve 23: −1.398
Curve 24: −1.523
Curve 25: −1.699
Curve 26: −1.854
Curve 27: −2.000
Curve 28: −2.097
Curve 29: −2.222
Curve 30: −2.301
Curve 31: −2.398
Curve 32: −2.523
Curve 33: −2.699
Curve 34: −3.398
Curve 35: no bromide

(Reproduced with permission from *Inorg. Chem. 5,* 1506 (1966))

MeA_2B_2, for [B] < [A]: MeB_4, $MeAB_3$, MeA_2B_2) and the following equations are valid:

$$\frac{(\bar{\varepsilon} - \varepsilon_{22})r^2}{\varepsilon_{40} - \bar{\varepsilon}} + \frac{1}{K_{31} K_{22}} + \frac{(\varepsilon_{31} - \bar{\varepsilon})r}{(\varepsilon_{40} - \bar{\varepsilon})} \cdot \frac{1}{K_{04}} \quad \text{(for } r > 1) \quad (8.71)$$

and

$$\frac{(\bar{\varepsilon} - \varepsilon_{04})r^2}{\varepsilon_{22} - \bar{\varepsilon}} = \frac{1}{K_{04} \cdot K_{13}} + \frac{(\varepsilon_{13} - \bar{\varepsilon})r}{(\varepsilon_{22} - \bar{\varepsilon})} \cdot \frac{1}{K_{04}} \text{ for } r > 1. \quad (8.72)$$

By plotting $(\bar{\varepsilon} - \varepsilon_{22})(\varepsilon_{40} - \bar{\varepsilon})^{-1} r^2$ versus $(\varepsilon_{31} - \bar{\varepsilon})(\varepsilon_{40} - \bar{\varepsilon})^{-1} r$ and $(\bar{\varepsilon} - \varepsilon_{04})(\varepsilon_{22} - \bar{\varepsilon})^{-1} r^2$ versus $(\varepsilon_{13} - \bar{\varepsilon})(\varepsilon_{22} - \bar{\varepsilon})^{-1} r$, respectively, choosing different arbitrary values for the unknown ε_{22}, two series of curves are obtained. When the value of ε_{22} is the real one, in both cases straight lines are obtained and the unknown constants can be calculated from the slopes and intercepts. The problem can be treated very simply by applying high-speed computers. The equilibrium constants determined by Srivastava and Newman at 25°C and $I = 4.5$ for the $Pd^{2+}-Cl^--Br^-$ and $Pd^{2+}-Br^--I^-$ systems are summarized in Table 8.2.

TABLE 8.2

Equilibrium constants of mixed halide complexes of palladium(II)

Reaction	$\log K$
$PdCl_4^{2-} + Br^- = PdCl_3Br^{2-} + Cl^-$	1.55 ± 0.05
$PdCl_3Br^{2-} + Br^- = PdCl_2Br_2^{2-} + Cl^-$	1.09 ± 0.07
$PdCl_2Br_2^{2-} + Br^- = PdClBr_3^{2-} + Cl^-$	0.95 ± 0.07
$PdClBr_3^{2-} + Br^- = PdBr_4^{2-} + Cl^-$	0.55 ± 0.05
$PdBr_4^{2-} + I^- = PdBr_3I^{2-} + Br^-$	2.75 ± 0.05
$PdBr_3I^{2-} + I^- = PdBr_2I_2^{2-} + Br^-$	3.00 ± 0.15
$PdBr_2I_2^{2-} + I^- = PdBrI_3^{2-} + Br^-$	1.70 ± 0.15
$PdBrI_3^{2-} + I^- = PdI_4^{2-} + Br^-$	0.80 ± 0.05

8.3.2 DISTRIBUTION OF THE COMPLEXES

The distribution of partial mole fractions of the various complexes is determined by the ratio r. The calculation can be made directly. Figure 8.11 shows the distribution of the different species in the $Pd^{2+}-Cl^--Br^-$ system.

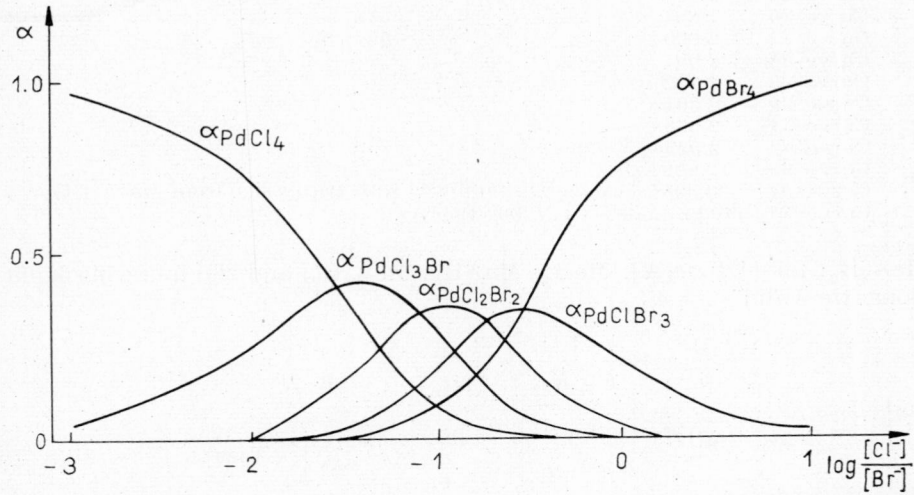

Fig. 8.11 Distribution of complexes in the Pd(II)–Cl–Br system as a function of log $[Cl^-]/[Br^-]$

8.4 Potentiometric study of mixed ligand complexes

If the conjugated acids of the ligands are weak the formation of mixed ligand complexes can be followed by pH titration. An exact general treatment is not possible, but the method is well applicable if certain conditions are fulfilled.

8.4.1 STUDY OF MIXED COMPLEXES OF LIGANDS OF VERY DIFFERENT BASICITY

Watters *et al.* [18, 19] have studied the mixed ligand complexes of oxalate and ethylenediamine. The experimental procedure consists of an acidimetric pH titration of the mixture containing integral ratios of the less basic ligand (oxalate in this case) to metal ion and an excess of the more basic ligand. The more basic ligand is selectively removed from the complex by the added acid and the mean number of more basic ligands bound is calculated according to the usual procedure. Mixed ligand complexes of 2,2'-dipyridyl and various more basic ligands were also investigated [20–22]. This method can be used when a strong complex of a multidentate ligand takes up another basic ligand. Thompson and Loraas [23] have studied the mixed ligand complexes of lanthanides with hydroxyethyl-ethylenediaminetriacetate and iminodiacetate. The application of high-speed computers is helpful even in these simple cases.

The constants for the Cd^{2+} (Zn^{2+})–oxalate–ethylenediamine system [24] are presented in a convenient form in Fig. 8.12.

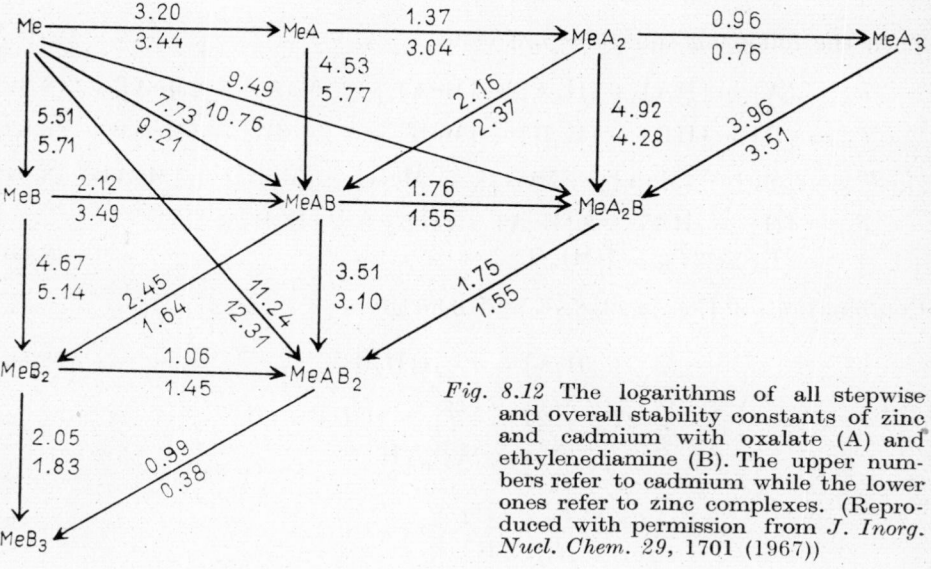

Fig. 8.12 The logarithms of all stepwise and overall stability constants of zinc and cadmium with oxalate (A) and ethylenediamine (B). The upper numbers refer to cadmium while the lower ones refer to zinc complexes. (Reproduced with permission from *J. Inorg. Nucl. Chem. 29*, 1701 (1967))

8.4.2 GENERAL TREATMENT OF THE MIXED COMPLEXES OF BASIC LIGANDS

The problem is much more difficult if the dissociation of both kinds of ligand has to be considered. Näsänen et al. [25, 26] have developed a method for the treatment of such systems and studied several mixed ligand complexes of copper(II). The method can be applied in cases where the ligands are bidentate and the co-ordination number of the central ion is four. The following equilibria must be considered.

For protonation of the ligands A and B:

$$K_{1A}^H = [HA][H]^{-1}[A]^{-1} \tag{8.73}$$

$$K_{12A}^H = [H_2A][H]^{-1}[HA]^{-1} \tag{8.74}$$

$$K_{1B}^H = [HB][H]^{-1}[B]^{-1} \tag{8.75}$$

$$K_{12B}^H = [H_2B][H]^{-1}[HB]^{-1}. \tag{8.76}$$

For the formation of parent complexes:

$$K_{1A} = [MA][H]^2[Me]^{-1}[H_2A]^{-1} \tag{8.77}$$

$$K_{2A} = [MeA_2][H]^2[MeA]^{-1}[H_2A]^{-1} \tag{8.78}$$

$$K_{1B} = [MeB][H]^2[Me]^{-1}[H_2B]^{-1} \tag{8.79}$$

$$K_{2B} = [MeB_2][H]^2[MeB]^{-1}[H_2B]^{-1} \tag{8.80}$$

For the formation of the mixed ligand complex:

$$\beta_{AB} = [MeAB][Me]^{-1}[A]^{-1}[B]^{-1}. \tag{8.81}$$

Then the following equations are valid:

$$T_A = [A] + [HA] + [H_2A] + [MeA] + 2[MeA_2] + [MeAB] \tag{8.82}$$

$$T_B = [B] + [HB] + [H_2B] + [MeB] + 2[MeB_2] + [MeAB] \tag{8.83}$$

$$T_M = [Me] + [MeA] + [MeA_2] + [MeB] + [MeB_2] + [MeAB] \tag{8.84}$$

$$T_H + [H] + [HA] + 2[H_2A] + [HB] + 2[H_2B] =$$
$$= 2T_A + 2T_B - [OH^-]. \tag{8.85}$$

Combination of Eqs. 8.73–8.76 and 8.85 gives

$$[H_2A] = a - b[H_2B] \tag{8.86}$$

where

$$a = \frac{2T_A + 2T_B - T_H + [OH] - [H]}{2 + K_{12B}^{H^{-1}}[H]^{-1}}. \tag{8.87}$$

$$b = \frac{2 + K_{12B}^{H^{-1}}[H]^{-1}}{2 + K_{12A}^{H^{-1}}[H]^{-1}}. \tag{8.88}$$

It follows then that

$$D_1[H_2B]^3 + D_2[H_2B]^2 + D_3[H_2B] + D_4 = 0. \tag{8.89}$$

The coefficients D_1, D_2, D_3 and D_4 can be calculated from

$$D_1 = -b^3 d_A f_A + b^2 d_B f_A + b d_A f_B - d_B f_B \tag{8.90}$$

$$D_2 = 3 ab^2 d_A f_A + b^2 d_A e_A - b^2 c_A f_A - b^2 c_B f_A - 2 abd_B f_A - ad_A f_B + \\ + c_A f_B - d_B e_B + c_B f_B \tag{8.91}$$

$$D_3 = -3 a^2 bd_A f_A + a^2 d_B f_B - 2 abd_A e_A + 2 abc_A f_B + 2 abc_B f_A - \\ - d_A b + bc_A e_A - d_B + c_B e_B \tag{8.92}$$

$$D_4 = a^3 d_A f_A + a^2 d_A e_A - a^2 c_A f_A - a^2 c_B f_A + ad_A - ac_A e_A - c_A + c_B \tag{8.93}$$

where the following abbreviations are used:

$$c_A = T_A - T_M; \quad c_B = T_B - T_M;$$

$$d_A = 1 + K_{12A}^{H^{-1}}[H]^{-1} + K_{1A}^{H^{-1}} K_{12A}^{H^{-1}}[H]^{-2}$$

$$d_B = 1 + K_{12B}^{-1}[H]^{-1} + K_{1B}^{H^{-1}} K_{12B}^{H^{-1}}[H]^{-2}; \quad e_A = K_{1A}[H]^{-2};$$

$$e_B = K_{1B}[H]^{-2}; \quad f_A = e_A K_{2A}[H]^{-2}; \quad f_B = e_B K_{2B}[H]^{-2}.$$

For the concentration of the uncomplexed metal ion we obtain

$$[M] = [2 T_M - (T'_A + T'_B)]/2 + e_A[H_2A] + e_B[H_2B] \tag{8.94}$$

where $T'_A = T_A - [H_2A] - [HA] - [A]$ and $T'_B = T_B - [H_2B] - [HB] - [B]$. Now the only unknown constant β_{AB} can be calculated. For the solution of Eq. 8.89 Näsänen used an electronic computer. Perrin, Sayce and Sharma [27], dealing with the same type of equilibria have developed a general computer programme SCOSUS (stability constant of single unknown species). Perrin and Sharma [28] have studied even more complicated systems involving the formation of three mixed ligand species: MeAB, MeA$_2$B and MeAB$_2$. The result of the pH titrations could be evaluated under certain conditions by the SCOSUS programme and in general by another programme SCOGS developed by Sayce [29].

An interesting case of mixed ligand complex formation is when the ligands are isomeric asymmetric molecules. In this case the stabilities of both the proton complexes and the parent metal complexes of the ligands are the same. The calculation method is therefore much simpler. Unfortunately, this extremely interesting problem has not attracted as much attention as it deserves. The data referring to the copper(II)–tartaric acid system [30] must be regarded with some reservation because differences were found even in the stability of the mono complex, depending on whether $(+)-$ or racemic tartaric acid was used. According to Bennett [31] the stability of the mixed ligand complex [Cu(l-asparagine)(d-asparagine)] is smaller than that of the non-mixed species. The stability of the analogous mixed ligand complexes of alanine and phenylalanine corresponds to the statistical considerations according to the studies of Simeon and Weber [32], while the NMR studies by McDonald and Phillips [33] indicate that the stability of the Co(d-histidine)(l-histidine) is higher than the stability of the non-mixed species. The thorough study of similar systems would be very important.

8.4.3 DISTRIBUTION OF COMPLEXES

The distribution of the different species cannot be given as simply as in the former cases. The distribution of the complexes depends on the total concentrations of the metal and the ligands, on their ratio and on the pH of the solution. The calculations are fairly complicated but the application of high-speed computers makes the procedure quite easy. The programmes developed by Perrin and Sayce [34] and by Sillén *et al.* [35] are particularly well suited for the treatment of systems containing mixed ligand complexes. Figure 8.13 illustrates the distribution of the different species at constant total concentration as a function of pH.

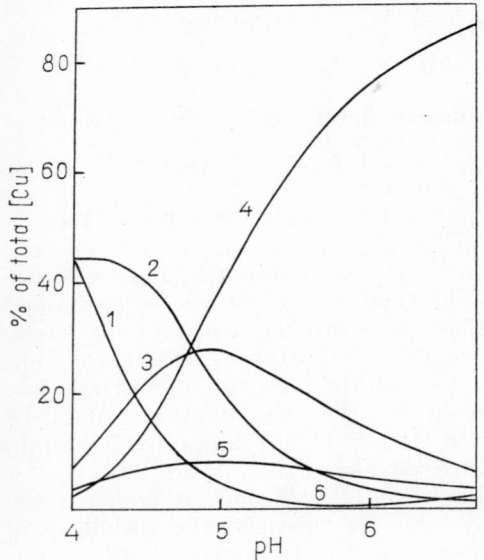

Fig. 8.13 Variation with pH of the composition of a solution of copper(II)ions (0.004 M), histamine (0.004 M), and serine (0.004 M).
Curve *1* — Cu²⁺;
Curve *2* — Cu (serine)⁺;
Curve *3* — Cu (histamine)⁺;
Curve *4* — — Cu (serine) (histamine);
Curve *5* — Cu (serine)₂;
Curve *6* — Cu (histamine)₂.
(Reproduced with permission from *J. Chem. Soc.* A 755 (1967))

8.5 Mixed ligand complexes of low co-ordination level

If the concentration of the ligands is low enough, only the first members of the series of successive complexes are formed. The number of mixed ligand complexes being much smaller in such cases, it is even possible that only one mixed ligand complex exists. The most thoroughly studied systems of this type are the mixed ligand complexes of iron(III) involving

ammonia and of several bidentate ligands by different nickel(II)–amino-polycarboxylate complexes does not result in the breaking of bonds between the central ion and the donor groups of the aminopolycarboxylate. Thermo-dynamic considerations seem to support this view. In the case of thorium even a terdentate ligand is taken up without displacing a donor group of the original aminopolycarboxylate, and the co-ordination sphere is expanded [53].

8.7 Formation of mixed ligand complexes by the expansion of the co-ordination sphere

The expansion of the co-ordination sphere occurs most frequently with square planar complexes by the uptake of two further ligands. This reaction is responsible for the difference of colour of certain cobalt(II) and nickel(II) complexes in different solvents. Only a few examples are mentioned here from the abundant literature of this problem. Sacconi, Lombardo and Paoletti [54] have studied the uptake of heterocyclic bases by biacetyl-bisbenzoylhydrazone-nickel(II) complex in benzene. The stability constant was calculated from the spectral changes caused by the variation of the base concentration. Császár [55] has studied by the same method the effect of substituents on the stability of mixed ligand complexes of nickel(II) with different derivatives of N-alkylsalicylaldimine and pyridine in chloro-form. Graddon and Watton [56] investigated the reaction of the bis(β-diketonato)copper(II) complex with heterocyclic bases in different non-aqueous solvents. The stability of these adducts depends on the nature of the β-diketone, the second ligand and the solvent.

Expansion of the co-ordination sphere may occur if the parent complex is tetrahedral. King, Körös and Nelson [57] have studied the equilibrium

$$\underset{\text{(tetrahedral)}}{CoL_2X_2 + 2L} = \underset{\text{(octahedral)}}{CoL_4X_2} \qquad (8.97)$$

in chloroform, where L = pyridine or 2-methylpyridine, and X = Cl^-, Br^-, I^-, OCN^-, SCN^- or $SeCN^-$.

The expansion of the co-ordination sphere occurs with the square planar dimethylglyoxime complexes of cobalt(II). The square planar parent complex can bind two halide or pseudohalide ions. Burger et $al.$ [58] have pointed out that the co-ordination of unidentate ligands can affect the strength of the metal ion–dimethylglyoxime bond in both directions, depending on the tendency of the ligand for back-co-ordination. In fact these reactions mean the displacement of very loosely bound water molecules.

An interesting example of increase in the co-ordination number is the reaction of $Ni(CN)_4^{2-}$ with halide and pseudohalide ions. The tetracyano complex is very stable and in the presence of high concentrations of cyanide [59], thiocyanate [60], iodide [60] and bromide [60] the following species are formed: $Ni(CN)_5^{3-}$, $Ni(CN)_4SCN^{3-}$, $Ni(CN)_4I^{3-}$, $Ni(CN)_4Br^{3-}$.

8.8 General methods for the study of the formation of mixed ligand complexes

The formation of any mixed ligand complex can be generally described by the equilibrium

$$\text{Me} + i\text{A} + j\text{B} = \text{MeA}_i\text{B}_j \tag{8.98}$$

Fronaeus [61] has developed a potentiometric method for the determination of the stability constant of this type of equilibrium by measuring the free concentration of one of the ligands. This method is the generalization of the method treated in Chap. 4.

The stability constants can be evaluated from the measurement of the concentration of the free metal ion as a function of the concentration of the ligands A and B. The concentration of the free metal can be determined potentiometrically [62] or polarographically [63]. For the evaluation of the constants Fridman *et al.* [62] described a graphical method, but in these complicated cases the application of electronic computers is very helpful.

Marcus [64] has developed liquid–liquid extraction methods for the evaluation of all the equilibrium constants for systems of co-ordination number 4, and applied them in the study of the mercury(II)–bromide–iodide system. In the first method the extraction of the central ion (in the three possible electrically neutral forms MeA$_2$, MeB$_2$ and MeAB) is measured as a function of the ratio of the free ligand concentrations. In the second method the stability constants are evaluated from the extraction of one

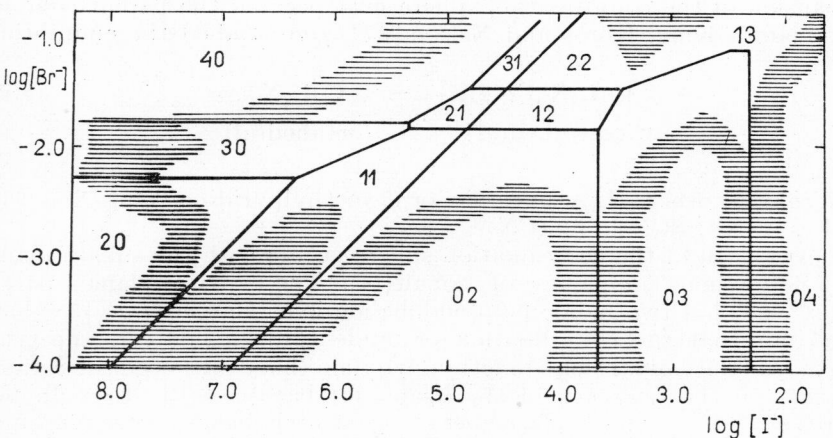

Fig. 8.14 Predominance area diagram for Hg(II)–Br$^-$–I$^-$ system. Areas correspond to predominance (>50% in shaded areas) of the complexes HgBr$_i$I$_j$ indicated by their indices (e.g. HgBr$_2$I is symbolized by 21) as a function of free bromide and free iodide concentration. (Reproduced with permission from *Acta Chem. Scand.* *11*, 619 (1957))

of the ligands (in the form of MeA_2 or MeB_2 and MeAB at different total concentrations of the metal and the ligands. Marcus introduced a convenient form for plotting the distribution of the different complexes as a function of the free ligand concentrations. The so-called predominance area diagram is shown in Fig. 8.14.

8.9 Mixed complex formation in the liquid–liquid extraction of metal ions

The formation of mixed ligand complexes in the liquid–liquid extraction of metal ions has already been treated in connection with the phenomenon of synergism (Section 5.5.3). It is self-evident that the conditions of the extraction of metal complexes favour the formation of mixed species. There are examples even for the existence of mixed ligand complexes containing three different ligands [65]. Here only some references are given for some particularly important papers [66–72].

8.10 Complexes with three different ligands

Although there is no available quantitative study of equilibria involving complexes containing three different ligands, there is some evidence for their formation. From a Raman study, Krishnan and Plane [73] have detected that chloride, thiocyanate and glycine are simultaneously coordinated to zinc(II) and to cadmium(II). An equilibrium analysis [74] of the complexometric method [75] recommended for the determination of thallium(III) in alkaline solution showed that the simultaneous coordination of tartrate, bromide and xylenol orange to thallium(III) has to be considered.

From the observation that the absorbance of the iron(III)–EDTA–H_2O_2–NH_3 system is higher than that of the iron(III)–EDTA–H_2O_2 system at the same pH obtained with borax buffer it was concluded that the mixed ligand complex contains EDTA, hydrogen peroxide and ammonia simultaneously [76]. This conclusion is not correct. The reason for the smaller absorbance in the absence of ammonia is the presence of borax; the formation of a peroxo derivative of the borate reduces the concentration of the perhydroxide ion and therefore also the concentration of the coloured iron(III)–EDTA–peroxide mixed ligand complex.

8.11 Factors determining the stability of mixed ligand complexes

Although there are some empirical relationships between the stabilities of mixed ligand complexes and the corresponding parent complexes, a general and satisfactory theoretical explanation is still lacking. Statistically the formation of the mixed ligand species is always favoured. Marcus pointed out that the statistical value of the equilibrium constant of the reaction expressed by Eq. 8.16 (K_{rij}) is

$$K_{\text{stat}} = \left(\frac{h!}{i!\,j!}\right)^h. \tag{8.99}$$

The statistical values for the different stability constants can be derived from Eq. 8.99 and the relationships given in Section 8.1.

The experimentally determined stability constants in most cases deviate from the statistically calculated constants. Because most of the earlier determined constants showed that the stability of the mixed species is preferential, it was thought that the extra stabilization is a general phenomenon.

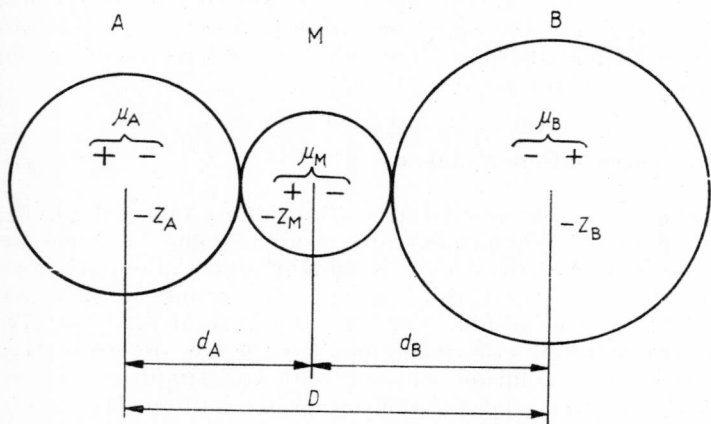

Fig. 8.15 The polarized ion model for the mixed ligand complex MeAB. For the meaning of symbols see the text

Theoretical considerations were therefore sought that would explain this high stability of the mixed ligand complexes. However, a survey of the available equilibrium constants clearly proves that deviations from the statistically expected stability occurs in both directions. On the basis of elementary electrostatic considerations Kida has pointed out that higher stability of mixed ligand complexes may be expected. The model used by Kida [78], however, is a very approximate one and it was considerably improved by Marcus and Eliezer [77, 79]. This polarized ion model for the mixed ligand complex MeAB is shown in Fig. 8.15.

The total electrostatic energy E_{el} for the formation of one molecule of the MeAB mixed ligand complex is given by

$$\frac{\varepsilon}{e^2}E_{\text{el}} = \frac{Z_A Z_{\text{Me}}}{d_A} + \frac{Z_B Z_{\text{Me}}}{d_B} - \frac{Z_A Z_B}{D} + \frac{Z_{\text{Me}}\,\mu_A}{d_A^2} + \frac{Z_{\text{Me}}\,\mu_B}{d_B^2} + \frac{Z_A\,\mu_{\text{Me}}}{d_A^2} -$$

$$-\frac{Z_B\,\mu_{Me}}{d_B^2} - \frac{Z_A\,\mu_B}{D^2} - \frac{Z_B\,\mu}{D^2} + \frac{2\,\mu_{Me}\,\mu_A}{d_A^3} - \frac{2\,\mu_{Me}\,\mu_B}{d_B^3} - \frac{2\,\mu_A\,\mu_B}{D^3} -$$

$$-\frac{\mu_A^2}{2\,\alpha_A} - \frac{\mu_{Me}^2}{2\,\alpha_{Me}} - \frac{\mu_B^2}{2\,\alpha_B} \tag{8.100}$$

where ε is the dielectric constant of the medium separating the constituents of the complex, e is the charge on an electron, and the terms in the first, second and third lines pertain to the ion-ion, ion-dipole and dipole-dipole interactions respectively, while those in the last line refer to the formation of induced dipoles. The dipole moment μ assumed at the centre of an ion is the product of its polarizability, α, and the field strength F at its centre. The field strengths at Me, A and B are given by expressions interconnected through the three dipole moments, and must be solved simultaneously. The values for the dipoles may be calculated from the values obtained for the field strengths and inserted into Eq. 8.100 to calculate the electrostatic energy of the mixed ligand complex. Similar calculations can be carried out for the parent complexes. An important feature of this treatment is that this model can explain both the increase and the decrease of the stability of the mixed ligand complex in comparison with the statistically calculated value. Unfortunately the available values for the dipole moments and polarizabilities are not correct enough for the calculations, and moreover it is impossible to give an unambiguous dielectric constant for the medium. Marcus and Eliezer assumed that the dielectric constant is equal to unity, but there are many arguments against such an assumption. Therefore the agreement of the theoretically calculated and experimentally found stability constants must be regarded with some reservation.

So, in most cases, we have to be content with empirical relationships. Such a relationship is shown by Eqs. 8.92 and 8.93, valid for the formation of some mixed ligand complexes involving aminopolycarboxylates and unidentate ligands. Jackobs and Margerum [80] have found that for many mixed ligand complexes of nickel(II) (NiLX) the free energy change of formation of the mixed ligand complex can be described by the equation

$$\Delta G_{NiLX} = (b\Delta G_{LN} + a\,\Delta G_{LC}) + (d\,\Delta G_{Ch} + z_1 z_2 \Delta G_Z + n\,\Delta G_{XN} + c\Delta G_{XC}). \tag{8.101}$$

The first two terms refer to the effect which groups co-ordinated to the nickel have on the second ligand; $b\Delta G_{LN}$ is for b bound nitrogen atoms and $a\Delta G_{LC}$ for a bound carboxylate groups. The remaining terms are for the interaction of the second ligand, X, with the complex NiL. The term $d\Delta G_{Ch}$ is for d chelate rings formed by X. The ion-ion electrostatic term is $z_1 z_2 \Delta G_Z$ where z_1 and z_2 are the charges of NiL and X, respectively. The last two terms, $n\Delta G_{XN}$ and $c\Delta G_{XC}$ refer to the free energy changes resulting from n amine bonds and c carboxylate bonds replacing co-ordinated water. No bond rupture of the metal ion and L ligand was assumed. A value was assigned to ΔG_{Ch} from Schwarzenbach's [81] calculations for the chelate of nickel(II).

The other free energy changes were calculated by computer from the stability constants found. As appears from Table 8.4 the empirical equation gives an excellent fit for the complexes studied.

TABLE 8.4

Comparison of calculated* and observed values of G_{NiLX}

Complex	$b \, \Delta G_{LN}$	$a \, \Delta G_{LC}$	$d \, \Delta G_{Ch}$	$z_1 z_2 \, \Delta G_Z$	$z_1 \, \Delta G_{DZ}$	$n \, \Delta G_{XN}$	G_{NiLX}, kcal/mole Calcd.	Obsd.
[Ni(EDTA) (NH$_3$)]$^{2-}$	1.10	−1.74	0	0	1.22	−2.68	−2.10	−1.85
[Ni(HEDTA) (NH$_3$)]$^-$	1.10	−1.74	0	0	0.61	−2.68	−2.71	−2.72
[Ni(NTA) (NH$_3$)]$^-$	0.55	−1.74	0	0	0.61	−2.68	−3.26	−3.44
[Ni(tetren) (NH$_3$)]$^{2+}$	2.75	0	0	0	−1.22	−2.68	−1.15	−1.14
[Ni(EDDA) (NH$_3$)]	1.10	−1.16	0	0	0	−2.68	−2.74	−2.72
[Ni(NTA) (ox)]$^{3-}$	0.55	−1.74	−3.82	2.06	0	0	−2.95	−2.95
[Ni(NTA) (en)]$^-$	0.55	−1.74	−3.82	0	0.61	−5.36	−9.76	−9.79
[Ni(NTA) (gly)]$^{2-}$	0.55	−1.74	−3.82	1.03	0	−2.68	−6.66	−6.65
[Ni(dien) (gly)]$^+$	1.65	0	−3.82	−2.06	0	−2.68	−6.91	−6.98
[Ni(dien) (ox)]	1.65	0	−3.82	−4.12	0	0	−6.29	−6.24

* Using $\Delta G_{LN} = 0.55$, $\Delta G_{LC} = -0.58$, $\Delta G_{Ch} = -3.82$, $\Delta G_Z = 1.03$, $\Delta G_{DZ} = 0.61$, and $\Delta G_{XN} = -2.68$ kcal/mole

HEDTA = hydroxyethylethylenediaminediacetate; NTA = nitrilotriacetate;

EDDA = ethylenediaminediacetate, tetren = triethylenetetramine, dien = diethylenetriamine; gly = glycine

Marcus, Eliezer and Zangen [79] found that for many complexes of the type MeAB, log K_r is a linear function of the difference of the logarithms of the second overall stability constant of the parent complexes (Fig. 8.16). As appears from Fig. 8.17 [82] this relationship is valid for several mixed ligand complexes of mercury(II), while it is quite inoperative for others. So it is not recommended to use this relationship in making assumptions for unknown K_r values.

8.12 The kinetic importance of mixed ligand complexes

It is self-evident that the different types of mixed ligand complex play an important role in various reactions, and only some characteristic examples are given here.

Margerum *et al.* [83–85] have pointed out in several careful studies that mixed ligand complexes are the intermediates in ligand displacement reactions. The mechanism of these reactions may be very complicated but the dependence of the rate on different conditions, especially pH, gives us

Fig. 8.16 Log K_r plotted
against the difference
of the logarithms
of the second stabil-
ity product of the
corresponding
parent complexes.
The mixed com-
plexes, in order of
increasing $\varDelta \log \beta_2$
are: $FeFSCN^+$,
$AgNH_3Br$,
$HgBrSCN$,
$CuIS_2O_3^{2-}$,
$FeSCNSO_4$,
$Cu(CH_3CO_2)CO_4^-$,
$UO_2(CH_3CO_2)SO_4^-$,
$AgIS_2O_3^{2-}$, $AgNH_3Cl$,
$FeSCNCl^+$, $HgClBr$,
$HgClSCN$,
$CuNH_3NO_2^+$,
$Ni(en)C_2O_4$,
$HgBrI$, $AgClS_2O_3^{2-}$,
$AgISCN^-$, $HgClI$,
$Cu(en)C_2O_4$,
$HgOHCN$,
$AgNH_3CN$,
$AgOHCN$,
$Hg(CH_3CO_2)CN$

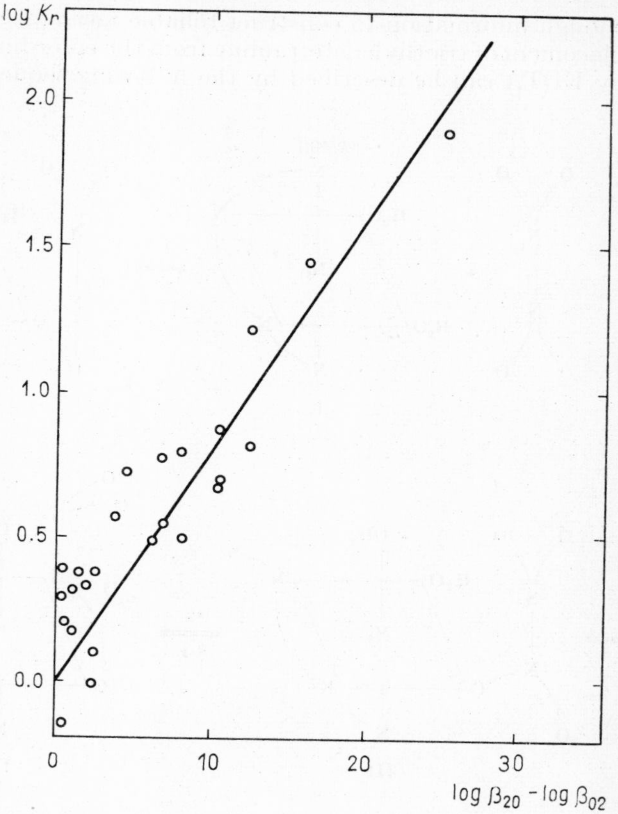

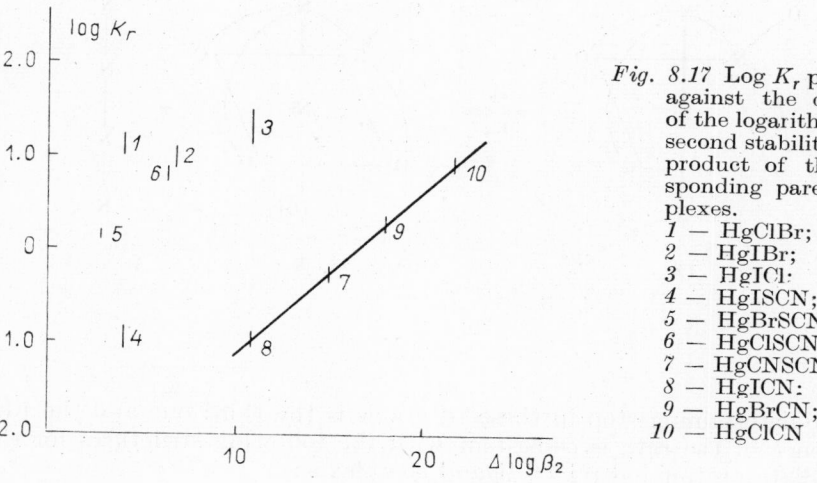

Fig. 8.17 Log K_r plotted
against the difference
of the logarithms of the
second stability
product of the corre-
sponding parent com-
plexes.
1 — $HgClBr$;
2 — $HgIBr$;
3 — $HgICl$:
4 — $HgISCN$;
5 — $HgBrSCN$;
6 — $HgClSCN$;
7 — $HgCNSCN$;
8 — $HgICN$:
9 — $HgBrCN$;
10 — $HgClCN$

enough information to construct reliable mechanisms. For example the displacement of triethylenetetramine from the co-ordination sphere of nickel(II)
by EDTA can be described by the following sequence of reactions:

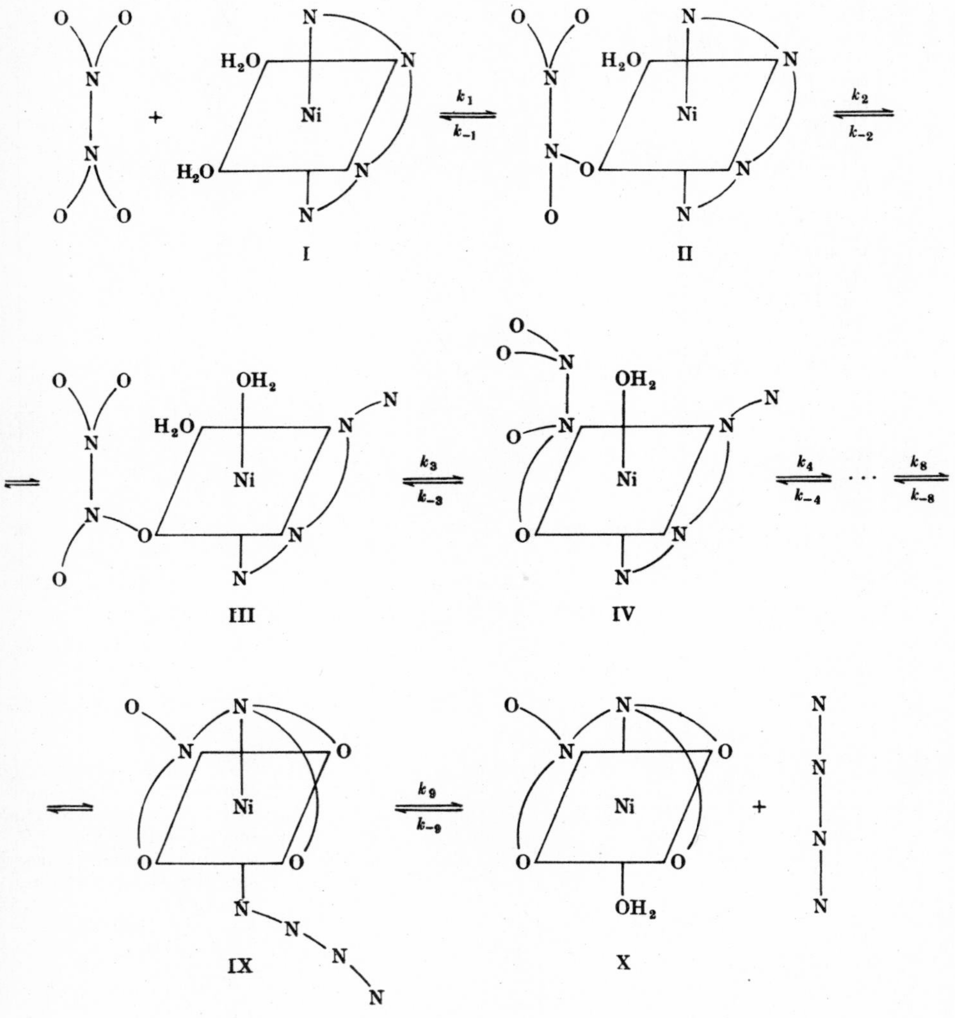

The rate-determining step in these reactions is the third one and the pH-
dependence of the rate is consistent with the following structures for the
intermediate protonated mixed ligand complexes:

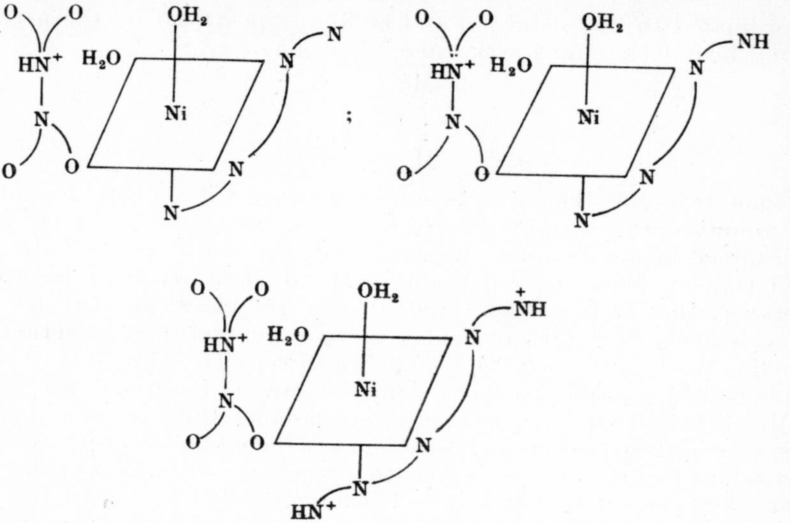

According to Margerum the stability constants of the series of mixed ligand complexes in Eq. 8.101 can be calculated from the stability constants of the parent metal ligand complexes corresponding to the segments co-ordinated. Thus, for the intermediate species III in Eq. 8.101 the stability constant of the polyamine segment is considered to be equal to that of the diethylene-triamine–nickel(II) complex, while the stability constant of the EDTA segment is assumed equal to the stability constant of the acetatonickel(II) complex. This is of course a fairly rough approximation.

Mixed ligand complexes are the intermediates in ligand-catalysed complex formation reactions [86, 87]. If the central ion Me reacts slowly with ligand A and rapidly with ligand B and if the complex MeB also reacts rapidly with ligand A, then ligand B catalyses the formation of the complex MeA. Schematically

$$Me + A \xrightarrow{k_1} MeA \qquad (8.102)$$

$$Me + B \xrightarrow{k_2} MeB \qquad (8.103)$$

$$MeB + A \xrightarrow{k_3} MeAB \xrightarrow{k'_3} MeA + B \qquad (8.104)$$

$$k_2, k_3, k'_3 > k_1. \qquad (8.105)$$

The formation of the mixed ligand complex, and its properties, are crucial from the point of view of the kinetic effect. The catalytic effect of hydroxide ion is quite general; recently several more selective ligand-catalytic effects have been found.

Mixed ligand complex formation is responsible for some catalytic effects in redox reactions, too. In connection with the activation of molecular hydrogen by silver(I) complexes Halpern [88] assumed that hydrogen is

co-ordinated to the metal ion and is doubly polarized by the permanently
bound ligand (X) and by the silver(I) itself

$$Me^{\delta +} - X^{\delta -}$$

$$H^{\delta -} \quad - H^{\delta +}$$

In some reactions the catalytic effect of the metal ion can be explained by
the simultaneous co-ordination of both reactants, that is, by the formation
of a mixed ligand complex. In these cases the metal ion mediates the electron transfer from the reductant to the oxidant. It must be mentioned,
however, that it is almost impossible to get clear-cut evidence for such
a mechanism. The effect of neutral salts on several redox reactions such as
$MnO_4^{2-}-MnO_4^-$ [89], $Fe(CN)_6^{4-}-Fe(CN)_6^{3-}$ [90], $SO_3^{2-}-Fe(CN)_6^{3-}$ [91] can be
explained by assuming a metal–ion bridging mechanism.

Mixed ligand complex formation is fundamentally important in various
enzyme-catalysed reactions. Most enzymes contain co-ordinatively bound
metal ions and during the enzymatic process co-ordination of the substrate
also takes place and thus a mixed ligand complex is formed.

According to Mildvan and Cohn [92] the mechanism of conversion of the
substrate complex into the product complex in the pyruvate kinase reaction
can be represented by the scheme shown by Fig. 8.18. Inhibitors are com-

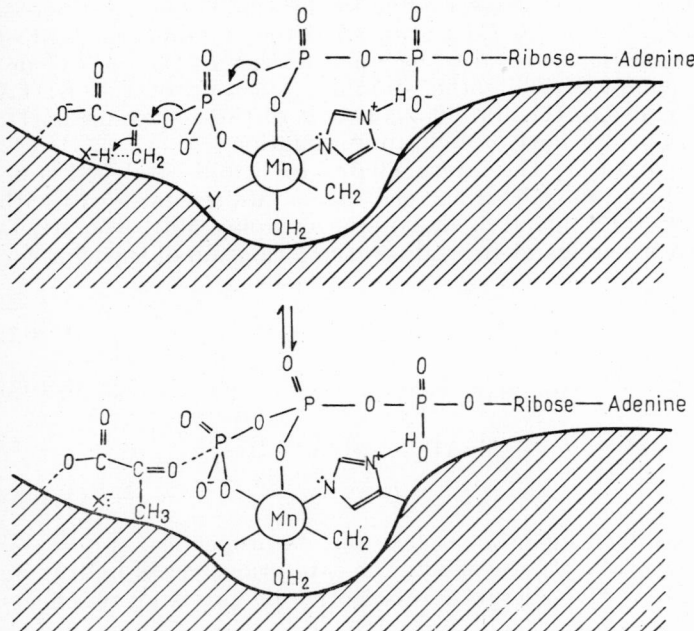

Fig. 8.18 Mechanism of conversion of the substrate complex to the product complex
in the pyruvate kinase reaction. (Reproduced with permission from *J. Biol.
Chem.* **241**, 1178 (1966))

pounds which are preferentially co-ordinated to the metal ion of the enzyme and therefore hinder the co-ordination of the substrate. The formation of such an enzyme–inhibitor–metal ion mixed ligand complex has been proved by Vallee, Coombs and Williams [93].

Mixed ligand complex formation plays an analogous role in the action of enzyme models [94]. The most thoroughly studied reaction type is trans-amination catalysed by metal ions [95–97]. The source of the catalytic action of metal ions is shown by Eq. 8.106.

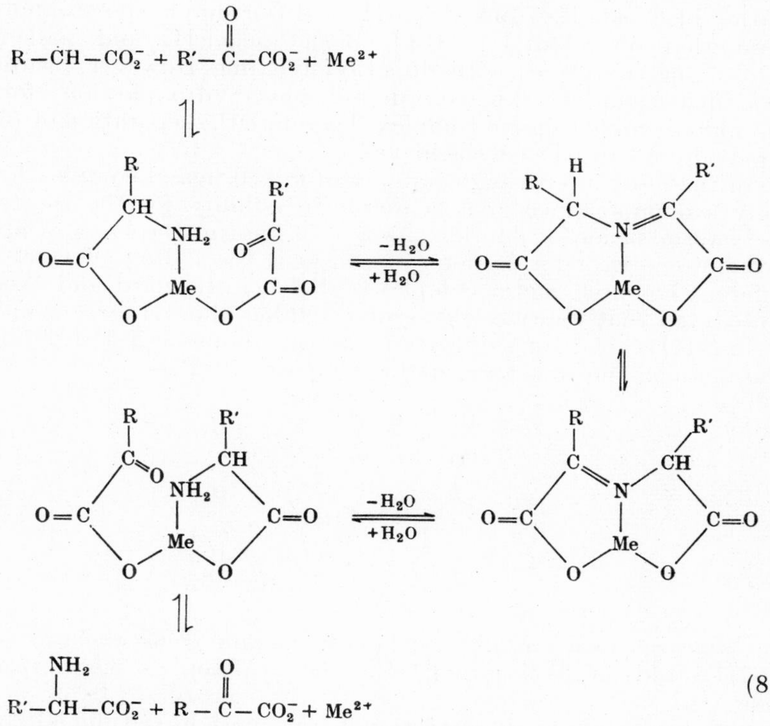

$$R-CH-CO_2^- + R'-\overset{\displaystyle O}{\overset{\|}{C}}-CO_2^- + Me^{2+}$$

$$(8.106)$$

Leussing *et al.* [98–100] thoroughly studied a number of systems of this type. When the stability constant of the mixed ligand species was found to be much greater than the statistically expected value, the formation of the metal complex of the Schiff's base was assumed.

8.13 Analytical application of mixed ligand complexes

In most analytical methods based on complex formation mixed ligand complexes are involved. This statement is equally valid for separations based on liquid–liquid extraction and for different titrimetric and gravimetric procedures. Therefore only a few examples are given here to illustrate the principles of the applications.

The tendencies of the different square-planar complexes to take up one or two further ligands to form a square pyramidal or a (distorted) octahedral complex are widely different. So the mixed ligand complex formation may contribute considerably to the selectivity of certain analytically important reactions. Among the bis(dimethylglyoximato) complexes of bivalent metal ions, that of nickel(II) does not show the slightest tendency towards expansion of the co-ordination sphere, while that of cobalt(II) can take up two ligands even as bulky as iodide. The formation of the anionic mixed ligand complex results in a dramatic increase in solubility, making possible the separation of cobalt(II) from nickel(II) and the simple spectrophotometric determination of cobalt(II) [101]. Both nickel(II) and palladium(II) form insoluble complexes with dimethylglyoxime. However, bis(dimethylglyoxime)palladium(II) can co-ordinate one hydroxide ion, forming a soluble anionic mixed ligand complex. Therefore the separation of nickel(II) from palladium can be easily achieved.

One of the most interesting examples of mixed ligand complex formation of analytical importance gives a unique possibility for the spectrophotometric determination of fluoride. As is well known, the basis of almost all spectrophotometric fluoride determinations is the fading effect of this ion on different coloured metal complexes. Belcher, Leonard and West [102] have observed that fluoride ion forms a soluble blue ternary complex with the cerium(III) chelate of 3[di(carboxymethyl)aminomethyl]-1,2-dihydroxy-anthraquinone. The structure of the complex [103] is:

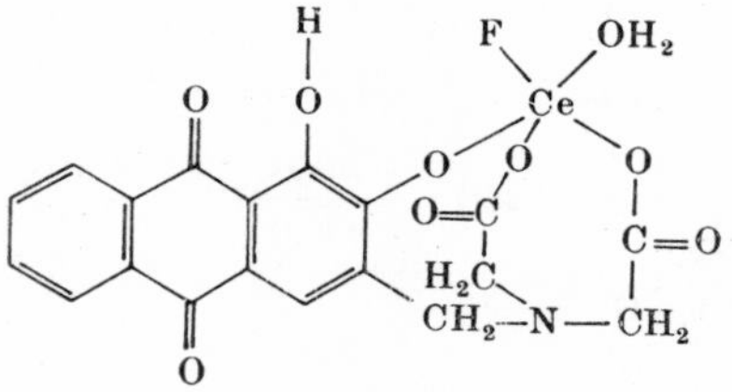

The same reaction was observed with the analogous chelates of lanthanum(III) and praseodymium(III). The reaction is very specific for fluoride ion.

Finally, it must be mentioned that many inert mixed ligand complexes may find analytical applications. Schilt et al. [104–105] studied the behaviour and analytical applications of different cyano-1,10-phenanthroline complexes of iron(II) and iron(III) as indicators and reagents in spectrophotometric analysis.

REFERENCES

1. BAILAR, J. C. JR. (Editor), *The Chemistry of the Coordination Compounds*, Reinhold, New York, 1956.
2. GRINBERG, A. A., *An Introduction to the Chemistry of Complex Compounds*, Pergamon, London, 1962.
3. JONES, M. M., *Elementary Coordination Chemistry*, Prentice Hall, Englewood Cliffs, 1964.
4. BASOLO, F. and PEARSON, R. G., *Mechanism of Inorganic Reactions*, Second Edition, Wiley, New York, 1967.
5. PREETZ, W. and BLASIUS, E., *Z. Anorg. Chem. 332*, 190 (1964).
6. SPIRO, T. and HUME, D. N., *J. Am. Chem. Soc. 83*, 4305 (1961).
7. HUHN, P. and BECK, M. T., *Acta Chim. Acad. Sci. Hung. 51*, 7 (1967).
8. BECK, M. T. and GAIZER, F., *Acta Chim. Acad. Sci. Hung. 41*, 423 (1964).
9. MARCUS, Y., *Acta Chem. Scand. 11*, 610 (1957).
10. BECK, M. T. and GAIZER, F., *J. Inorg. Nucl. Chem. 26*, 1775 (1964).
11. GAIZER, F., MURAY, L. and BECK, M. T., unpublished results.
12. GAIZER, F. and BECK, M. T., *J. Inorg. Nucl. Chem. 28*, 503 (1966).
13. DE WITT, R. and WATTERS, J. I., *J. Am. Chem. Soc. 76*, 3810 (1954).
14. NEWMAN, L. and HUME, D. N., *J. Am. Chem. Soc. 79*, 4571 (1957).
15. NEWMAN, L. and HUME, D. N., *J. Am. Chem. Soc. 79*, 4581 (1957).
16. SRIVASTAVA, S. C. and NEWMAN, L., *Inorg. Chem. 5*, 1506 (1966).
17. SRIVASTAVA, S. C. and NEWMAN, L., *Inorg. Chem. 6*, 762 (1967).
18. WATTERS, J. I., *J. Am. Chem. Soc. 81*, 1560 (1959).
19. WATTERS, J. I. and DE WITT, R., *J. Am. Chem. Soc. 82*, 1333 (1960).
20. L'HEUREUX, G. A. and MARTELL, A. E., *J. Inorg. Nucl. Chem. 28*, 481 (1966).
21. SIGEL, H. and GRIESSER, R., *Helv. Chim. Acta 50*, 1842 (1967).
22. SIGEL, H., BECKER, K. and McCORMICK, D. B., *Biochim. Acta, 148*, 655 (1967).
23. THOMPSON, L. C. and LORAAS, J. A., *Inorg. Chem. 2*, 89 (1963).
24. KANEMURA, Y. and WATTERS, J. I., *J. Inorg. Nucl. Chem. 29*, 1701 (1967).
25. NÄSÄNEN, R., MERILÄINEN, P. and LUKKARI, S., *Acta Chem. Scand. 16*, 2384 (1962).
26. NÄSÄNEN, R. and KOSKINEN, M., *Suomen Kemistilehti B40*, 23 (1967).
27. PERRIN, D. D., SAYCE, I. G. and SHARMA, V. S., *J. Chem. Soc. A* 1755 (1967).
28. PERRIN, D. D. and SHARMA, V. S., *J. Chem. Soc. A* 446 (1968).
29. SAYCE, I. G., *Talanta 15*, 1397 (1968).
30. FRONAEUS, S., *Dissertation*, Lund, 1948.
31. BENNETT, W. E., *J. Am. Chem. Soc. 81*, 246 (1959).
32. SIMEON, V. and WEBER, O. A., *Croatica Chem. Acta, 38*, 161 (1966).
33. McDONALD, C. C. and PHILLIPS, W. D., *J. Am. Chem. Soc. 85*, 3736 (1963).
34. PERRIN, D. D. and SAYCE, I. G., *Talanta 14*, 833 (1967).
35. INGRI, N., KAKOLOWICZ, W., SILLÉN, L. G. and WARNQUIST, B., *Talanta 14*, 1261 (1967).
36. RABINOWITCH, E. and STOCKMAYER, W. H., *J. Am. Chem. Soc. 64*, 335 (1942).
37. LISTER, M. W. and RIVINGTON, D. E., *Can. J. Chem. 33*, 1572 (1955).
38. LISTER, M. W. and RIVINGTON, D. E., *Can. J. Chem. 33*, 1591 (1955).
39. LISTER, M. W. and RIVINGTON, D. E., *Can. J. Chem. 33*, 1603 (1955).
40. YALMAN, R. G., *J. Am. Chem. Soc. 83*, 4142 (1961).
41. SMITH, T. D., *J. Inorg. Nucl. Chem. 11*, 314 (1959).
42. SCHWARZENBACH, G. and HELLER, J., *Helv. Chim. Acta 31*, 576 (1951).
43. VANDEGAAR, J., CHABEREK, S. and FROST, A. E., *J. Inorg. Nucl. Chem. 11*, 210 (1959).
44. RINGBOM, A., SIITONEN, S. and SAXÉN, B., *Anal. Chim. Acta 16*, 541 (1957).
45. BECK, M. and CSISZÁR, B., *Magy. Kém. Foly. 66*, 259 (1960).
46. KOCHANY, G. L. and TIMNICK, A., *J. Am. Chem. Soc. 83*, 2777 (1961).
47. SCHWARZENBACH, G., *Helv. Chim. Acta 32*, 839 (1949).
48. JØRGENSEN, C. K., *Acta Chem. Scand. 10*, 887 (1956).
49. MARGERUM, D. W., BYDALEK, T. J. and BISHOP, J. J., *J. Am. Chem. Soc. 83*, 1791 (1961).

50. BECK, M. T. and BJERRUM, J., unpublished results.
51. BHAT, T. R., RADHAMMA, D. and SHANKAR, J., J. Inorg. Nucl. Chem. 27, 2641 (1965).
52. JACKOBS, N. E. and MARGERUM, D. W., Inorg. Chem. 6, 2038 (1967).
53. CAREY, G. H., BOGUCKI, R. F. and MARTELL, A. E., Inorg. Chem. 3, 1388 (1964).
54. SACCONI, L., LOMBARDO, G. and PAOLETTI, P., J. Inorg. Nucl. Chem. 8, 217 (1958).
55. CSÁSZÁR, J., Magy. Kém. Foly. 68, 440 (1962).
56. GRADDON, D. P. and WATTON, E. C., J. Inorg. Nucl. Chem. 21, 49 (1961).
57. KING, H. C. A., KŐRÖS, E. and NELSON, S. M., J. Chem. Soc. 5449 (1963).
58. BURGER, K., RUFF, I. and RUFF, F., Proc. Symp. Coord. Chem. Tihany, Akadémiai Kiadó, Budapest, 1965, p. 205.
59. MCCULLOUGH, R. L., JONES, L. H. and PENNEMAN, R. A., J. Inorg. Nucl. Chem. 13, 286 (1960).
60. BECK, M. T. and BJERRUM, J., Acta Chem. Scand. 16, 2050 (1962).
61. FRONAEUS, S., Acta Chem. Scand. 4, 72 (1959).
62. FRIDMAN, YA. D., SARBAYEV, J. S. and SOROTSEN, R. I., Zh. Neorgan. Khim. 5, 795 (1960).
63. SCHAAP, W. B. and MCMASTERS, D. L., J. Am. Chem. Soc. 83, 4699 (1961).
64. MARCUS, Y., Acta Chem. Scand. 11, 811 (1957).
65. IRVING, H., Proc. Symp. Coord. Chem. Tihany, Hungary 1965, Akadémiai Kiadó, Budapest, 1966, p. 219.
66. SEKINE T. and DYRSSEN, D., J. Inorg. Nucl. Chem. 26, 1727 (1964).
67. DYRSSEN, D. and PETKOVIC, D., Acta Chem. Scand. 19, 653 (1965).
68. LI, N. C., WANG, S. M. and WALKER, W. R., J. Inorg. Nucl. Chem. 27, 2263 (1965).
69. CASEY, R. J., FARDY, J. J. and WALKER, W. R., J. Inorg. Nucl. Chem. 29, 1139 (1967).
70. DEPTULA, C. and MINC, S., J. Inorg. Nucl. Chem. 29, 221 (1967).
71. SEKINE, T. and DYRSSEN, D., J. Inorg. Nucl. Chem. 29, 1481 (1967).
72. ZOLOTOV, Y. A., SERYAKOVA, I. V. and VOROBYEVA, G. A., Talanta 14, 737 (1967).
73. KRISHNAN, K. and PLANE, R. A., Inorg. Chem. 6, 55 (1967).
74. SZABÓ, A. and BECK, M. T., to be published.
75. STRELOW, W. E. and VON TOERIEN, F. S., Anal. Chim. Acta 36, 189 (1966).
76. POEDER, B. C., BOEF, G. and FRANSWA, C. E. M., Anal. Chim. Acta 27, 339 (1962).
77. MARCUS, Y. and ELIEZER, I., J. Phys. Chem. 66, 1661 (1962).
78. KIDA, S., Bull. Chem. Soc. Japan 34, 962 (1961).
79. MARCUS, Y., ELIEZER, I. and ZANGEN, M., Proc. Symp. Coord. Chem. Tihany, Akadémiai Kiadó, Budapest, 1965, p. 409.
80. JACKOBS, N. E. and MARGERUM, D. W., Inorg. Chem. 6, 2038 (1967).
81. SCHWARZENBACH, G., Helv. Chim. Acta 35, 2344 (1952).
82. BECK, M. T. in ref. 79, pp. 4, 7.
83. RORABACHER, D. B. and MARGERUM, D. W., Inorg. Chem. 3, 382 (1964).
84. JANES, D. L. and MARGERUM, D. W., Inorg. Chem. 5, 1135 (1966).
85. CARR, J. D., LIBBY, R. A. and MARGERUM, D. W., Inorg. Chem. 6, 1083 (1967).
86. BECK, M. T., J. Inorg. Nucl. Chem. 15, 250 (1960).
87. BECK, M. T., Record. Chem. Progr. 27, 37 (1966).
88. HALPERN, J., J. Phys. Chem. 63, 398 (1959).
89. GIERSTEN, L. A. and WAHL, A. C., J. Am. Chem. Soc. 81, 1572 (1959).
90. WAHL, A. C., Z. Electrochem. 64, 90 (1960).
91. SWINEHART, J. H., J. Inorg. Nucl. Chem. 27, 2313 (1967).
92. MILDVAN, A. S. and COHN, M., J. Biol. Chem. 241, 1178 (1966).
93. VALLEE, B. L., COOMBS, T. L. and WILLIAMS, R. J. P., J. Am. Chem. Soc. 80, 397 (1958).
94. BRESLOW, R. and CHIPMAN, D., J. Am. Chem. Soc. 87, 4195 (1965).
95. METZLER, D. E., IKAWA, M. and SNELL, E. E., J. Am. Chem. Soc. 76, 648 (1954).
96. LONGENECKER, J. B. and SNELL, E. E., J. Am. Chem. Soc. 79, 142 (1957).
97. MATSUO, Y., J. Am. Chem. Soc. 79, 2016 (1957).

98. LEUSSING, D. L., *Talanta 11*, 189 (1964).
99. LEUSSING, D. L. and HANNA, E. M., *J. Am. Chem. Soc. 88*, 693, 696 (1966).
100. LEUSSING, D. L. and STANFIELD, C. K., *J. Am. Chem. Soc. 88*, 5726 (1966).
101. BURGER, K. and RUFF, I., *Acta Chim. Acad. Sci. Hung. 45*, 77 (1965).
102. BELCHER, R., LEONARD, M. A. and WEST, T. S., *Talanta 2*, 92 (1959).
103. LEONARD, M. A. and WEST, T. S., *J. Chem. Soc.* 4477 (1960).
104. SCHILT, A. A. and CRESWELL, A. M., *Talanta 13*, 911 (1966).
105. SCHILT, A. A. and FRITSCH, K., *J. Inorg. Nucl. Chem. 28*, 2677 (1966).

Chapter 9

COMPLEXES
OF THE OUTER-SPHERE TYPE

In the condensed phase the co-ordination sphere of the central ion is usually saturated, i.e. the maximum number of donor groups is bound to the metal centre. The co-ordination number of a given metal ion is not necessarily constant; its value depends on different factors, primarily on the size and electronic structure of the ligand(s). Under certain conditions an expansion of the co-ordination sphere may occur in the presence of suitable ligands. For example it is well known that the square-planar complexes of copper(II) can usually take up two further ligands, forming a distorted octahedron. The complexes in which some of the donor atoms are bound at short (frequently exceptionally short) distances whereas others are much more weakly bound at larger distances, are termed *anisotropic* complexes by Jørgensen [1]. Even the expansion of the co-ordination sphere does not mean that a co-ordinatively fully saturated complex has no affinity at all for other species present in the solution. This type of reaction was already recognized by Werner [2] and now there is abundant evidence for the formation of the so-called *outer-sphere* complexes. In the following, this term always refers to species in which anionic or electrically neutral ligands are attached to a co-ordinatively saturated complex. In these complexes there is no direct bonding between the central ion and the ligand in the outer sphere. It appears that the inner-sphere and the outer-sphere complexes represent two extreme types and the anisotropic complexes are intermediates between them. Of course there are no sharp frontiers between these types, and a clear-cut distinction between inner- and outer-sphere complexes is only seldom possible [3]. The treatment is much easier if the inner-sphere complex is inert. The changes in the outer sphere are very rapid and therefore can be studied quite separately from the slow reactions involving changes in the inner sphere. The results of such studies can also be considered in the treatment of the analogous labile systems, because it is reasonable to suppose that the stability and structure of the outer sphere are not considerably influenced by the inertness of the inner sphere.

9.1 Detection of outer-sphere complex formation and determination of stability constants

When in a solution only equilibria involving changes in the outer sphere have to be considered, i.e. in systems where the inner-sphere complexes are inert, the same experimental and calculation methods can be applied as in the case of the study of successive complex formation. Optical rotatory dispersion measurement can be used even more extensively. Whilst a change in the optical rotatory power can be used to study equilibria of labile complexes only if the ligand is optically active, in the case of inert complexes outer-sphere complex formation can be followed polarimetrically if the inert complex itself is asymmetric.

However, even in these simple systems some special problems arise. It may be expected that in outer-sphere complex formation, if specific forces are involved at all, these must be much less pronounced than in the case of direct interactions of metal ions with ligands. Therefore, the principle of using a constant ionic medium [4] must be applied very cautiously. The favourite non-complex-forming ligands, nitrate and perchlorate, are not as innocent as in the case of inner-sphere complex formation. In the study of outer-sphere complexes of tris(phenanthroline)cobalt(III), fluoride ion could be used as a swamping ion [5].

The situation is much more complicated if the simultaneous formation of inner- and outer-sphere complexes occurs. The usual equilibrium analysis cannot distinguish between the corresponding outer- and inner-sphere complexes $Me(H_2O)_6L$ and $Me(H_2O)_5L$. Consequently the stability constant of the first complex, obtained by the usual methods (β_1), is a composite constant:

$$\beta_1 = \frac{[Me(H_2O)_5L] + [Me(H_2O)_6L]}{[Me(H_2O)_6][L]} = \beta_{10} + \beta_{01}. \qquad (9.1)$$

Analogously it can be written in general that

$$\beta_n = \sum_{i=0}^{n} \beta_{i(n-i)} \qquad (9.2)$$

where the second subscript refers to the number of ligands bound in the outer sphere.

9.1.1 Experimental methods in the study of inert complexes

Polarography. The effect of different anions on the shift of the half-wave potential and the decrease of the diffusion current was observed by Laitinen *et al.* [6] in the polarographic reduction of $Co(NH_3)_6^{3+}$. Later, both effects were quantitatively studied for the evaluation of stability constants [7, 8]. Methods were developed for the determination of the stability constants of outer-sphere complexes from the kinetics of polarographic reduction [9] and from the chronopotentiometric transition time [10]. These latter methods have the advantage that they can be applied to both reversible and irreversible electrode processes.

Conductivity measurements can be used under suitable experimental conditions to obtain quantitative information on the stability of certain outer-sphere complexes [11]. Recently a continuous variation approach was applied in acetone solution [12].

Solubility measurements were made in testing the validity of the Debye-Hückel theory. Davies [13] pointed out that the deviations from the theoretically calculated solubilities can be explained by the formation of outer-sphere complexes. This simple technique can provide very exact and reliable constants. Solubility experiments by Larsson indicated [14] the formation of outer-sphere complexes involving neutral ligands.

Liquid–liquid extraction has received much less attention than it deserves in the determination of stability constants of outer-sphere type complexes. Larsson [15] showed by extraction experiments that $Co(NH_3)_6^{3+}$ can take up as many as four salicylate ligands to form a negatively charged species. Another important observation is that different ligands (salicylate and cyclohexane) are simultaneously bound in the outer sphere.

Ion-exchange measurements [16, 17] indicated the formation of the following anionic species: $Co(NH_3)_6Cl_4^-$, $Co(NH_3)_6Br_4^-$, $Co(NH_3)_6(SO_4)_3^{3-}$ and $Co(NH_3)_6(S_2O_3)_3^{3-}$.

Measurements of pH can be used in the case of the uptake of a basic ligand in the outer sphere. Papers by Peacock and James [18] and by Archer, East and Monk [19] contain a lot of data referring to $Co(NH_3)_6^{3+}$ and $Co(NH_3)_5Cl^{2+}$ and different organic ligands.

Nuclear magnetic resonance methods are well suited for the study of weak interactions. Alei [20] studied the association between $Cr(H_2O)_6^{3+}$ and perchlorate ion (in water) by measuring the effect of the paramagnetic ion on the chemical shift of ^{17}O nuclei in enriched water as a function of Cr(III) and ClO_4^- concentration. Stengle and Langford [21] developed another NMR technique. They studied the effect of a paramagnetic ion on the relaxation time of nearby nuclei. The association of some chromium(III) complexes with PF_6^- and F^- ions was observed. The interaction of the solvent (deuterated dimethylsulphoxide–D_2O mixture) and the hydrogen atoms attached to the nitrogen atoms in the $Co(en)_3^{3+}$ complex was also proved by NMR studies [22].

Spectrophotometric studies are perhaps the most popular in studying the formation of outer-sphere complexes. The effect of anionic ligands on the spectra of some cobalt(III) complexes was observed by Kiss and Czeglédy [23]; the first quantitative study was made by Linhard [24]. Taube and Posey [25] observed a remarkable change in the molar absorptivities of the outer-sphere complexes with temperature. This possibility must be considered in the evaluation of spectrophotometric data obtained at different temperatures. Recently Ogino and Saito [26] elaborated a method to follow the association of the asymmetric inert complexes with an optically active ligand. The method requires the comparative study of racemic complex–ligand and one of the enantiomorph–ligand systems.

Polarimetry and study of circular dichroism are amongst the most powerful techniques for studying outer-sphere complexes. After observations and qualitative studies by Werner [27], Jaeger [28] and Kirschner *et al.* [29, 30], Larsson [31–34] applied these methods to obtain stability data. Besides equilibrium data these methods give important information on the structure of the outer sphere and on the mode of the bonding therein [35].

Reaction kinetic studies can be applied because of the large difference in the rate of the reactions in the inner and in the outer sphere [25]. The outer-sphere complex formation can be regarded as a fast reaction preceding the slow reactions in the inner sphere. The stability constant of the outer-sphere complex can be obtained by measuring the rate as a function of the concentration of the ligand. Recently, a number of such studies have been made, but sometimes the data are regarded as "in no sense thermodynamic equilibrium constants" [36]. This view seems to be overcautious, especially when the value obtained from the kinetic measurements agrees well with the independently determined constants.

9.1.2 DATA REFERRING TO INERT COMPLEXES

Recently, a nearly complete list of the available equilibrium constants referring to inert inner-sphere complexes has been published [37]. It appears from these data that there are sometimes very large differences between the published values, which frequently cannot be explained by the differences in the experimental conditions. Some values determined by polarimetry and circular dichroism [31–34] indicate the formation of a whole series of successive outer-sphere complexes, while the data obtained by spectrophotometry can be fully explained by the formation of one complex [38]. According to Bjerrum [39] the source of discrepancy is that both the dextro and laevo forms of $Co(en)_3^{3+}$ exist as equilibrium mixtures of four conformations, each with its own optical rotation, and these equilibria are disturbed by the formation of outer-sphere complexes. It would be very desirable to make comparative studies with different asymmetric complexes.

The data for the outer-sphere complexes of anions of dibasic organic acids [18, 19] seem to indicate that these ligands can occupy two sites in the outer co-ordination sphere.

9.1.3 STUDY OF LABILE COMPLEXES

In systems where the inner-sphere complexes are labile the main problem is the distinction between inner- and outer-sphere type complexes. Although many approaches to this problem have been suggested, only the *relaxation techniques* developed in the last few years by Eigen *et al.* [40] can be regarded as reliable. That the formation of an MeL complex occurs in successive steps is reflected in the fact that a single relaxation time cannot be assigned to each step. Eigen and Tamm [41] describe the formation of the sulphato complexes of bivalent metal ions by the following scheme:

$$Me_{aq}^{2+} + SO_{4aq}^{2-} \rightleftharpoons \left[Me^{2+} \; O \underset{H}{\overset{H}{\diamond}} O \; SO_4^{2-} \right]_{aq} \rightleftharpoons$$

$$\rightleftharpoons \left[Me^{2+} \; O \underset{H}{\overset{H}{<}} SO_4 \right]_{aq} \rightleftharpoons MeSO_{4aq} . \qquad (9.3)$$

The first step represents in fact, a series of diffusion-controlled processes which, however, can be regarded as a single step. The second step is the formation of the outer-sphere complex which is followed by the transformation to the inner-sphere complex. Similar schemes are valid for other pairs of metal ions and ligands. The analysis of the relaxation spectra gives information on the distribution of the corresponding inner- and outer-sphere complexes. Unfortunately only a few systems have been studied, but it is expected that these methods will be frequently applied in the future. The results show that in the case of all the systems where different oxyanions were used as ligands, a considerable proportion of the mono-complexes exist as outer-sphere complexes; in many systems the outer-sphere complexes even predominate.

The possibility of spectroscopic distinction between inner- and outer-sphere co-ordination has been discussed in several papers [1, 42], but the result is far from unambiguous. The effects produced by outer-sphere complex formation may be disguised by simultaneous inner-sphere association.

A number of authors have dealt with distinguishing between the two types of complex formation by analysis of thermodynamic data [43–48]. It seems quite evident that both types of complex are formed simultaneously but neither of the suggested methods can tell us the distribution of these complexes, or can indicate which type predominates.

9.2 Stability of outer-sphere complexes

The fundamental reason for the interaction of the co-ordinatively saturated complexes with further ligands, resulting in outer-sphere complexes, is undoubtedly the electrostatic attraction between the inner-sphere complex and the ligand. This is reflected in the fact that so far there is no evidence for the formation of an outer-sphere complex involving an uncharged inner-sphere complex. Bjerrum [49] treated the problem from purely electrostatic considerations and defined a distance between the oppositely charged ions within which they are regarded as being associated. This distance, q, is arbitrarily chosen so that the work needed to separate the ions is four times the mean kinetic energy per degree of freedom:

$$q = \frac{z_+ \, z_- \, e^2}{2 \, \varepsilon \, kT} \tag{9.4}$$

where z_+ and z_- are the charges of the ions, e is the charge of the electron, ε is the dielectric constant of the medium, k is the Boltzmann constant and T is the absolute temperature. Then the association constant is given by

$$K = \frac{4 \, N}{1000} \int_a^q \exp \left\{ - \frac{z_+ \, z_- \, e^2}{\varepsilon \, kT} \right\} r^2 \, \mathrm{d}r \tag{9.5}$$

where a is the radius of the central ion and r is the distance between the centres of the two ions. It follows that the higher the charge of the ions and the smaller the dielectric constant of the medium, the greater will be the tendency for the formation of complexes. Most of the experimentally determined stability constants support this view. However, the following observations seem to be in conflict with the simple electrostatic explanation.

(*i*) Formation of anionic species, by overcompensation of the original positive charge by the uptake of several anions in the outer sphere. Evidence was obtained by anion-exchange experiments, liquid–liquid extraction, optical rotation studies, by conductivity and reaction kinetic measurements.

(*ii*) Sometimes there is no evidence for complex formation when it would be expected from consideration of the charges of ions, e.g. Larsson [50] found no interaction between $Co(en)_3^{3+}$ and $Fe(CN)_6^{3-}$. Furthermore, in many cases the stability of the iodo-complex is much greater than that of the corresponding fluoro-complex.

(*iii*) The fact that the stability constants of many outer-sphere complexes are bigger that those of the corresponding inner-sphere complexes.

Considering that ligands in the outer sphere may be bound by hydrogen bonds to the ligands of the inner sphere, this last-mentioned fact is easily understandable.

The other facts, mentioned under (*i*) and (*ii*), suggest that besides the simple electrostatic interaction charge-transfer also occurs and this contributes to the stabilization of the outer-sphere complexes. A certain degree of charge-transfer is indicated by different spectroscopic observations [24, 51, 52]. The charge-transfer bands of $Co(NH_3)_5 X^{2+}$ (X=F, Cl, Br, I) are gradually shifted towards shorter wavelengths in the order $I^- > Br^- > Cl^- > F^-$, clearly indicating that the share of the central ion in controlling the electron pairs coming from halide ions changes in the same order. The fact that the same phenomenon was observed with $Co(NH_3)_6^{3+}$ and halide ions, suggests that the electronic interaction is transmitted by the NH_3 molecules of the inner sphere. The comparative study of inert complexes containing ligands of widely different ability to mediate electronic effects (e.g. ethylenediamine and 2,2'-dipyridyl) would throw more light on this interesting and important phenomenon.

9.3 Kinetic role of outer-sphere complexes

Taube and Posey [25] found that the rate of the reaction

$$Co(NH_3)_5H_2O^{3+} + SO_4^{2-} \rightarrow Co(NH_3)_5SO_4^+ H_2O$$

s almost independent of the concentration of the sulphate ion over a wide range. They explained this fact by postulating the initial formation of an outer-sphere type complex, the rearrangement of which gives the final products of the reaction. As mentioned previously this behaviour is quite general [53] and from a kinetic study both the specific rate constant and the stability constant of the outer-sphere complex can be evaluated.

It is reasonable to expect that the S_N2 mechanism involving a pre-equilibrium between the reactants occurs in the case of substitution reactions of four-co-ordinated square-planar complexes, too. The kinetic studies indicate such pre-equilibria [54], but these association products may be regarded as anisotropic complexes.

Outer-sphere complexes may play an important role in redox reactions between a cationic complex and an uncharged or anionic ligand. However, no quantitative study has so far been made in this direction.

Outer-sphere complex formation and its kinetic effects may be responsible for the somewhat enigmatic role of certain inert complexes in catalytic reactions [55]. The catalytic effects of complexes are generally associated with transient changes in the inner co-ordination sphere. However, if the catalysed reaction is faster than the substitution reaction of the catalyst complex, this explanation cannot be correct. In these cases the catalytic effect is due to the formation of outer-sphere complexes. The enormous repulsion forces which hinder the reaction between two anionic reactants, dramatically decrease if one of the reactants is attached to a co-ordinatively saturated cationic complex. The catalytic effect of $Co(NH_3)_6^{3+}$ in the reaction between persulphate and iodide ions [56] can be explained by such an effect.

Another possibility is electron mediation by the inert complex. Infrared spectroscopic studies indicate [57] an electronic interaction between ligands occupying *trans* positions. As already mentioned there is convincing evidence for an electronic interaction between the central metal ion and the ligands in the outer sphere; therefore the inner-sphere complex itself can act as electronic mediator. It may be expected that the more polarizable the inner-sphere ligands and the central ion, the more efficient is the inert complex in transferring the electrons of the outer-sphere ligands. A condition of such a mechanism is the simultaneous co-ordination of the reactants in the outer-sphere. There is evidence for the preferred formation of the mixed ligand outer-sphere complex $Fe(phen)_3S_2O_8I^-$ [58]. It follows, however, from the nature of the phenomenon that it is very difficult to find definitive experimental evidence for such a mechanism.

9.4 Analytical applications of outer-sphere complexes

Although there are only a few examples of the deliberate analytical applications of outer-sphere complexes, it is evident that in most of the procedures based on solvent extraction this type of complex is involved.

The different outer-sphere complexes of tris(phenanthroline)iron(II) have received the greatest attention. Margerum and Banks [59] utilized the extractability of the perchlorate of this complex into nitrobenzene for the determination of iron traces. It can be inferred from the interfering effect of cobalt(II) and copper(II) that the phenanthroline complexes of these ions also form extractable outer-sphere complexes with perchlorate. Vydra and Přibil [60] achieved a selective extraction of the tris(phenanthroline)iron(II) complex with iodide as counter-ion and chloroform and nitrobenzene as extractants.

The same complex was applied to the determination of a number of anions. In these cases tris(phenanthroline)iron(II) complex is added in excess and the outer-sphere complex formed is extracted. A number of methods were developed by Yamamoto et al. [61–63]. Trichloroacetate can be determined with nitrobenzene as extractant in the presence of acetate [61]. Moderate excess of mono- and dichloroacetate does not interfere. Archer and Doolittle developed methods for the determination of hexafluorophosphate [64] and hexafluoroarsenate [65] ions, using n-butyronitrile as extractant, while Magee et al. [66] elaborated spectrophotometric and two-phase titrimetric methods for the determination of gold in the form of $AuCl_4^-$ and $AuBr_4^-$ ions.

Dagnall and West [67] thoroughly studied the analytical applications of an interesting type of complex. Further studies are necessary to decide whether the complexes involved are mixed ligand anisotropic or outer-sphere complexes. They observed that while the colour of the silver complex of Bromopyrogallol Red (BPR) is yellow, in the presence of 1,10-phenanthroline a blue complex of the composition $[Ag(phen)_2]_2BPR$ is formed. This reaction makes possible a very sensitive and selective determination of silver(I) in aqueous solution. The extraction of the complex into nitrobenzene is even more advantageous [68]. The complex can be applied to the titrimetric [69] or spectrophotometric [70] determination of cyanide.

It was found that Rose Bengal Extra [tetrachloro(P)tetraiodo(R)-fluorescein] and the more easily available Erythrosin [tetraiodo(R)fluorescein] form analogous complexes with the phenanthroline complexes of bivalent metal ions. Spectrophotometric [71] and spectrofluorimetric [72] methods were recommended for the determination of copper(II).

REFERENCES

1. JØRGENSEN, C. K., *Proc. Symp. Coord. Chem. Tihany, Hungary*, p. 11. Akadémiai Kiadó, Budapest, 1965.
2. WERNER, A., *Neuere Anschauungen auf dem Gebiete der anorganischen Chemie*, 3rd Ed. Vieweg, Braunschweig, 1913.
3. YONEDA, H., *Bull. Chem. Soc. Japan 29*, 68 (1956).
4. BIEDERMANN, G. and SILLÉN, L. G., *Arkiv Kemi 5*, 425 (1952).
5. ILCHEVA, L. and BECK, M. T. to be published.
6. LAITINEN, H. A., BAILAR, J. C., JR., HOLTZCLAW, H. F. and QUAGLIANO, J. V., *J. Am. Chem. Soc. 70*, 2999 (1948).
7. LAITINEN, H. A. and GRIEB, M. W., *J. Am. Chem. Soc. 77*, 5201 (1955).
8. TANAKA, N., OGINO, K. and SATO, G., *Bull. Chem. Soc. Japan 39*, 366 (1966).
9. VLČEK, A. A., *Progr. Inorganic Chemistry 5*, 268 (1963).
10. TANAKA, N. and YAMADA, A., *Z. Anal. Chem. 224*, 117 (1967).
11. DAVIES, C. W., *Ion Association*, p. 12. Butterworths, London, 1962.
12. HUGHES, M. N. and TOBE, M. L., *J. Chem. Soc.* 1204 (1965).
13. DAVIES, C. W., *J. Chem. Soc.* 2421 (1930).
14. LARSSON, R., *Acta Chem. Scand. 12*, 708 (1958).
15. LARSSON, R., *Acta Chem. Scand. 16*, 2305 (1962).
16. LARSSON, R., *Acta Chem. Scand. 14*, 697 (1960).
17. LARSSON, R. and TOBIASON, I., *Acta Chem. Scand. 16*, 1919 (1962).
18. PEACOCK, J. M. and JAMES, J. C., *J. Chem. Soc.* 2233 (1951).
19. ARCHER, D. W., EAST, D. A. and MONK, C. B., *J. Chem. Soc.* 720 (1965).
20. ALEI, M., *Inorg. Chem. 3*, 44 (1964).
21. STENGLE, T. R. and LANGFORD, C. H., *J. Phys. Chem. 69*, 3299 (1965).
22. FUNG, B. M., *J. Am. Chem. Soc. 89*, 5788 (1967).
23. KISS, A. and CZEGLÉDY, D., *Z. Anorg. Allgem. Chem. 239*, 27 (1938).
24. LINHARD, M., *Z. Elektrochem. 50*, 224 (1944).
25. TAUBE, H. and POSEY, F. A., *J. Am. Chem. Soc. 75*, 1463 (1953).
26. OGINO, K. and SAITO, U., *Bull. Chem. Soc. Japan, 40*, 826 (1967).
27. WERNER, A., *Ber. 45*, 121 (1912).
28. JAEGER, F. M., *Z. Krist. 55*, 209 (1915).
29. ALBINAK, M. J., BHATNAGAR, D. C., KIRSCHNER, S. and SONNESSA, A. J., *Can. J. Chem. 39*, 2360 (1961).
30. KIRSCHNER, S. and BHATNAGAR, D. C., *Proc. Symp. Coord. Chem. Tihany, Hungary*, p. 23. Akadémiai Kiadó, Budapest, 1965.
31. LARSSON, R., *Acta Chem. Scand. 16*, 2267 (1962).
32. LARSSON, R. and JOHANSSON, L., *Proc. Symp. Coord. Chem. Tihany, Hungary*, p. 31. Akadémiai Kiadó, Budapest, 1965.
33. LARSSON, R. and NORMAN, B., *J. Inorg. Nucl. Chem. 28*, 1291 (1966).
34. LARSSON, R., MASON, S. F. and NORMAN, B. J., *J. Chem. Soc. A* 301 (1966).
35. MASON, S. F. and NORMAN, B. J., *J. Chem. Soc. A* 307 (1966).
36. LANGFORD, C. H. and MUIR, W. R., *J. Am. Chem. Soc. 89*, 3141 (1967).
37. BECK, M. T., *Coord. Chem. Rev. 3*, 91 (1968).
38. OLSEN, I. and BJERRUM, J., *Acta Chem. Scand. 21*, 1112 (1967).
39. BJERRUM, J., *Werner Centennial Symposium*, p. 178. G. B. Kaufmann (ed.), *American Chemical Society*, Washington, 1967.
40. EIGEN, M. and DE MAEYER, L., *Investigation of Rates and Mechanisms of Reactions*, Part II, S. L. Friess, E. S. Lewis and A. Weissberger (eds.), 2nd Ed., p. 895. Interscience, New York, 1963.
41. EIGEN, M. and TAMM, K., *Z. Elektrochem. 66*, 107 (1962).
42. SMITHSON, J. M. and WILLIAMS, R. J. P., *J. Chem. Soc.* 457 (1958).
43. ROSSOTTI, F. J. C., *Modern Coordination Chemistry*, (edited by J. Lewis and R. G. Wilkins), p. 21. Interscience, New York, 1960.
44. CHOPPIN, G. R and STRAZIK, W. F., *Inorg. Chem. 4*, 1250 (1965).
45. DUNCAN, J. F. and KEPERT, D. L., *The Structure of Electrolytic Solutions*, W. J. Hamer (ed.), p. 380. Wiley, New York, 1959.
46. MANNING, P. G., *Can. J. Chem. 43*, 2911 (1965).

47. MANNING, P. G., *Can. J. Chem. 43*, 3258 (1965).
48. MANNING, P. G., *Can. J. Chem. 43*, 3476 (1965).
49. BJERRUM, N., *Kgl. Danske Vid. Selsk. Math. Fys. Medd.*, 7, 1 (1926).
60. LARSSON, R., *Acta Chem. Scand. 21*, 257 (1967).
51. LINHARD, M., SIEBERT, H. and WEIGEL, M., *Z. Anorg. Allgem. Chem. 278*, 287 (1955).
52. BAKER, W. A., JR. and PHILLIPS, M. G., *Inorg. Chem. 4*, 915 (1965).
53. TOBE, M. L., *Record Chem. Progr. 27*, 79 (1966).
54. BASOLO, F. and PEARSON, R., *Mechanisms of Inorganic Reactions*, 2nd Ed. p. 375. Wiley, New York, 1967.
55. BECK, M. T., *Record Chem. Progr. 27*, 37 (1966).
56. INDELLI, A. and AMIS, E. S., *J. Am. Chem. Soc. 82*, 332 (1960).
57. CHATT, J., DUNCANSON, L. A., SHAW, B. L. and VENANZI, L. M., *Disc. Faraday Soc. 26*, 131 (1958).
58. NIKOLASEV, V. and BECK, M. T., to be published.
59. MARGERUM, D. W. and BANKS, C. V., *Anal. Chem. 26*, 200 (1954).
60. VYDRA, F. and PŘIBIL, R., *Talanta 3*, 72 (1959).
61. YAMAMOTO, Y., KUMAMARU, T. and UEMURA, Y., *Anal. Chim. Acta 39*, 51 (1967).
62. YAMAMOTO, Y. and KINUWAKI, S., *Bull. Chem. Soc. Japan 37*, 434 (1964).
63. YAMAMOTO, Y., KOTSUJI, K. and TANAKA, S., *Bull. Chem. Soc. Japan 38*, 499 (1965).
64. ARCHER, V. S. and DOOLITTLE, F. G., *Anal. Chem. 39*, 371 (1967).
65. ARCHER, V. S. and DOOLITTLE, F. G., *Talanta 14*, 921 (1967).
66. NASOURI, F. G., SHAHINE, S. A-F. and MAGEE, R. J., *Anal. Chim. Acta 36*, 346 (1966).
67. DAGNALL, R. M. and WEST, T. S., *Talanta 11*, 1533 (1964).
68. DAGNALL, R. M. and WEST, T. S., *Talanta 11*, 1627 (1964).
69. DAGNALL, R. M., EL-GHAMRY, M. T. and WEST, T. S., *Talanta 13*, 1667 (1966).
70. DAGNALL, R. M., EL-GHAMRY, M. T. and WEST, T. S., *Talanta 15*, 107 (1968).
71. BAILEY, B. W., DAGNALL, R. M. and WEST, T. S., *Talanta 13*, 753 (1966).
72. BAILEY, B. W., DAGNALL, R. M. and WEST, T. S., *Talanta 13*, 1661 (1966).

Chapter 10
POLYNUCLEAR COMPLEXES

Complexes having more than one central ion are termed polynuclear complexes. The formation of polynuclear species causes many experimental and theoretical difficulties in the quantitative study of complex systems. Owing to these difficulties the results are usually much more ambiguous than the results referring to the formation of the parent complexes. It is sometimes very difficult even to get unambiguous evidence for the existence of a particular species. As Sillén, whose theoretical and experimental studies are fundamentally important in the field of the chemistry of polyions, wrote in his paper [1]: "The more I work with polyions, the more I think we must be cautious with claims to have proved the existence of any particular species."

There are many examples of the formation of well defined polynuclear inert complexes. Knowledge of them is important from the viewpoint of equilibrium studies because evidently analogous structures exist in the case of polynuclear complexes which are in equilibrium with their constituents. It may be mentioned that from the structural point of view the metal complexes of oxy-anions may also be regarded as polynuclear complexes, the oxy-anions themselves being complexes. Usually this has no consequence for equilibrium studies, but in the formation of heteropoly acids the formation of complexes of just this type must be considered.

Between the central ions bridging groups are situated. The only exceptions are the ions containing a metal–metal bond, the most common example being Hg_2^{2+}.

Because no dissociation of mercury(I) ion occurs in solution, the treatment of the complexes of the mercury(I) ion is the same as that of the mononuclear complexes.

The formation of polynuclear complexes is due to the fact that a ligand bound to a metal ion still has the ability to donate another pair of electrons. This is the case when the co-ordinated donor atom still has unshared pairs of electrons, or the number of donor groups in the ligand exceeds the maximum co-ordination number of the central ion, or the steric arrangement of the donor atoms of the multidentate ligand does not make it possible for all the donor atoms to be co-ordinated to the same central ion.

The bridging group may be *monatomic*, *oligoatomic* or *polyatomic*. In complexes containing a monatomic bridge the same donor atom is co-ordinated to two central ions. For example: $[Ag–I–Ag]^+$, $[Zn–\overset{\underset{|}{H}}{O}–Zn]^{3+}$. Only donor atoms having more than one unshared pair of electrons can function as monatomic bridges. This is a necessary but not sufficient condition. For example the water molecule cannot be a bridging ligand, although there are two pairs of unshared electrons on its oxygen atom. In the oligoatomic bridges the donor atoms co-ordinated to the metal ions are neighbours or are separated by only one further atom. For example:

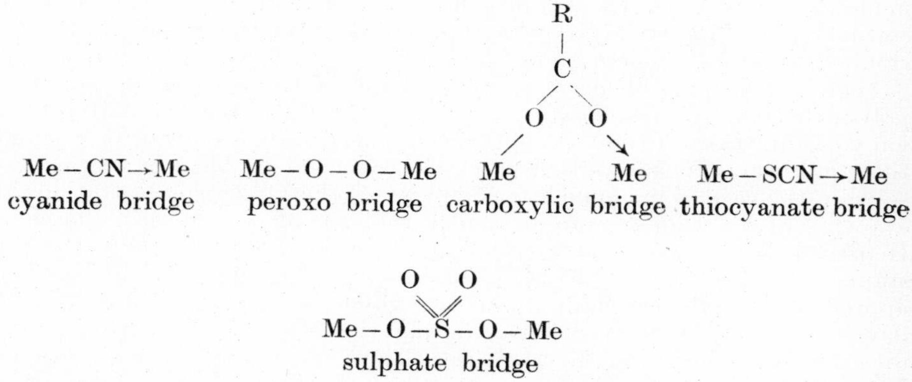

$$
\begin{array}{cccc}
& & \underset{\nearrow\ \ \searrow}{\overset{\overset{\displaystyle R}{\overset{|}{\underset{|}{C}}}}{O\ \ \ \ O}} & \\
& & Me\qquad\qquad Me & \\
Me-CN\rightarrow Me & Me-O-O-Me & & Me-SCN\rightarrow Me \\
\text{cyanide bridge} & \text{peroxo bridge} & \text{carboxylic bridge} & \text{thiocyanate bridge}
\end{array}
$$

$$
\overset{\overset{\displaystyle O\quad O}{\diagdown\!\!\diagup}}{Me-O-S-O-Me}
$$
$$
\text{sulphate bridge}
$$

The polyatomic bridging ligands may be extremely various. In this case the donor atoms co-ordinated to different metal ions are separated by some, in certain cases by many, other atoms. Only one example is given here, some others will be mentioned later:

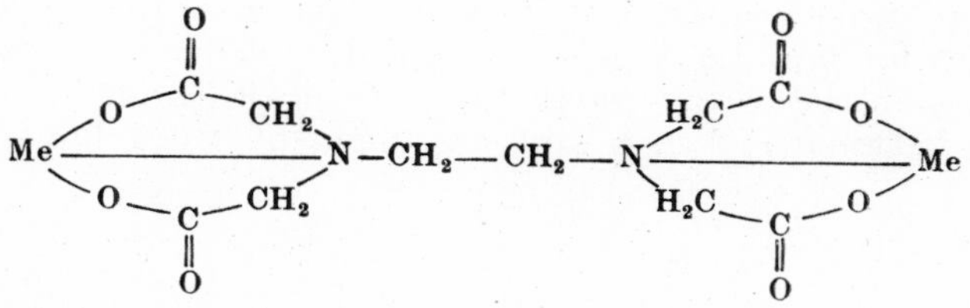

If the central metal ions are identical the complex may be termed *homopolynuclear*, otherwise *heteropolynuclear*. The term *homonuclear* is sometimes used when the nuclearity of all the complexes in the equilibrium system is the same.

The central ions may be connected by more than one bridge, and the bridging atoms are not necessarily the same or even of the same type.

10.1 Some general problems in the calculation of stability constants of polynuclear complexes

In a complex system containing polynuclear complexes the general formula for the binary species is Me_pL_q and the stability product is given as

$$\beta_{pq} = \frac{[Me_p L_q]}{[Me]^p[L]^q} \tag{10.1}$$

The average number of ligands bound per metal ion $(\bar{n})$ is as an important a quantity in the evaluation of the stability constants as in the case of mononuclear systems. Its meaning, however, is a bit different; it cannot be considered as an average co-ordination number. For this reason, a different symbol, usually Z, is sometimes used for it.

There are two problems in the evaluation of the stability constants of polynuclear complexes. First, to calculate the free ligand and free metal ion concentrations if they are not directly measurable; second, to calculate the equilibrium constants themselves. This second problem is very easy to solve in the case of mononuclear complexes if the "free" concentrations are known, but much more complicated in the case of polynuclear complexes. Hedström [2] showed that in principle it is possible to evaluate the constants by a graphical procedure analogous to Leden's method, but this approach has only very limited practical value. Hedström [2] and Sillén [3] developed some methods for calculating the "free" concentrations from experimentally measurable quantities. These ingenious methods are not treated here, because it became quite clear that the stability constants of polynuclear species in the general case can be obtained by high-speed electronic computers. Sillén et al. [4–7] developed special programmes for the study of these systems.

We must briefly treat, however, Sillén's "core + links" hypothesis [8, 9]. According to this, not all complexes of composition Me_pL_q exist, but only species consisting of a "core" and of a certain number of "links". That is, the formula of the complexes can be given as $Me_s(L_tMe)_n$ and $L_r(L_tMe)_n$ where r, s and t are constants, the value of n may be any integer which gives integer values for p and q; r gives the number of ligands, s the number of the metal ions in the core, t equals the number of ligands in the links. The relationship between r, s, t, p and q is given by the following equations:

$$r = -ts \tag{10.2}$$

$$p = tq + r = t(q - s) \tag{10.3}$$

The hypothesis does not require that all complexes described by Eq. 10.3 must exist, but it excludes the existence of complexes of different composition. The hypothesis was applied to the treatment of the hydrolysis of many metal ions [10] and to the formation of silver(I) complexes of N, N'-bis(2-hydroxyethyl)dithio-oxamide [11, 12]. In the latter and some other cases it is assumed that several sets of different "core + links" complexes are formed.

The validity of the "core and links" hypothesis for a given system can be judged from the curves $\bar{n}(\log [L]T_{Me})$ referring to different total metal ion concentrations. Coinciding curves indicate the formation of mononuclear complexes only. If the curves are parallel, and the horizontal spacing between two curves, $\Delta \log [L]$, is proportional to the difference between the two $\log T_{Me}$ values, $\Delta \log T_{Me}$, a series of "core + links" complexes is formed. The spacing gives the value of t:

$$\frac{\Delta \log T_{Me}}{\Delta \log [L]} = \text{constant} = t. \tag{10.4}$$

Although the "core + links" complexes occur probably less frequently than it was first thought [13], it is always worth-while to consider the possibility of their formation.

10.2 Polynuclear complexes with monatomic bridges

The following ligands may function as monatomic bridges: F^-, Cl^-, Br^-, I^-, O^{2-}, HO^-, S^{2-}, HS^-, Se^{2-}, Te^{2-}. There are some other ligands which act as bridging ligand in the solid state, such as P^{3-}, As^{3-}, Sb^{3-}. The formation of polynuclear complexes is very common in the hydrolysis of metal ions. These complexes contain O^{2-} and OH^- bridges and will be treated in detail. There are only very few examples of the formation of polynuclear complexes containing fluoride bridges. Interestingly enough, such complexes of composition $NiSnF_n^{4-n}$ ($n = 1, 2, 3$) exist in the nickel(II)–tin(II)–fluoride system [14], although the stability of the nickel(II) fluoride complex is fairly small. The other more polarizable monatomic ligands form polynuclear complexes with different class (b) metal ions.

10.2.1 FORMATION OF POLYNUCLEAR COMPLEXES IN THE HYDROLYSIS OF METAL IONS

The well known phenomenon that solutions of salts of bivalent and ter-valent metal ions are acidic is due to the hydrolysis of these metal ions. For a long time only the formation of mononuclear hydrolysis products was considered, but Niels Bjerrum [15] assumed that in the hydrolysis of chromium(III) polynuclear complexes, e.g. $Cr_2(OH)_2^{4+}$, are also formed. In the last two decades the hydrolysis of different metal ions was extensively investigated, first of all by Sillén's school. It became evident that though almost every metal ion behaves in a different manner, the formation of only mononuclear hydroxo-complexes may be regarded more as the exception than the rule. The formation of polynuclear hydroxo-complexes can be considered both with cations and such oxy-anions as MoO_4^{2-}, VO_3^{3-}, etc. These oxy-anions exist in the simple form MO_4^{n-} only in alkaline solution, poly-nuclear species being formed on acidification. In the case of metal ions the situation is the opposite. They exist in acidic solutions as aquo-complexes

and different mono- and polynuclear oxo- and hydroxo-complexes are
formed on adding alkali. There is, however, a common feature of the hydrolysis of cations and oxy-anions. The greatest degree of complexity occurs
at an intermediate pH. The cations in strongly alkaline solutions are trans-
formed into mononuclear hydroxo-complexes, and the molecular weight
of the polynuclear species derived from an oxy-anion decreases [16] in the
presence of a high concentration of strong acid. If a much wider range of
pH were attainable, there would not be a qualitative difference between
the behaviour of cations and oxy-anions.

Most of the quantitative information is obtained by potentiometric
studies. The reader may consult three magnificent reviews by Sillén [1, 17,
18] and the papers quoted therein. For the details of the experimental
technique, the recent papers by Biedermann et al. [19, 20] and by Baes [21]
are suggested for study.

In the study of such complicated reactions as the hydrolysis of metal ions
and oxy-anions it is particularly important to apply as many experimental
approaches as possible. The following methods have been used: spectrophotometry [22, 23], Raman spectroscopy [16, 24–26], ultracentrifugation
[16, 24, 25, 27–32], light-scattering [31, 33, 34], coagulation studies [35–37].
The light-scattering and ultracentrifugation studies provide information on
the molecular weight; from the coagulation experiments (determination of
the critical concentration of a given cation at a certain pH to produce coagulation of a suitable sol) the charge of the predominant complex ion can be
approximately judged. The comparative study of the vibrational spectrum
of solutions and solid complexes makes it possible to assign a certain structure to the species in the solution. It must be mentioned that one always
must be very cautious in deducing the existence of a particular species in
the solution, from X-ray data referring to solid complexes.

10.2.2 POLYNUCLEAR COMPLEXES WITH MONATOMIC BRIDGES OTHER THAN OXYGEN

The formation of polynuclear species was shown by Leden [38] in the
case of the silver(I)–halide system. Such complexes may be formed when
halide is added to silver(I) solutions, and when the silver(I) halide is dissolved in excess of halide ions. It is plausible that the reaction of silver(I)
halide complexes with excess of silver(I) also leads to the formation of polynuclear complexes. Hellwig [39] was the first to observe that the solubility
of silver(I) halides in silver(I) nitrate solutions exhibits a minimum at
a certain silver(I) nitrate concentration. It was assumed that the increasing
solubility is due to the formation of complexes such as Ag_3I^{2+}. Complexes
of this type, in which there is apparently an anionic centre and the metal
ions seem to form a shell around this centre, were considered by Werner [40]
to be a special group of compounds and were recently termed metallo-
complexes by Bergerhoff [41]. They may be considered most reasonably as
polynuclear complexes having a monatomic bridging ligand. It must be
borne in mind that in solution other ligands (water, anions) are also co-ordinated to metal ions.

Lieser [42] determined the solubility of silver(I) halides as a function of the concentration of silver(I) nitrate and from the results he calculated the constants given in Table 10.1.

TABLE 10.1

Stability constants of complexes of the type Ag_nX^{n-1} *at 20°C*

Reaction	Stability constant		
	X = Cl	X = Br	X = I
$Ag^+ + X^- = AgX$	2×10^3	—	—
$2\,Ag^+ + X^- = Ag_2X^+$	4×10^4	1.14×10^7	1.0×10^{11}
$3\,Ag^+ + X^- = Ag_3X^{2+}$	9.1×10^4	1.35×10^8	4.16×10^{13}

Similar solubility behaviour was found in the case of mercury(II) iodide, and Yatsimirskii and Shutov [43] could prove the existence of the Hg_2I^{3+} ion. It is expected that analogous behaviour will be found in the case of other polarizable anions and cations.

Lieser [44] studied the solubility of the sulphide, selenide and telluride of silver(I) as a function of silver(I) nitrate concentration. The large increase in the solubility of these compounds with increasing silver(I) concentration suggests that, in the polynuclear complexes formed, the degree of nuclearity is fairly high. In the case of the telluride the analysis of the solubility and conductivity data show that the predominant complex is Ag_8Te^{6+}. Lieser could prepare a compound of the composition $Ag_6Te(NO_3)_6$. The composition of this complex is a bit puzzling from the theoretical point of view.

10.3 Polynuclear complexes with oligoatomic bridges

At present there are only a few known examples of complexes of this type and this field seems to be very promising for equilibrium studies. Hein and Daniel [45] determined the solubility of silver(I) permanganate in solutions of silver(I) nitrate of different concentrations. Just as in case of the halides a solubility minimum can be observed and the increasing solubility is caused by the formation of complexes $Ag_nMnO_4^{n-1}$. Hein and Daniel calculated some equilibrium constants, but these cannot be regarded as reliable. Similar solubility behaviour was found with silver(I) sulphate and silver(I) chromate [46]. It seems probable that in the latter case the following complex is formed

$$\begin{bmatrix} Ag-O & & O-Ag \\ & \diagdown\;\diagup & \\ & Cr & \\ & \diagup\;\diagdown & \\ Ag-O & & O-Ag \end{bmatrix}^{2+}$$

Hellwig [39] found the solubility behaviour of the pseudohalides of silver(I) quite analogous to that of the halides. Considering that complexes of the composition Ag_3CN^{2+} and Ag_3SCN^{2+} exist in solution, one must assume that these ligands may function as monatomic and oligoatomic bridges.

The metal "salts" of different cyano-complexes can be regarded as complexes of this type. These complexes are in fact polynuclear species, but their small solubility makes equilibrium studies difficult.

The co-ordinative unsaturation of mercury(II) cyanide is the source of its interaction with hexacyanoferrate(II). A solid compound of composition $K_4Fe(CN)_6 \cdot 3Hg(CN)_2 \cdot 4H_2O$ was prepared by Kane [47] as early as 1840, and several analogous compounds were prepared and analysed by Strömholm [48] in 1913. A spectrophotometric study indicated a weak interaction in solution [49], but apparently there is no interaction between mercury(II) cyanide and hexacyanoferrate(III) and hexacyanocobaltate(III). Further studies are necessary to decide whether charge-transfer contributes to the stabilization of the polynuclear complex.

10.4 Polynuclear complexes with multidentate ligands

It may appear plausible to distinguish two types of multidentate ligand, depending on whether the number of the donor groups or their steric situation is the source of polynuclear complex formation. Considering, however, that the co-ordination numbers of different metal ions are different, and that in most cases both factors must be considered, this classification cannot be rigid. Nevertheless, two ligands representing the two extreme types may be mentioned. Triethylenetetraminehexa-acetate has ten functional groups, more than enough to saturate the co-ordination sphere of most metal ions:

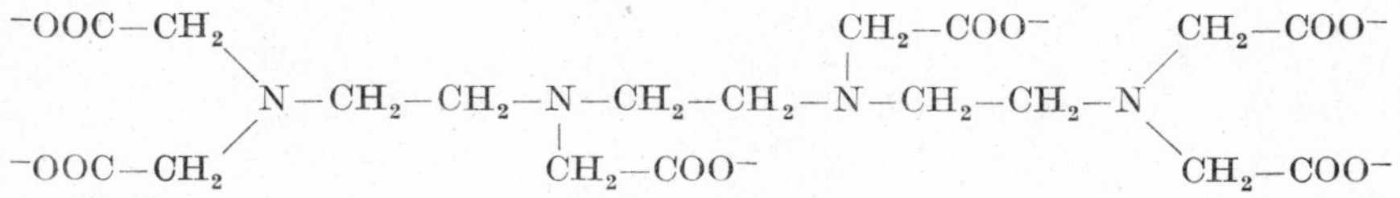

$$^-OOC-CH_2 \quad\quad CH_2-COO^- \quad CH_2-COO^-$$
$$N-CH_2-CH_2-N-CH_2-CH_2-N-CH_2-CH_2-N$$
$$^-OOC-CH_2 \quad\quad CH_2-COO^- \quad\quad CH_2-COO^-$$

Dithio-oxamide has only four donor groups

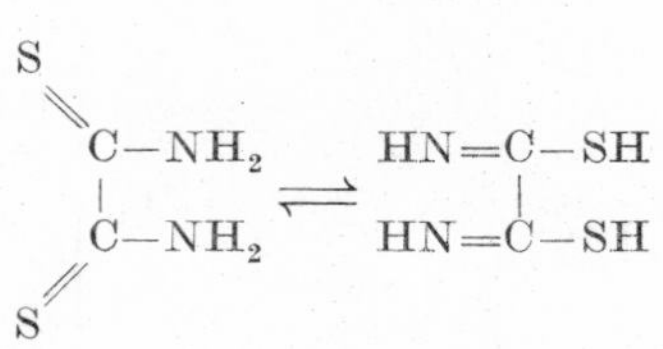

$$\begin{array}{c} S \\ \| \\ C-NH_2 \\ | \\ C-NH_2 \\ \| \\ S \end{array} \rightleftharpoons \begin{array}{c} HN=C-SH \\ | \\ HN=C-SH \end{array}$$

but their steric arrangement prevents more than two being attached to the same metal ion, and therefore polynuclear complexes of extremely low solubility are formed:

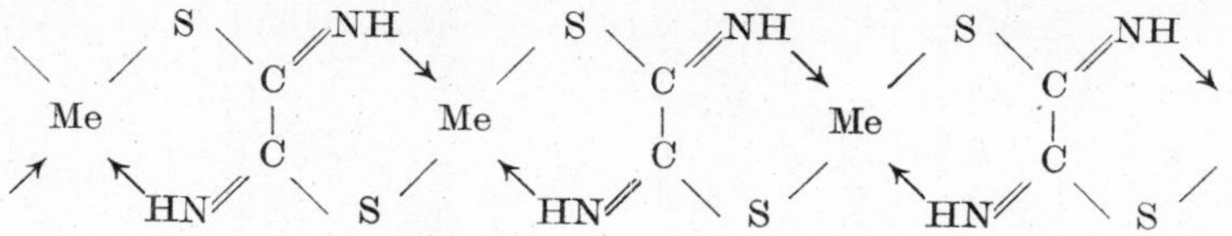

Most of the studies referring to polynuclear complexes of multidentate ligands are concerned with polyaminopolycarboxylic acids. The existence of such complexes was demonstrated by spectrophotometric [50], polarographic [51], potentiometric [52–57] and NMR [58, 59] studies. For the calculation of stability constants the potentiometric data are the most suitable. In some cases well defined solid compounds were prepared. The nuclearity of the complexes is different, but usually only bi- and ternuclear complexes are formed. With ethylenediaminetetra-acetate several metal ions and oxyanions form binuclear complexes, e.g. U(IV) [50], Tl(III) [59], Sn(II) [60], Mo(VI) [51]. In many cases the formation of heterobinuclear complexes was observed [61, 62]. Schmid and Reilley [61] demonstrated by potentiometric experiments the interaction of the mercury(II)–EDTA complex with various bivalent metal ions. The equilibrium constants defined by

$$K_{\text{MeY Hg(OH)}_2} = \frac{[\text{MeY Hg(OH)}_2]}{[\text{HgY}][\text{Me}][\text{OH}]^2} \tag{10.5}$$

are summarized in Table 10.2.

TABLE 10.2

Stability constants of several heterobinuclear complexes of EDTA

Me^{2+}	$\log K_{\text{MeYHg(OH)}_2}$	$\log K_{\text{MeY}}$	$\Delta \log K$
Ca^{2+}	10.7	13.3	2.6
Mn^{2+}	13.8	16.1	2.3
Cd^{2+}	16.4	18.7	2.3
Zn^{2+}	16.4	19.1	2.3

It appears from these data that the stability of the binuclear complex is roughly proportional to the stability of the EDTA complex of Me^{2+} ions. Schmid and Reilley on the basis of a very careful analysis of the experimental data and the comparison of several ligands suggest that the complex has the structure shown in Fig. 10.1. It appears that for the dissociation represented by Eq. 10.5 a considerable rearrangement of the co-ordination sphere of the mercury(II) is necessary.

It may be expected that on increase in the number of methylene groups (x) separating the two iminodiacetate groups of EDTA the probability of the formation of the binuclear complex increases, while the stability of the

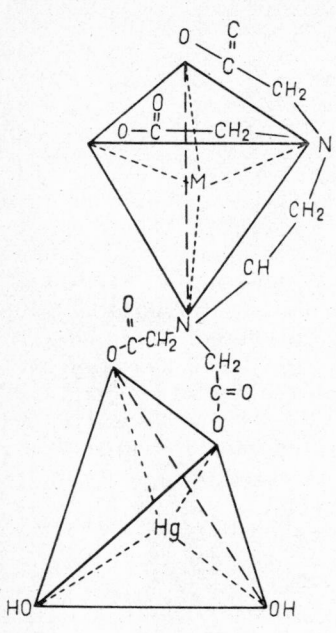

Fig. 10.1 Structure of MeEDTAHg(OH)$_2^{2-}$ complex. (Reproduced with permission from *J. Am. Chem. Soc. 78*, 5519 (1956))

mononuclear complex decreases. The results obtained by Schwarzenbach and Ackermann [52] are in agreement with this assumption and illustrate that the greater the value of x, the more independent is the behaviour of the two iminodiacetate groups (Table 10.3).

TABLE 10.3

*The effect of the number of methylene groups (x) separating
two iminodiacetate groups on the stability
of the mono- and binuclear calcium chelates*

x	$\log K_{CaY}$	$\log K_{Ca_2Y}$
2	10.5	—
3	7.1	0.7
4	5.2	2.0
5	4.6	2.6

The stability of the polynuclear complexes increases with increasing number of functional groups. This is well illustrated by studies referring to the complexes of diethylenetriaminepenta-acetate [53], and particularly of triethylenetetraminehexa-acetate [56]. The stability constants determined by Bohigian and Martell [56] are summarized in Table 10.4.

TABLE 10.4

Stability constants of complexes of triethylenetetraminehexa-acetate

Equilibrium	log K	Equilibrium	log K
$Mg^{2+} + H_2L^{4-} \rightleftharpoons MgH_2L^{2-}$	2.81	$CaL^{4-} + Ca^{2+} \rightleftharpoons Ca_2L^{2-}$	4.32
$Mg^{2+} + HL^{5-} \rightleftharpoons MgHL^{2-}$	7.55	$Ca_2L^{2-} + Ca^{2+} \rightleftharpoons Ca_3L$	4.00
$Mg^{2+} + L^{6-} \rightleftharpoons MgL^{4-}$	8.43	$2\,Cu^{2+} + L^{6-} \rightleftharpoons Cu_2L^{2-}$	27.6
$MgL^{4-} + Mg^{2+} \rightleftharpoons Mg_2L^{2-}$	5.5	$2\,Co^{2+} + L^{6-} \rightleftharpoons Co_2L^{2-}$	28.0
$Mg_2L^{2-} + Mg^{2+} \rightleftharpoons Mg_3L$	5.32	$2\,Ni^{2+} + L^{6-} \rightleftharpoons Ni_2L^{2-}$	29.5
$Ca^{2+} + H_2L^{4-} \rightleftharpoons CaH_2L^{2-}$	3.76	$La^{3+} + L^{6-} \rightleftharpoons LaL^{3-}$	23.1
$Ca^{2+} + HL^{5-} \rightleftharpoons CaHL^{3-}$	8.23	$2\,La^{3+} + L^{6-} \rightleftharpoons La_2L$	26.9
$Ca^{2+} + L^{6-} \rightleftharpoons CaL^{4-}$	9.89	$Th^{4+} + L^{6-} \rightleftharpoons ThL^{2-}$	27

The experimental findings are in accordance with the following structures for the mono-, bi- and ternuclear chelates:

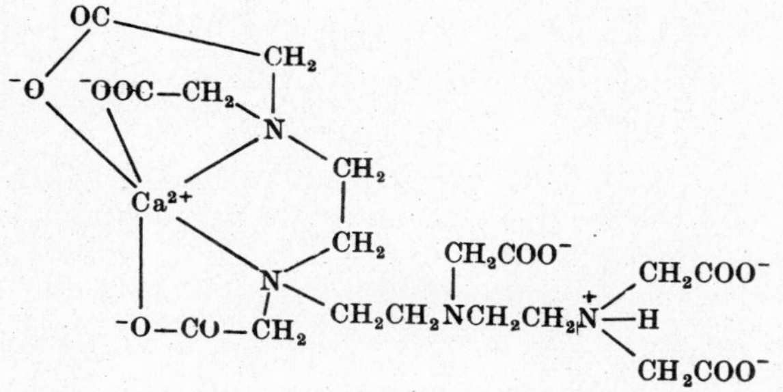

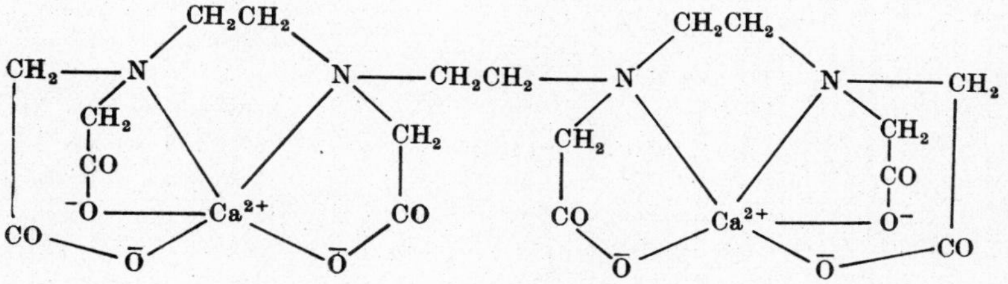

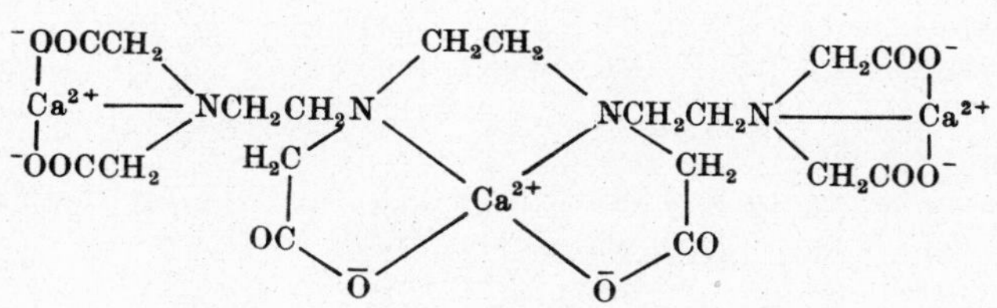

Polynuclear complexes may be formed in the hydrolysis of certain metal chelates. In these cases either the multidentate ligand or hydroxo groups or both of them may function as bridges [63–66]. Investigations by Zompa and Bogucki [57] furnish a particularly interesting example. They studied the copper(II) complex of tetrakis(aminomethyl)methane:

$$\begin{array}{ccc}
H_2N-CH_2 & & CH_2-NH_2 \\
& C & \\
H_2N-CH_2 & & CH_2-NH_2
\end{array}$$

Two of the four amine groups are sterically available for the same copper(II) ion of square–planar co-ordination. With Cu(II) : ligand ratios of 1 : 2 and 2 : 1 stable mono- and binuclear chelates are formed at low pH. At higher pH the binuclear one hydrolyses to produce an insoluble polymer

$$\begin{array}{c}
HO \\
\diagdown \\
Cu \\
\diagup \\
HO
\end{array}
\begin{array}{c}
NH_2-CH_2 \\
\diagup \\
\\
NH_2-CH_2
\end{array}
\begin{array}{c}
NH_2 \\
\diagdown \\
C \\
\diagup \\
NH_2
\end{array}
\begin{array}{c}
\diagdown \\
Cu \\
\diagup
\end{array}
\left[
\begin{array}{c}
OH \\
\diagup \\
Cu \\
\diagdown \\
OH
\end{array}
\begin{array}{c}
NH_2-CH_2 \\
\diagdown \\
\\
NH_2-CH_2
\end{array}
\begin{array}{c}
CH_2-NH_2 \\
\diagdown \\
C \\
\diagup \\
CH_2-NH_2
\end{array}
\begin{array}{c}
\diagdown \\
Cu \\
\diagup
\end{array}
\right]^{(2n-2)+}_{n-1}
\begin{array}{c}
OH \\
\diagup \\
\\
\diagdown \\
OH
\end{array}$$

in which the value of n is in the range 10–20. Zompa and Bogucki attempted to reduce the nuclearity of the complex by adding ethylenediamine to act as a terminal group. Although quantitative results could not be obtained, a decrease in the nuclearity was shown by the formation of a soluble chelate.

10.5 Polynuclear complexes involving central ions of the same atomic number but of different oxidation state

It was observed already by Werner [67] that the colour of complexes containing both platinum(II) and platinum(IV) is much deeper than that of the corresponding mononuclear complexes of these metal ions. He drew attention to the parallelism between this phenomenon and the behaviour of the hydroquinone–quinone system. The most common example is provided by the hydroxides of iron. Iron(II) hydroxide is greenish white, iron(III) hydroxide is rust brown, the colour of iron(II)–iron(III) hydroxide is black. As pointed out by Wells [68] the same atomic number is not a necessary condition that the formation of the polynuclear complex should be accompanied by an unusually deep colour. He found that the colour of $Cs_2Ag(I)Au(III)Cl_6$ was much deeper than might be expected without a special interaction. The close similarity in the electronic configuration of the two central ions makes possible an interaction of this kind.

After some sporadic observations Davidson et al. [69–72] studied thoroughly such interactions in solutions. It is very likely that all of these polynuclear complexes are binuclear ones, and although they are strongly absorbing species, their thermodynamic stability is rather low, therefore at present no reliable stability constants have been determined. Interaction was found with the following pairs of ions in aqueous hydrochloric acid solution: Sb(III)–Sb(V) [69], Sn(II)–Sn(IV) [70], Cu(I)–Cu(II) [71] and Fe(II)–Fe(III) [72]. The specific interaction between iron(II) and iron(III) appears from Fig. 10.2.

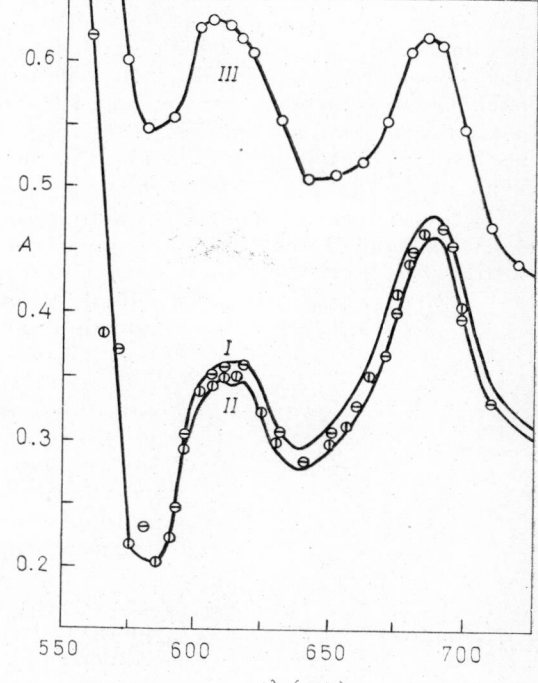

Fig. 10.2 Effect of different salts on the spectrum of iron(II) in 12 M HCl.
Curve *I* — 0.88 M FeCl$_3$;
Curve *II* — 0.88 M FeCl$_3$ + 0.13 M MeCl$_2$ (Me = = Mn, Mg, Zn);
Curve *III* — 0.88 M FeCl$_3$ + + 0.13 M FeCl$_2$. (Reproduced with permission from *J. Am. Chem. Soc.* 72, 5557 (1950))

No interaction was found between the corresponding aquo-cations, i.e. for the formation of these complexes chloride ions or other bridging ligands should be present. Similarly no interaction was found either between Fe(CN)$_6^{4-}$ and Fe(CN)$_6^{3-}$ [73] or between tris(phenanthroline)iron(II) and tris(phenanthroline)iron(III) complexes [74].

The formation of complexes of this type can be observed in some redox reactions. Thiomalic acid is oxidized by copper(II) and the product is of intense violet colour. It follows from the stoichiometry of the reaction that the complex formed is a pentanuclear one, containing one copper(II) and four copper(I) central ions [75].

In the oxidation of the aminopolycarboxylic complexes of cobalt(II) by chromium(VI) a fairly stable intermediate of very strong colour is formed. This intermediate is probably a polynuclear complex containing cobalt(III) and chromium(V) centres [76].

10.6 Polymerization of metal chelates

In all cases discussed so far, the formation of the polynuclear species involved displacement and/or condensation reactions. True polymerization equilibria occur with neutral four-co-ordinated complexes in non-aqueous solution. The driving force of such reactions is that polymerization increases the co-ordination number, the bridging atom being shared by two central ions. Cotton and Fackler [77] observed that the magnetic moment and absorption spectrum of bis-(2,6-dimethyl-3,5-heptanediono)nickel(II) complex dissolved in toluene depend on both the temperature and concentration. The monomer is diamagnetic; the change in the absorbance and in the magnetic susceptibility can be quantitatively explained by the formation of a trimeric complex similar to the bis(2,4-pentanediono) complex which has been accurately determined and is shown schematically in Fig. 10.3. It seems likely that the trimer has the same structure in solution.

However, in the case of the bis(ketophosphonyl)cobalt(II) complex, where the formation of the trimeric complex in solution was also observed by Cotton et al. [78] the structure is quite different. All six chelate rings are closed around the terminal cobalt atoms and none around the cobalt ion in the central position.

The stability of the trimeric bis(2,4-pentanediono)nickel(II) complex is much greater, dissociation becoming appreciable only at temperatures near and above 200°. The polymerization is hindered by bulky substituents. Similar association and steric hindrance of the association were found with bis(N-alkyl)salicylaldimine by Ferguson [79] and by Holm [80]. Only the complex of the methyl derivative exhibits a tendency for association.

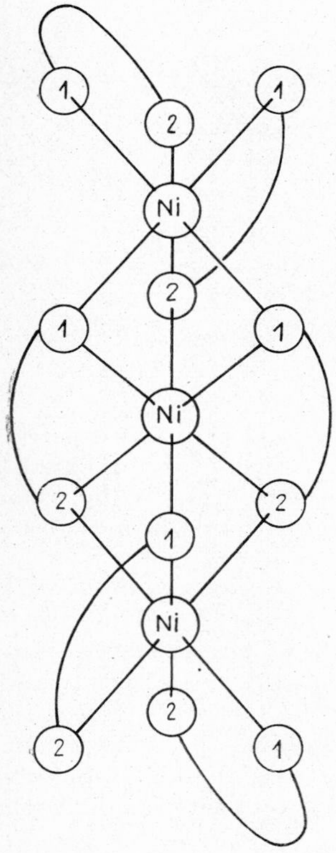

Fig. 10.3 Schematic structure of the bis(2,4-pentanediono) nickel(II) trimer. 1 and 2 represent the oxygen atoms of the ligand molecule

10.7 Distribution of polynuclear complexes

The distribution of polynuclear complexes can be calculated from the relevant equilibria. The application of computers is very convenient, and the programme Haltafall can be applied [81]. Evidently, the distribution of the species is not independent of the total metal ion concentration as in the case of mononuclear complexes. It may be said generally that the fraction of the polynuclear species increases with increasing metal ion concentration. This tendency is shown in Fig. 10.4, referring to the distribution of species in the copper(II)–2,7-diaminosuberic acid system [54].

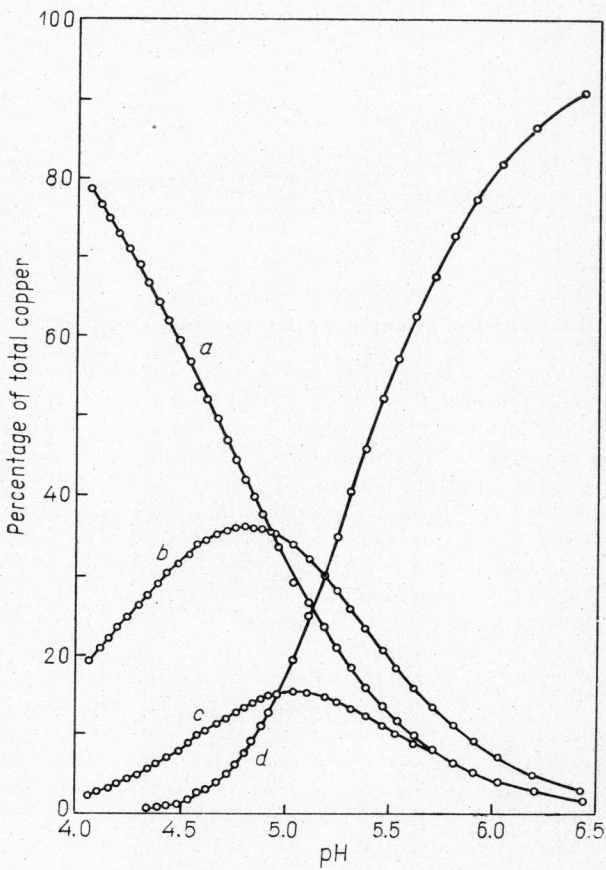

Fig. 10.4 (a) Distribution of complexes for copper(II)–2,7-diaminosuberate (DAS) system as a function of pH. $T_{Cu} = 5 \times 10^4$ M, $T_{DAS} = 5.034 \times 10^{-4}$ M.
Curve a — Cu^{2+}; Curve b — Cu^+LRLH; Curve c — Cu^+LRLCu^+;
Curve d — $(CuLRL)_2$ cycle complex. R represents the skeleton of the ligand while L stands for each pair (amino and carboxylic) of donor groups

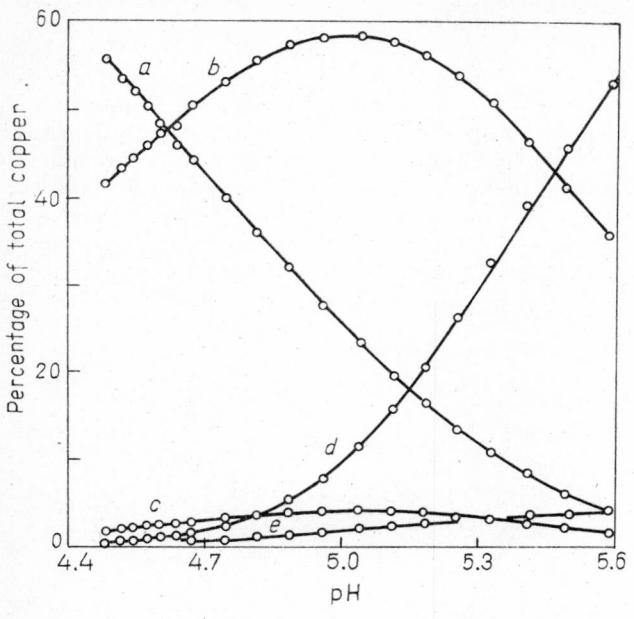

(b) Distribution of complexes for copper(II)–2,5-diaminosuberate system as a function of pH. $T_{Cu} = 10^{-4}$ M, $T_{DAS} = 4.984 \times 10^{-4}$ M.
Curve
a — Cu^{2+};
Curve
b — $HLRLCu^+$;
Curve
c — $^+CuLRLCu^+$;
Curve
d — $(CuLRL)_2$;
Curve
e — $Cu(LRLH)_2$.
(Reproduced with permission from *Inorg. Chem.* 2, 839 (1963))

10.8 Kinetic importance of polynuclear complexes

Polynuclear complexes, usually binuclear, are formed in the reaction of certain metal complexes with another central ion. If the structure of the ligand makes it possible a binuclear complex or a whole series of complexes is formed as intermediate. The stability of such binuclear species was calculated by Bydalek and Margerum [82] in a similar way to that treated already in connection with mixed ligand complexes.

Polynuclear complexes may play a role in the formation of complexes. It is well known that the rate of formation of the EDTA complex of chromium(III) increases with increasing pH. It was observed further that if the initial pH of the solutions is greater than 2, the rate decreases if aged solutions are used instead of freshly prepared ones. The reason is that above pH 2 a binuclear species is formed slowly and this complex reacts much more slowly with EDTA than the mononuclear one does [83].

The formation of polynuclear species may be important in catalytic reactions too. The source of the catalysis of aquation of halide complexes by metal ions is the affinity of these catalyst ions for the bound halide. Aluminium(III), thorium(IV), iron(III) and other class (*a*) metal ions [84–86] accelerate the aquation of fluoro-complexes, whereas lead(II), mercury(II), silver(I) and other class (*b*) metal ions catalyse that of the complexes of heavier halides [87–89]. It is likely that binuclear complexes are intermediates in these reactions.

Binuclear complexes themselves may be catalysts of certain reactions. Jonassen and Ramanujam [90] even assume that such species are responsible for catalytic effects in many systems. We feel that there is no general rule in this respect. In complex systems exhibiting catalytic properties, where both mononuclear and binuclear species exist, only careful experiments can decide which form is the more effective catalyst. Two characteristic examples are mentioned. As is well known, in the presence of excess of EDTA the catalytic effect of iron(III) on the decomposition of hydrogen peroxide is completely inhibited. However, if iron(III) is in excess the activity is even greater than in the absence of EDTA [91]. The probable explanation is that in a binuclear complex two iron(III) ions exert a concerted action on the same peroxide molecule. On the other hand in the case of oxygen-carrying cobalt(II) complexes catalysing autoxidation of ascorbic acid, the experiments indicated that a mononuclear oxygen-carrying species was the active catalyst; the common binuclear complex was much less efficient [92]. Recently independent rate studies on the formation of oxygen-carrying complexes have proved the existence of the mononuclear oxygen-carrying species [93].

REFERENCES

1. SILLÉN, L. G., *Proc. Welch Foundation Conf. Chem. Res.*, 1964, p. 187.
2. HEDSTRÖM, B., *Acta Chem. Scand. 9*, 613 (1955).
3. SILLÉN, L. G., *Acta Chem. Scand. 15*, 1981 (1961).
4. SILLÉN, L. G., *Acta Chem. Scand. 23*, 159 (1964).
5. INGRI, N. and SILLÉN, L. G., *Acta Chem. Scand. 16*, 173 (1962).
6. SILLÉN, L. G., *Acta Chem. Scand. 18*, 1085 (1964).
7. INGRI, N. and SILLÉN, L. G., *Arkiv Kemi 23*, 97 (1964).
8. SILLÉN, L. G., *Acta Chem. Scand. 8*, 299 (1954).
9. SILLÉN, L. G., *Acta Chem. Scand. 8*, 318 (1954).
10. HIETANEN, S. and SILLÉN, L. G., *Acta Chem. Scand. 8*, 1607 (1954).
11. VAN POUCKE, L. C., HERMAN, M. A. and EECKHAUT, Z., *Anal. Chim. Acta 40*, 55 (1968).
12. VAN POUCKE, L. C. and HERMAN, M. A., *Anal. Chim. Acta 41*, 1 (1968).
13. SILLÉN, L. G., *Acta Chem. Scand. 16*, 1051 (1962).
14. RAU, R. L. and BAILAR, J. C., JR., *J. Electrochem. Soc. 107*, 745 (1960).
15. BJERRUM, N., *Thesis*, Copenhagen, 1908.
16. AVESTON, J., ANACKER, E. W. and JOHNSON, J. S., *Inorg. Chem. 3*, 735 (1964).
17. SILLÉN, L. G., *Quart. Rev. 13*, 146 (1959).
18. SILLÉN, L. G., *Proc. Symp. Coord. Chem. Tihany, Hungary*, p. 123. Akadémiai Kiadó, Budapest, 1965.
19. BIEDERMANN, G. and NEWMAN, L., *Arkiv Kemi 22*, 303 (1964).
20. BIEDERMANN, G. and CIAVATTA, L., *Arkiv Kemi 22*, 253 (1964).
21. BAES, C. F., JR., *Inorg. Chem. 4*, 588 (1965).
22. KING, E. L. and PANDOW, M. L., *J. Am. Chem. Soc. 74*, 1966 (1952).
23. OFFNER, H. G. and SKOOG, D. A., *Anal. Chem. 38*, 1520 (1966).
24. AVESTON, T., *Inorg. Chem. 3*, 981 (1964).
25. AVESTON, J. and JOHNSON, J. S., *Inorg. Chem. 3*, 1051 (1964).
26. MARONI, V. A. and SPIRO, T. G., *J. Am. Chem. Soc. 88*, 1410 (1966).
27. JOHNSON, S. J., KRAUS, K. A. and SCATCHARD, G., *J. Phys. Chem. 58*, 1034 (1954).
28. HOLMBERG, R. W., KRAUS, K. A. and JOHNSON, I. S., *J. Am. Chem. Soc. 78*, 5506 (1956).
29. JOHNSON, I. S. and KRAUS, K. A., *J. Am. Chem. Soc. 78*, 3937 (1956).

30. HENTZ, F. C., JR. and JOHNSON, J. S., *Inorg. Chem.* 5, 1337 (1966).
31. NELSON, W. H. and TOBIAS, R. S., *Inorg. Chem.* 3, 653 (1964).
32. AVESTON, J., *J. Chem. Soc.* A 1599 (1966).
33. NELSON, W. H. and TOBIAS, R. S., *Can. J. Chem.* 42, 731 (1964).
34. HENTZ, F. C., JR. and TYREE, S. Y., JR., *Inorg. Chem.* 4, 873 (1965).
35. MATIJEVIC, E. and JANAUER, G. E., *J. Colloid Interfacial Sci.* 21, 197 (1966).
36. MATIJEVIC, E. and STRYKER, L. J., *J. Colloid Interfacial Sci.* 22, 68 (1966).
37. MATIJEVIC, E., MATHAI, K. G. and KERKER, M., *J. Phys. Chem.* 66, 1799 (1966).
38. LEDEN, I., *Proc. Symp. Coord. Chem. Copenhagen*, p. 77. Danish Chemical Society, Copenhagen, 1954.
39. HELLWIG, K., *Z. Anorg. Chem.* 25, 157 (1900).
40. WERNER, A., *Neuere Anschauungen auf den Gebiete der anorganischen Chemie*, 3rd Ed., p. 300. Vieweg, Braunschweig, 1913.
41. BERGERHOFF, G., *Angew. Chem. Intern. Ed.* 3, 686 (1964).
42. LIESER, K. H., *Z. Anorg. Allgem. Chem.* 304, 296 (1960).
43. YATSIMIRSKII, K. B. and SHUTOV, A. A., *Zh. Fiz. Khim.* 26, 842 (1952).
44. LIESER, K. H., *Z. Anorg. Allgem. Chem.* 305, 255 (1960).
45. HEIN, F. and DANIEL, W., *Z. Anorg. Allgem. Chem.* 234, 155 (1957).
46. BECK, M. T., unpublished results.
47. KANE, R., *J. Prakt. Chem.* 19, 405 (1840).
48. STRÖMHOLM, D., *Z. Anorg. Chem.* 84, 208 (1913).
49. BECK, M. T., unpublished results.
50. ERMOLAYEV, N. P. and KROT, N. N., *Zh. Neorgan. Khim.* 8, 2447 (1963).
51. PECSOK, R. L. and SAWYER, D. T., *J. Am. Chem. Soc.* 78, 5496 (1956).
52. SCHWARZENBACH, G. and ACKERMANN, H., *Helv. Chim. Acta* 31, 1029 (1948).
53. CHABEREK, S., FROST, A. E., DORAN, M. A. and BICKNELL, N. J., *J. Inorg. Nucl. Chem.* 11, 184 (1959).
54. HAWKINS, C. J. and PERRIN, D. D., *Inorg. Chem.* 2, 839 (1963).
55. UHLIG, E., *Z. Anorg. Allgem. Chem.* 320, 296 (1963).
56. BOHIGIAN, T. A. JR. and MARTELL, A. E., *Inorg. Chem.* 4, 1264 (1965).
57. ZOMPA, L. J. and BOGUCKI, R. F., *J. Am. Chem. Soc.* 88, 5186 (1966).
58. CHAN, S. I., KULA, R. J. and SAWYER, D. T., *J. Am. Chem. Soc.* 86, 377 (1964).
59. DÓZSA, L., SZABÓ, A. and BECK, M. T., to be published.
60. LANGER, H. G., *J. Inorg. Nucl. Chem.* 26, 767 (1964).
61. SCHMID, R. W. and REILLEY, C. N., *J. Am. Chem. Soc.* 78, 5513 (1956).
62. BENNETT, M. C. and WISE, W. S., *Trans. Faraday Soc.* 52, 696 (1956).
63. FELDMAN, I., HAVILL, J. R. and NEUMAN, W. F., *J. Am. Chem. Soc.* 76, 4726 (1954).
64. FELDMAN, I., NORTH, C. A. and HUNTER, H. B., *J. Phys. Chem.* 64, 1224 (1960).
65. GUSTAFSON, R. L., RICHARD, C. and MARTELL, A. E., *J. Am. Chem. Soc.* 82, 1526 (1960).
66. BOGUCKI, R. F. and MARTELL, A. E., *J. Am. Chem. Soc.* 80, 4170 (1958).
67. WERNER, A., *Z. Anorg. Chem.* 12, 53 (1896).
68. WELLS, H. L., *Am. J. Sci. 3*, 315 (1922).
69. WHITNEY, J. E. and DAVIDSON, N., *J. Am. Chem. Soc.* 69, 2076 (1947).
70. WHITNEY, J. E. and DAVIDSON, N., *J. Am. Chem. Soc.* 71, 3809 (1949).
71. McCONNELL, H. and DAVIDSON, N., *J. Am. Chem. Soc.* 72, 3168 (1950).
72. McCONNELL, H. and DAVIDSON, N., *J. Am. Chem. Soc.* 72, 5557 (1950).
73. IBERS, J., unpublished results, quoted in ref. 71.
74. SERES, I. and BECK, M. T., unpublished results.
75. KLOTZ, I. M., CZERLINSKI, G. H. and FIESS, H. A., *J. Am. Chem. Soc.* 80, 2920 (1958).
76. BECK, M. T., SERES, I. and BÁRDI, I., *Acta Chim. Acad. Sci. Hung. 41*, 231 (1964).
77. COTTON, F. A. and FACKLER, J. P. Jr., *J. Am. Chem. Soc. 83*, 2818 (1961).
78. BISHOP, J. J., COTTON, F. A., EISS, R. and HUGEL, R. P., *Nature 214*, 1111 (1967).
79. FERGUSON, J., *Spectrochim. Acta 17*, 316 (1961).
80. HOLM, R. H., *J. Am. Chem. Soc. 83*, 4683 (1961).
81. INGRI, N., KAKOŁOWICZ, W., SILLÉN, L. G. and WARNQVIST, B., *Talanta 14*, 1261 (1967).
82. BYDALEK, T. J. and MARGERUM, D. W., *Inorg. Chem.* 2, 678 (1963).

83. GÉHER, J. and BECK, M. T. to be published.
84. VLČEK, A. A., *Nature 197*, 786 (1963).
85. JOHNSON, W. L. and JONES, M. M., *Inorg. Chem. 5*, 1345 (1966).
86. BECK, M. T. and DÓZSA, L., *J. Am. Chem. Soc. 89*, 5713 (1967).
87. BRÖNSTED, J. N. and LIVINGSTON, R., *J. Am. Chem. Soc. 49*, 435 (1927).
88. ELVING, P. J. and ZEMEL, B., *J. Am. Chem. Soc. 79*, 5855 (1957).
89. HIGGINSON, W. C. E. and HILL, M. P., *J. Chem. Soc.* 1620 (1959).
90. JONASSEN, H. B. and RAMANUJAM, V. V., *J. Phys. Chem. 63*, 411 (1959).
91. BECK, M. T., GÖRÖG, S. and KISS, Z., *Acta Chim. Acad. Sci. Hung. 42*, 321 (1964).
92. BECK, M. T., *Record Chem. Progr. 27*, 37 (1966).
93. SIMPLICIO, J. and WILKINS, R. G., *J. Am. Chem. Soc. 89*, 6092 (1967).

Chapter 11

FACTORS INFLUENCING AND DETERMINING THE STABILITY CONSTANTS OF METAL COMPLEXES

It is of vital importance to find relationships between the stability constants of complexes and the characteristic properties of their constituents. Such relationships may give information on the nature of chemical bonding in complexes and make possible the estimation of unknown stability constants. The stability constants are determined by some atomic properties of the metal ions and ligands and by external conditions. It is not easy to separate the internal and external effects because the external factors may influence the atomic properties of the constituents of the complex; nevertheless this distinction is helpful in the following systematic treatment.

A prerequisite for the establishment of valid relationships is that the stability data considered must be reliable. Therefore we have first to deal with the critical evaluation of stability constants.

11.1 Critical evaluation of stability constants

When the stability constants for a given complex system are available, two questions arise.

(i) Is it possible to describe the properties of the given system?

(ii) Are there other sets of stability constants which describe the system equally well?

While it is relatively easy to answer the first question, the possibility of alternative formation schemes and sets of stability constants is fairly difficult to exclude. The greater the number of complex species, the more difficult it is to find an unambiguous set of equilibrium constants. The different methods are not equally sensitive to the concentration of different species, and it is therefore possible that different properties of the system can be quantitatively explained by postulating different numbers of species. Therefore it is not a necessary aim of equilibrium studies to find the least number of constants by which a certain property of the system can be described. However, it would be even more dangerous if the only criterion for reliability were "the best fit" of the calculated curve with the experimentally found one. Only very careful consideration can decide which species should be considered in a given system.

In this century, and particularly in the last two decades, a vast amount of stability data has been determined and published. The first, and especially the second, edition of 'Stability Constants' offers a wonderful collection of equilibrium data on metal complexes [1]. In the second edition, data referring to complexes of 1029 organic and 80 inorganic ligands are listed. The aim of the compilers being to give a complete collection of published equilibrium data, the reliability of the tabulated values was not considered, although sometimes the doubt of the compilers is expressed. In comparing data obtained by different authors for the same system under approximately the same conditions, one frequently finds great differences. As an example a few 'constants' referring to the stability of the tetracyanonickel(II) complex are collected in Table 11.1.

TABLE 11.1

Stability constant of the tetracyanonickel(II) complex

Method	$\log \beta_4$	Ref.
Electrode potential measurement	12.5	[2]
Polarography	15.5	[3]
Thermodynamic calculation	22	[4]
Polarography	>24	[5]
Toxicity measurement	30	[6]
Spectrophotometry	31.1	[7]
Spectrophotometry	31.5	[8]

This is the most extreme example, but unfortunately there are many others in which there are differences in the order of magnitude of the constants. It is interesting that toxicity measurements provided the first approximately correct value for the stability constant of the tetracyanonickel(II) complex, but in the absence of other data it would have been too brave to declare this value to be the reliable one.

The reliability of a stability constant is determined by the following factors: (a) adequacy of the experimental method; (b) the exactness of the experimental work; (c) consideration of all relevant equilibria; (d) the calculation method; (e) the reliability of the auxiliary data used.

The source of the greatest error is evidently the inadequacy of the experimental method applied. In all cases some measurement dependent on concentration gives the necessary data for the calculation. If the relationship between the measured quantity and concentration is not well defined, or is not known, then on using such data one cannot expect to obtain constants with physical meaning. For example, from measuring the electrode potential one can calculate the concentration of the solvated metal ion in question, assuming the validity of the Nernst equation. However, this equation is not valid if the electrode reaction is irreversible. The reason for the abnormally small values of β_4 for tetracyanonickel(II) obtained potentiometrically and polarographically is that the reaction $Ni^{2+} + 2e \rightarrow Ni^0$ is not strictly reversible. Hence Hume and Kolthoff [5] correctly considered their value as the lower limit. *If the chemical evidence suggests that the experimental method used is not suitable to provide reliable data for the concentration of the different species, the data must not be considered for any further calculation.*

Even the most careful experimental work cannot result in sound data if an inadequate method is used. However, even an ideally suitable experimental method will provide only poor data if the experimental work is not satisfactorily performed. It is necessary to work with well defined conditions: temperature, ionic strength and ionic composition, presence and amount of organic solvents, etc. Sometimes even minute amounts of foreign ions can seriously interfere, so the purity of materials has to be checked. In the case of autoxidizable organic ligands it is particularly important to determine whether the compound is contaminated by oxidation products. The reproducibility of the results, the scatter of points on the curves, and most explicitly an exact calculation of the error, give indications of the quality of the experimental work.

In the data provided by good experimental work with adequate methods, all the equilibria contributing substantially to the distribution of different species are reflected. As was shown in the previous chapters, well founded calculation methods have been developed for practically all types of simple and complicated complex equilibria. In many cases the consideration of all the equilibria occurring would lead to unsurmountable difficulties and some simplification is necessary to perform the calculations. The danger here is oversimplification. Mostly in older papers, but sometimes even nowadays, one meets the habit of taking into consideration only the stepwise formation of mononuclear binary complexes and of disregarding such types of equilibria as protonation of metal complexes, formation of polynuclear species and mixed ligand complexes and the co-ordination of ligands in the outer sphere. This practice evidently may result in erroneous stability constants. For example the stability constant of the iron(III)-EDTA complex was determined spectrophotometrically by following the effect of increasing acid concentration on the decomposition of the metal chelate [9]. The value obtained is smaller by nearly two orders of magnitude than that determined by electrometric measurements at higher pH [10]. This big difference cannot be accounted for by the difference in the ionic strengths. The reason for the discrepancy is twofold [11]. First, in acidic medium stepwise protonation of the iron chelate occurs; secondly, at lower pH the formation of positively charged EDTA species has to be taken into account. This example clearly shows that it is necessary — at least potentially — to consider the basic character of all the donor groups of the chelate-forming ligands.

To illustrate the importance of side-reactions, the determination of the stability constant of the bismuth(III)–1,2-diaminocyclohexanetetra-acetate complex may be mentioned. The constant was determined polarographically, by following the competition between copper(II) and bismuth(III) in an acetate-buffered medium at pH 4.3 [12]. The value obtained is less by seven orders of magnitude than the true one [13], evidently because the formation of acetate- and hydroxo-complexes of the two metal ions was not considered.

The most frequent error in the evaluation of experimental data is the *abuse* of Bjerrum's 'half value' method. Although Bjerrum himself stressed [14] the limitations of this approximation, it is so rapid and simple that many authors cannot resist the temptation to apply it even when the system does not meet the necessary requirements.

An excellent method was developed by Nagypál and Gergely [15] to decide whether the assumption that only mononuclear binary complexes are formed is correct or not. They introduced certain "reduced formation functions" which must be symmetrical if the stepwise formation of parent complexes occurs, whatever the shape of the original formation curve.

It does not need any further explanation that the reliability of all the auxiliary constants used in the evaluation of the stability constants is crucial. The consequence of choosing an erroneous acidic dissociation constant for the ligand was thoroughly treated by Cabani [16]. Unfortunately there are examples where the value of good work was decreased by use of poor auxiliary data.

11.2 Effect of external factors on the stability constants

In discussing the effect of external factors one must bear in mind that usually the change of one external factor also results in changes in the others, and therefore their effects cannot be studied separately and independently.

11.2.1 EFFECT OF PRESSURE

In practically all cases complex equilibria are studied under normal pressure. However, the importance of studies of the effect of pressure on the equilibria is not exclusively theoretical. The pressure at the bottom of oceans varies between 100 and 1000 atmospheres, so the effect of pressure must be considered in dealing, for example, with equilibria in sea-water. The increase of pressure promotes reactions resulting in decrease of volume and inhibits those resulting in volume increases. In general the increase of pressure leads to the enhancement of the dissociation of weak electrolytes, including complexes [17–20]. It was found that the stability constant of $FeCl^{2+}$ decreases by a factor of at least 20 when the pressure is increased from 1 to 2000 atm, and in the case of $FeNO_3^{2+}$ there is a decrease of about 20% in the stability constant on going from 1 to 3500 atm [20].

The effect of pressure on the magnetic properties of nickel(II) chelates of different salicylaldimines and aminotroponeimineates, has been studied and the equilibrium constants for the planar and tetrahedral transformation were calculated from the magnetic data [21]. The logarithms of the equilibrium constants vary linearly with the pressure, but the slope may be positive or negative, depending on the ligand.

11.2.2 EFFECT OF TEMPERATURE

The following relationship is valid between the free energy of formation of a complex and its stability constant:

$$\Delta G = -2.303 \, RT \log K \, . \tag{11.1}$$

On the other hand the free energy is given by the sum of an enthalpy and an entropy term:

$$\Delta G = \Delta H - T \Delta S \, . \tag{11.2}$$

That is, complex formation is favoured by negative enthalpy and by positive entropy changes. It is very difficult to predict the contribution of these terms because the solvation of the constituents of the complex also must be considered. According to Williams [22] the entropy term is usually favourable when the ligand is anionic, and is generally unfavourable in the case of neutral ligands.

It follows from Eqs. 11.1 and 11.2 that the enthalpy and the entropy terms cannot be calculated from stability data referring to a single temperature. From the dependence of the stability constant on temperature, it is possible in principle to evaluate both the enthalpy and the entropy terms, but this procedure is not very exact. It is more advisable to determine the thermodynamic functions by additional calorimetric experiments. Recently a fairly large amount of such data has been collected.

In certain cases there is a simple relationship between the entropy and enthalpy of formation for a series of complexes, e.g. Kennedy and Lister [23] found that for the complexes $CoCl^+$, $NiCl^+$, $CuCl^+$, $CoBr^+$, $NiBr^+$ and $CuBr^+$, ΔH and ΔS are related by the equation

$$\Delta H = 254 \Delta S + 201 \tag{11.3}$$

(ΔH is expressed in calories, ΔS in cal/deg, and the two constants have the units of deg (254) and calories (201) respectively).

The temperature influences all other external factors; its effect is therefore very complicated, even if the result can be given by a simple approximately valid relationship.

11.2.3 EFFECT OF DIELECTRIC CONSTANT

Because at least one of the constituents is charged and the other is also charged or has a dipole moment, the dielectric constant of the medium evidently strongly affects the stability constants. However, one must always distinguish the macroscopic dielectric constant of the medium and the dielectric constant of the layer between the metal ion and the ligands. The latter is much smaller than the macroscopically measured dielectric constant, but its exact value is usually not available.

According to Van Uitert et al. [24] there is a linear relationship between the stability constant of certain acetylacetonate complexes and the mole fraction of dioxan used in the solvent to change the dielectric constant of the medium. It appears that this relationship is generally valid for oxygen-donor ligands [25]. Irving and Rossotti [25] pointed out that if the stability constants of complexes of the same ligand with different metal ions, referring to aqueous solution, are plotted versus the corresponding stability constants referring to a water–organic solvent mixture, a straight line is obtained (Fig. 11.1). The intercept of the straight line is the difference between the acidic dissociation constants of the ligand in the two media.

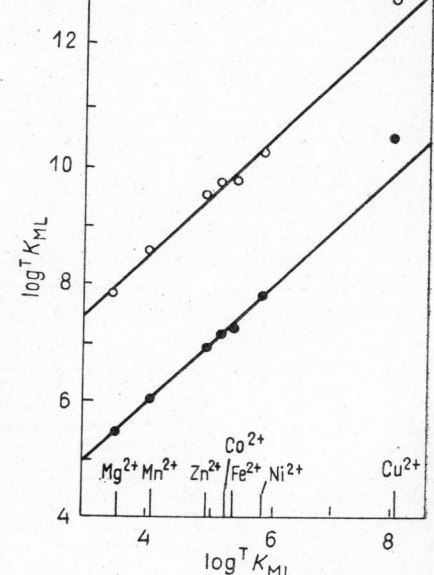

Fig. 11.1 Log K of several metal acetyl-acetonate complexes at 30 °C in 75% (O) and in 50% (●) aqueous dioxan plotted versus log K in water. Slopes of the straight lines are unity, the intercepts are equal to the differences of logarithms of acidic dissociation constants of acetylacetone determined in water and in the mixed solvent. (Reproduced with permission from *Acta Chem. Scand. 10*, 72 (1956))

11.2.4 EFFECT OF IONIC STRENGTH AND IONIC MEDIUM

The effect of different ions may vary considerably, but the general effect is to change the activity coefficients. The stability constants obtained at different ionic strengths can be extrapolated to zero ionic strength. Table 11.2 shows the effect of ionic strength on the stability constant of the zinc(II) acetylacetonate complex [26].

TABLE 11.2

The effect of ionic strength on the stability of the zinc(II) acetylacetonate complex

T_{Zn} M	I	$\log K_1$	$\log K_1^*$	$\log K_2$	$\log K_2^*$
2×10^{-3}	0.0098	4.90	5.02	3.77	3.82
5×10^{-3}	0.024	4.82	4.99	3.72	3.80
1×10^{-2}	0.049	4.74	4.95	3.68	3.78
2×10^{-3}	0.10	4.70	4.96	3.62	3.76
2.5×10^{-2}	0.122	4.70	4.98	3.70	3.83
2×10^{-3}	0.20	4.66	4.98	3.67	3.83
5×10^{-2}	0.25	4.64	4.98	3.68	3.84
2×10^{-3}	0.30	4.61	4.96	3.63	3.81
1×10^{-1}	0.487	4.61	5.00	3.80	3.98

K_n = concentration constant; K_n^* = thermodynamic constant

The effect of different ions on the activity coefficients may differ widely. In this case the principle of using a constant ionic medium is not valid and reliable constants can be obtained only by considering the activity coefficients [27, 28].

In moderately concentrated solutions the effect of 'neutral salts' on the water concentration and on the dielectric constant of the solution must also be considered.

Many experiences show that the stability constant depends on the nature of the counter-ion of the complex-forming anion used, even when these cations do not form complexes. According to Fialkov and Spivakovskii [29], in the cadmium(II)–chloride system even the co-ordination number depends on which alkali chloride was used as chloride source. Tate and Jones [30] potentiometrically determined the stability constant of the $CdNO_3^+$ ion, using different nitrates. Their data — summarized in Table 11.3 — show that even in this simple system there are large differences in the values, depending on the counter-ion. Therefore, with a complicated system it is by no means certain that it is necessary to assume a change in the maximum co-ordination number in order to explain the experimental data.

TABLE 11.3

Effect of different cations on the stability of $CdNO_3^+$ *at 25°C*

Salt	K_1	Salt	K_1
$Sr(NO_3)_2$	1.063	$Nd(NO_3)_3$	0.550
$Ca(NO_3)_2$	0.904	$Al(NO_3)_3$	0.539
$Mg(NO_3)_2$	0.817	$La(NO_3)_3$	0.533
$NaNO_3$	0.747	$LiNO_3$	0.142

11.3 Relationships between the properties of central ions and ligands and the stability constants of complexes

The values of stability constants under given experimental conditions are determined by the properties of the central ions and the ligands. In looking for relationships of this kind one must always bear in mind that the characteristic atomic properties are interdependent. The most characteristic relationships are treated in this Chapter; relationships referring to some particular types of complex (mixed ligand, outer-sphere, polynuclear complexes) are discussed in the corresponding Chapters.

11.3.1 RELATIONSHIPS BETWEEN THE PROPERTIES OF THE CENTRAL ION AND THE STABILITY OF COMPLEXES

11.3.1.1 The ionic radius

If the interaction of the metal ion and the ligand were purely electrostatic, the stability constants for complexes of metal ions of the same charge should be inversely proportional to metal ion radii. For ions of similar electronic configuration this relationship may be approximately valid, but it completely fails when metal ions of different groups of the periodic system are compared. For example the radii of sodium and copper(I) ions or the radii of calcium and cadmium ions are approximately the same, but the stabilities of their complexes with the same ligands are vastly different. This indicates the importance of the electronic configuration.

When metal ions of different charge are compared the *ionic potential* (the ratio of the charge to the radius of the ion) must be considered instead of the ionic radius.

11.3.1.2 The ionization potential and electronegativity

In complex formation, electrons, lost in the ionization process, are gained from the donor atoms; the ionization potential can therefore be regarded as a direct measure of the electron affinity of the metal ion and a correlation can be expected between the stability of the complex and the ionization potential of the metal ion. Although this comparison is only a rough approximation owing to the change in the electronic configuration during the formation of complexes, fairly good linear relationships have been found [31–33].

Van Panthaleon van Eck [34] found that for many metal ions the following relationship is valid

$$\log K_1 = p(I - q) \tag{11.4}$$

where I is the ionization potential (for the reaction $Me \rightarrow Me^{m+} + me$ in the gas phase), and p and q are constants depending only on the ligand and

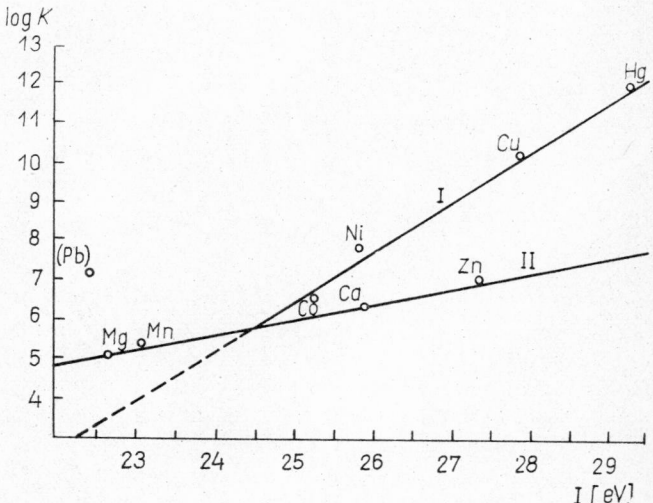

Fig. 11.2 Stability constant of different metal complexes of glycine as a function of the ionization potential of the central ion. The stability constants are dimensionless quantities, the concentrations being expressed in mole fractions. (Reproduced with permission from *Rec. Trav. Chim.* 72, 50 (1953))

the experimental conditions (medium, temperature, etc) but independent of the metal ion; p depends on the number of donor groups and the polarizability of the ligand, while q is characteristic of the donor groups. For certain bidentate ligands metal complexes can be divided into two groups, each group having a characteristic p and a characteristic q value (Fig. 11.2). From a comparison of different systems van Panthaleon van Eck concluded that in the case of the type II complexes (greater p and q values) the charged group of the ligand is co-ordinated (in case of glycine the carboxylic group), while with type I complexes either the uncharged group is co-ordinated or chelation occurs. This explanation is not very convincing, because all experience seems to prove that chelation is a general phenomenon in the case of these complexes, independent of the nature of the metal ion.

Many authors have established relationships between the electronegativities of metal ions and the stability constants of the complexes [35–39]. Figure 11.3 shows the stability constants of complexes of glycine with bivalent metal ions as a function of the electronegativities of the central ions.

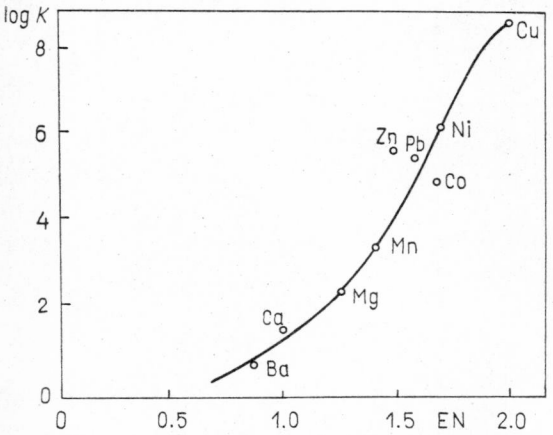

Fig. 11.3 Stability constants of different metal glycine complexes as a function of the electronegativities of the metal ions (Reproduced with permission from *Nature 174*, 887 (1954))

It is interesting that although the ionization potentials and the electronegativities of metal ions are related quantities, all the points fall on the same curve, while two straight lines were obtained in Fig. 11.2.

Important conclusions can be drawn from the comparison of the dependence of the successive formation constants of metal complexes on the ionization potential of the central ions. Figure 11.4 shows such graphs referring to the ethylenediamine complexes of bivalent transition metal ions.

It is clear that good linear relationships are valid for the mono and bis complexes, but the tris complexes of zinc(II) and copper(II) strongly deviate from the general trend. The most likely reason for this behaviour is that the copper(II) complexes prefer the square planar configuration to the octahedral one, and there is therefore a drop in the value of successive stability constants after co-ordination number four has been achieved. On the other hand, while the configuration of bis complexes of all metal ions except zinc(II) are octahedral [in the case of copper(II) distorted octahedral], two

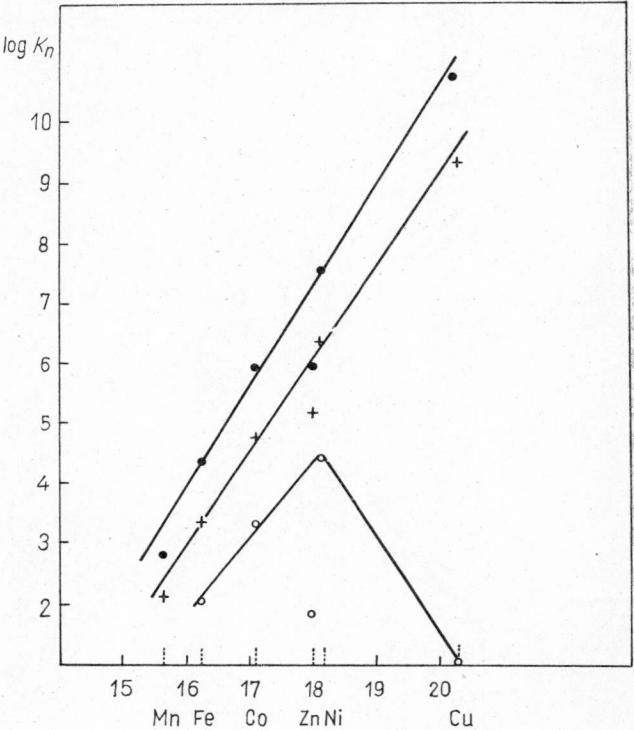

Fig. 11.4 Correlation between the successive stability constants of the ethylenediamine complexes and the ionization potentials of transition metal ions

co-ordination sites being occupied by water molecules, the $Zn(en)_2^{2+}$ complex is most probably tetrahedral even in aqueous solution. The necessity for rearrangement of the co-ordination sphere in the uptake of the third ligand molecule is reflected in the lower value of K_3.

11.3.1.3 The electronic configuration of the central ion

The complex-forming abilities of the transition metal ions are frequently characterized by stability orders. The stability order established by Irving and Williams [40, 41] is particularly important because it is valid for most nitrogen and oxygen donor ligands, irrespective of the nature of the ligand.

Some examples are given in Fig. 11.5. This order $Mn^{2+} < Fe^{2+} < Co^{2+} < < Ni^{2+} < Cu^{2+} > Zn^{2+}$ was rationalized by Irving and Williams by comparison of the ionic radii and the second ionization potential I_2 of the metal ions in question (Table 11.4).

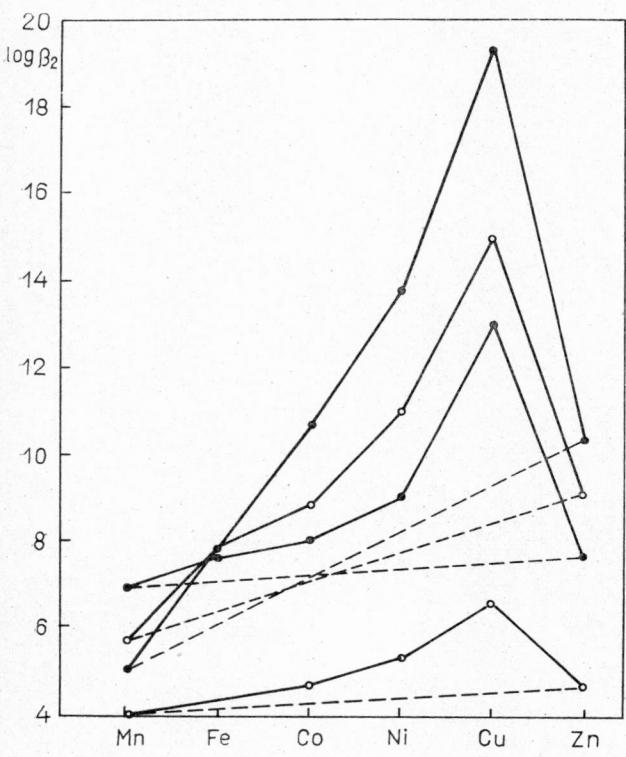

Fig. 11.5 Log β_2 of complexes of several ligands as a function of the atomic number of bivalent transition metal ions. ○ salicylaldehyde; ◑ glycine; ● ethylenediamine; ◓ oxalate. Dashed lines are linear interpolation between Mn^{2+} and Zn^{2+}

TABLE 11.4

Ionic radii, ionization potentials and hydration heats of some bivalent metal ions

	Mg^{2+}	Mn^{2+}	Fe^{2+}	Co^{2+}	Ni^{2+}	Cu^{2+}	Zn^{2+}	Cd^{2+}	Ca^{2+}
Ionic radii, Å	0.66	0.78	0.76	0.74	0.73	0.72	0.72	0.96	0.99
I_2, kcal/mole	525.7	534.7	558.6	586	599	648.6	633.5	599.9	417.5
H, kcal/mole	464.0	444.7	467.9	497	507	507.2	491.5	436.5	381.9

According to the crystal field theory the general validity of the Irving–Williams order is a consequence of crystal field stabilization [42, 43]. For all configurations other than d^0 (Ca^{2+}, Sc^{3+}), d^5 (Mn^{2+}, Fe^{3+}) and d^{10} (Zn^{2+}) the splitting of the d electrons between two energy levels by the electrical field of the ligands lowers the total energy of the system. This decrease in energy is termed the crystal field stabilization energy (CFSE). That is, no stabilization occurs in the case of Mn^{2+} and Zn^{2+} complexes, and the stabilization is of maximum value for Ni^{2+} and Cu^{2+} complexes. The crystal field stabilization energy can be calculated from spectroscopic data. For many systems these values are in good agreement with the energy values derived from the deviation of the experimentally found stability constants from the values obtained by linear interpolation between the Mn^{2+} and Zn^{2+} complexes.

The considerations above are valid for high-spin complexes only. With many ligands — termed strong field ligands — spin coupling occurs. Deviations from the Irving–Williams order occur with complexes of this kind.

The situation is more complicated in the case of lanthanide complexes. While the Irving–Williams order is uniformly valid for all weak field ligands, in the case of the complexes of the rare earth ions three groups can be distinguished [44]. Three representative examples are shown in Fig. 11.6. As can be seen, a difference in the trend of the stability constants is exhibited after gadolinium. The decrease of the stability constant at gadolinium is not yet sufficiently explained.

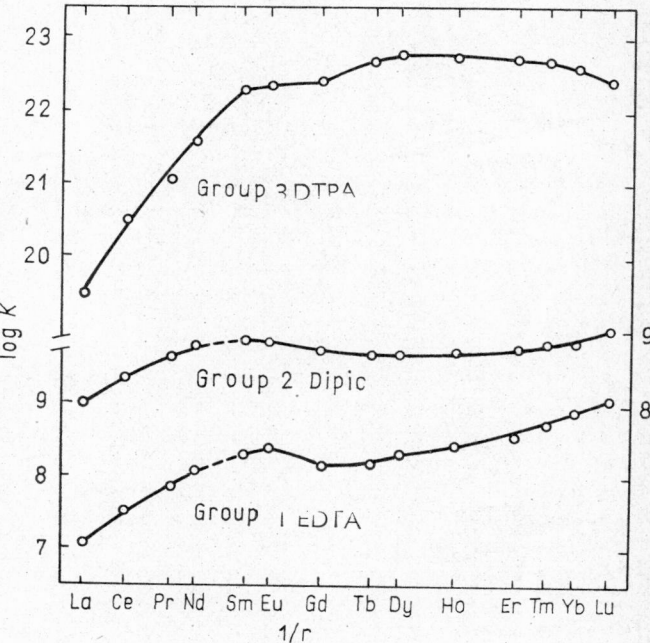

Fig. 11.6 Typical plots of stability constants of complexes of tervalent lanthanide ions against the reciprocal of the ionic radius of the corresponding metal ion.
EDTA = ethylenediamine-N,N'-tetraacetic acid;
Dipic = dipicolinic acid; DTPA = diethylenetriamine, N,N,N',N',N''-pentaacetic acid.
(Reproduced with permission from *Chem. Rev. 65*, 1 (1965))

11.3.1.4 Classification of the central ions according to the periodic table

There is a strict correlation between the electronic structure of ions and the position of the corresponding elements in the periodic table. As is shown by Fig. 11.7, the periodic table offers a very convenient classification of the acceptor ions [45]. The basis of this classification is the relative tendency

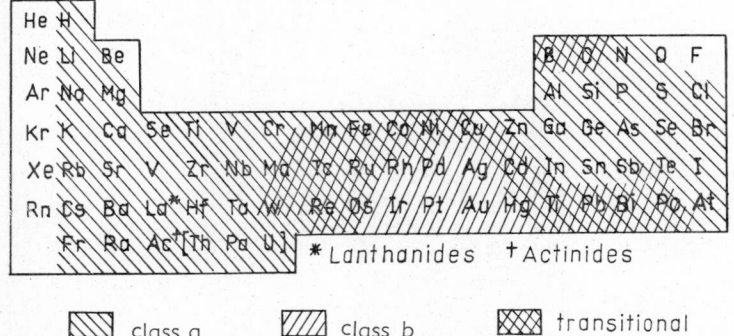

Fig. 11.7 The distribution of acceptor ions in the periodic table. (Reproduced with permission from Quart. Rev. 12, 265 (1958))

of the acceptor ions towards complex formation with different donor atoms. For class (a) ions the following orders are valid

$$F \gg Cl > Br > I$$
$$O \gg S > Se > Te$$
$$N \gg P > As > Sb$$

while for class (b) ions just the opposite is true. Class (c) ions do not exhibit such great differences in the tendency towards complex formation. This classification of acceptor ions was offered independently by Ahrland and Larsson [46] and by Carleson and Irving [47]. The characteristic difference between the relative stabilities of different metal ions was earlier observed and roughly explained by Sidgwick [48] and by Grinberg [49]. Evidently the bond in the complexes of class (a) ions is mainly electrovalent, while there is considerable covalency in the complexes of the more polarizing class (b) ions. From simple electrostatic considerations, the fluoro-complex should be the most stable halide complex of a particular metal ion, the ionic radius of fluoride being the smallest. The reversed order for class (b) ions, according to Ahrland, Chatt and Davies [45] is caused by the availability of electrons from ($n - 1$) d orbitals of the metal for dative π-bonding.

Pearson [50] extended the classification to donor atoms, and introduced the new terms "hard" and "soft" acids (acceptors) and bases (donors). Although this concept became popular [51], its generalization must be treated with some reservation. As was convincingly pointed out by Williams and Hale [52], at present the class (a) and class (b) behaviour cannot be fully explained and use of the terms "hard" and "soft" may lead to confusion.

11.3.1.5 Comparison of the stability constants of complexes of different ligands with two central ions

Irving and Rossotti [25] pointed out that there is a roughly linear relationship between the stability constants of complexes of two metal ions with a series of related ligands. Figure 11.8 shows this relationship for the complexes of copper(II) and nickel(II) with 56 different ligands.

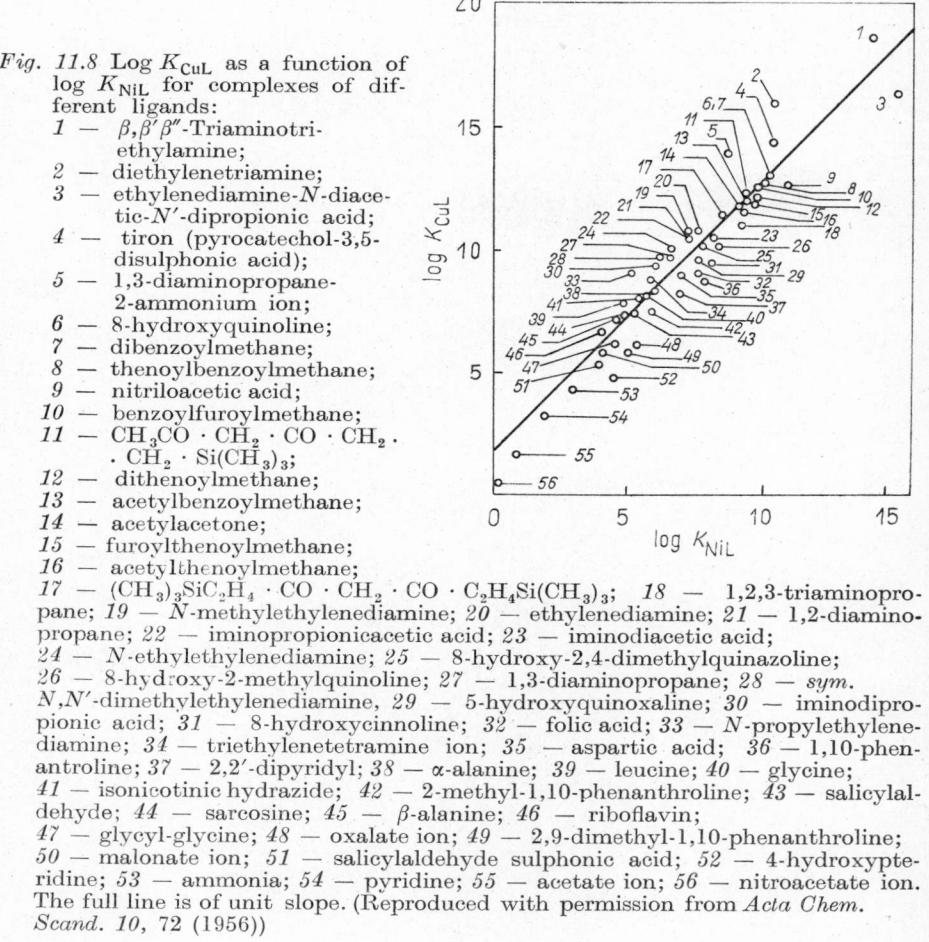

Fig. 11.8 Log K_{CuL} as a function of log K_{NiL} for complexes of different ligands:

1 — $\beta,\beta'\beta''$-Triaminotriethylamine;
2 — diethylenetriamine;
3 — ethylenediamine-N-diacetic-N'-dipropionic acid;
4 — tiron (pyrocatechol-3,5-disulphonic acid);
5 — 1,3-diaminopropane-2-ammonium ion;
6 — 8-hydroxyquinoline;
7 — dibenzoylmethane;
8 — thenoylbenzoylmethane;
9 — nitriloacetic acid;
10 — benzoylfuroylmethane;
11 — $CH_3CO \cdot CH_2 \cdot CO \cdot CH_2 \cdot CH_2 \cdot Si(CH_3)_3$;
12 — dithenoylmethane;
13 — acetylbenzoylmethane;
14 — acetylacetone;
15 — furoylthenoylmethane;
16 — acetylthenoylmethane;
17 — $(CH_3)_3SiC_2H_4 \cdot CO \cdot CH_2 \cdot CO \cdot C_2H_4Si(CH_3)_3$; 18 — 1,2,3-triaminopropane; 19 — N-methylethylenediamine; 20 — ethylenediamine; 21 — 1,2-diaminopropane; 22 — iminopropionicacetic acid; 23 — iminodiacetic acid; 24 — N-ethylethylenediamine; 25 — 8-hydroxy-2,4-dimethylquinazoline; 26 — 8-hydroxy-2-methylquinoline; 27 — 1,3-diaminopropane; 28 — *sym.* N,N'-dimethylethylenediamine, 29 — 5-hydroxyquinoxaline; 30 — iminodipropionic acid; 31 — 8-hydroxycinnoline; 32 — folic acid; 33 — N-propylethylenediamine; 34 — triethylenetetramine ion; 35 — aspartic acid; 36 — 1,10-phenantroline; 37 — 2,2'-dipyridyl; 38 — α-alanine; 39 — leucine; 40 — glycine; 41 — isonicotinic hydrazide; 42 — 2-methyl-1,10-phenanthroline; 43 — salicylaldehyde; 44 — sarcosine; 45 — β-alanine; 46 — riboflavin; 47 — glycyl-glycine; 48 — oxalate ion; 49 — 2,9-dimethyl-1,10-phenanthroline; 50 — malonate ion; 51 — salicylaldehyde sulphonic acid; 52 — 4-hydroxypteridine; 53 — ammonia; 54 — pyridine; 55 — acetate ion; 56 — nitroacetate ion. The full line is of unit slope. (Reproduced with permission from *Acta Chem. Scand. 10,* 72 (1956))

Freiser, Fernando and Cheney [53] compared the stability constants of nickel(II) and zinc(II) complexes with 72 ligands. For 42 ligands (aminoacids, salicylaldehyde, β-diketones, etc) the following relationship was found to be valid:

$$\log \beta_2^{Ni} = 1.13 \log \beta_2^{Zn}. \tag{11.5}$$

The ligands for which Eq. 11.5 is not valid can be classified into two distinct groups. For both groups the same linear relationships are valid, but the intercept is not zero. For ligands containing two oxygen donor atoms and forming five-membered rings the intercept is -2.0, while for ligands containing two nitrogen donor atoms and also forming five-membered rings the intercept is $+2.0$.

Yatsimirskii [54] compared the stability constants of the complexes of Mg^{2+} and Zn^+. These two metal ions are of the same radius and charge, but of different electronic configuration. The difference in the stability constants is characteristic of the bonding. In Table 11.5 the differences of the stability constants ($\Delta \log K_{ZnMg} = \log K_{Zn} - \log K_{Mg}$) are summarized for some unidentate ligands.

TABLE 11.5

The value of $\Delta \log K_{ZnMg}$ for several unidentate ligands

Ligand	$\Delta \log K_{ZnMg}$	Ligand	$\Delta \log K_{ZzMg}$
F^-	-0.5	$P_2O_7^{4-}$	0.7
H_2O	0.0	OH^-	2.3
SO_4^{2-}	0.0	NH_3	2.6
$S_2O_3^{2-}$	0.6	NR_3	2.8
$RCOO^-$	0.6		

The negative value obtained for fluoride seems to indicate that the bond in the fluoro-complexes is more ionic than in the aquo-complexes.

Table 11.6 shows the $\Delta \log K_{ZnMg}$ values for several oxygen and nitrogen donor ligands. It appears from the data that the stability difference is

TABLE 11.6

The value of $\Delta \log K_{ZnMg}$ for several O- and N-containing ligands

Ligand	$\Delta \log K_{ZnMg}$	Ligand	$\Delta \log K_{ZnMg} - 0.6\, n_0$
$CH_3CO_2^-$	0.6	$C_2H_4(NH_2)_2$	2.7
$C_2H_5CO_2^-$	0.5	$CH_3CHNHCO_2^-$	3.2
$C_3H_7CO_2^-$	0.5	$HNAc_2^{2-}$	2.1
$1/2\ C_2O_4^{2-}$	0.7	$CH_3NAc_2^{3-}$	3.0
$1/2\ CH_2(CO_2)_2^{2-}$	0.5	NAc_3^{3-}	3.2
$1/2\ C(COCH_3)_2^{2-}$	0.7	$(CH_2NAc_2)_2^{4-}$	3.1
$1/4\ (C_2O_4^{2-})_2$	0.7	H_2C-CH_2	2.5
$1/4\ [C(COCH_3)_2^{2-}]_2$	0.7	$\begin{array}{c} \mid \quad\quad \mid \\ H_2C \quad CO_2^- \\ \diagdown \quad \diagup \\ NH \end{array}$	
Average	0.6 ± 0.1	Average	2.8 ± 0.3

$Ac = -CH_2COO^-$

a simple function of the number of donor N and O atoms (n_N and n_O):

$$\Delta \log K_{ZnMg} = 2.8 n_N + 0.6_{n_0} \tag{11.6}$$

Yatsimirskii claimed that this relationship can be used to find out which potential donor atoms are involved in the complex formation. This is indicated by the data of Table 11.7. With oxalate and malonate $\Delta \log K_{ZnMg} = 1.2$, showing that both carboxylic groups are involved in complex formation, while the value 0.6 for the succinate and glutarate complexes indicates that no chelate formation occurs with these ligands. It is worth mentioning that for hydroxycarboxylic acids $\Delta \log K_{ZnMg}$ falls between the values corresponding to the co-ordination of one and two oxygen atoms. This may be due to a weak bond between the undissociated alcoholic hydroxy group and the metal ion.

TABLE 11.7

Mode of co-ordination with several ligands

Ligand	$\Delta \log K_{ZnMg}$	Co-ordination
$C_2O_4^{2-}$	1.3	2 O
$CH_2(CO_2)_2^{2-}$	1.1	2 O
$(CH_2CO_2)_2^{2-}$	0.6	O
$CH_2(CH_2CO_2)_2^{2-}$	0.5	O
$CH_2OHCO_2^-$	1.0	2 O
$CH_3CHOHCO_2^-$	0.9	2 O
$CH_2CO_2^-$		2 O
$\mid$	1.3	
$CHOHCO_2^-$		
$C_4H_4O_6^{2-}$	1.3	2 O
$H_2N(CH_2)_2NAc_2^{2-}$	7.4	2 N, 2 O
NAc_3^{3-}	5.0	N, 3 O
NAc_2Prop^{-3}	4.8	N, 3 O
$NAc_2Prop_2^{-3}$	4.2	N, 2 O
$(CH_2NAc)_2^{4-}$	7.6	2 N, 4 O
$(CH_2NAcProp)_2^{4-}$	7.6	2 N, 4 O
$(CH_2NProp_2)_2^{4-}$	6.0	2 N, 2 O

$Ac = -CH_2COO^-$; $Prop = -CH_2CH_2COO^-$

There are two arguments against this reasoning. First, in calculating the stability constants certain assumptions about the mode of co-ordination might have already been made; secondly it is not guaranteed that the mode of co-ordination is exactly the same with the two metal ions. For the location of the binding sites the analysis of equilibrium data must be used very cautiously. For this purpose the differential proton relaxation studies introduced by Dillon and Rossotti [55] offer a much more reliable approach.

11.3.2 Correlations between the properties of a ligand and the stability of its metal complexes

11.3.2.1 Nature of the donor atom

The number of monatomic ligands is much smaller than that of the poly-atomic ligands in which the donor atoms are the constituents of the functional groups. The halides act in almost all cases as monatomic ligands; in most of the organic ligands O, N and S act as donor atoms. Evidently the radius and charge — or the dipole character — of the donor atom are very important from the point of view of electrostatic interaction. At present our kowledge of the effect of the electron structure of the donor atom is fairly incomplete. The heavier donor atoms and ligands containing bonding pairs of electrons can act as acceptors for d electrons of metal ions with full or almost full d orbitals, using their vacant p orbitals (or π^* antibonding orbitals) for the purpose. This phenomenon, the so-called *back-co-ordination* or *back-donation*, helps us to understand the complex formation tendency of different ligands and the class (a) and class (b) behaviour of metal ions.

Just as, in the case of metal ions, correlations were found between the ionization potential and the stability of complexes, van Panthaleon van Eck pointed out that a relationship exists between the electron affinity (E in eV) of the ligand and the logarithm of the stability constant of the complex for the halide complexes of Cd^{2+} and Hg^{2+}:

$$\log K = a(E - b) \tag{11.7}$$

where a and b are constants depending on the metal ion, although no chemical meaning can be assigned to them. The relationship is not valid for the fluoro-complexes.

11.3.2.2 Correlations between the basicity of the ligand and the stability of complexes

Considering that in most cases the formation of complexes means a competition between metal ions and protons, it is reasonable to expect that there is some correlation between the stability constant of the complex and the acidic dissociation constant of the conjugate acid of the ligand. First Larsson [56] found that there is a linear relationship between the corresponding constants for the complexes of silver(I) with organic amines. Subsequently similar correlations were found in many complex systems [25, 57–60]. Such comparisons can be made to predict the behaviour of ligands when the linear relationship can be thermodynamically rationalized [25] but refer only to the first complex formed, MeL. Figure 11.9 shows the limitation of the validity of the linear relationship for the complexes of the derivatives of 8-hydroxyquinoline.

It is clear from the figure that deviation from the linear relationship is exhibited when the substitution occurs at a position adjacent to the donor

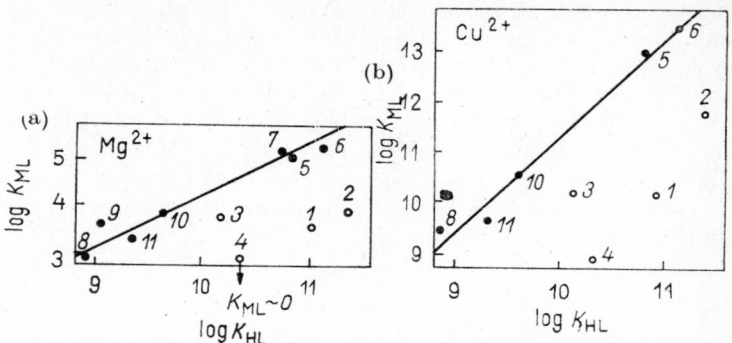

Fig. 11.9 Log K_{ML} as a function of log K_{HL} for Mg^{2+} (a) and Cu^{2+} (b) complexes of 8-hydroxyquinoline derivatives in 50% v/v aqueous dioxan (0.3 M $NaClO_4$) at 20°C. Open points refer to ligands with a substituent adjacent to the chelating nitrogen atom. *1* — 2-methyl-8-hydroxyquinoline; *2* — 1,2,3,4-tetrahydro-10-hydroxyacridine; *3* — 8-hydroxy-2,4-dimethylquinazoline; *4* — 4,8-dihydroxy-4-metyl-2-phenylquinazoline. Full points refer to the following ligands; *5* — 8-hydroxyquinoline; *6* — 5-methyl-8-hydroxyquinoline; *7* — 6-methyl-8-hydroxyquinoline; *8* — 8-hydroxycinnoline; *9* — 8-hydroxy-4-methylcinnoline; *10* — 8-hydroxyquinazoline; *11* — 5-hydroxyquinoxaline. (Reproduced with permission from *Acta Chem. Scand. 10*, 72 (1965))

atom. Only complexes of ligands containing the same donor atoms can be compared. A linear relationship cannot be expected, for example, in the case of the stability constants of the metal complexes of 2,2'-dipyridyl and its reduction products and the corresponding acidic dissociation constants [61] (Table 11.8).

TABLE 11.8

Acidic dissociation constants of 2,2'-dipyridyl and its reduction products and the stability constants of their copper(II) complexes

Ligand	K_1	K_2	β_2
2,2'-dipyridyl	1.59	4.27×10^{-5}	4.57×10^{13}
2,2'-pyridyl piperidyl	10^{-2}	7.95×10^{-10}	1.59×10^{12}
2,2'-dipyperidyl	10^{-7}	6.9×10^{-11}	8.69×10^{17}

In the case of multidentate ligands the situation is more complicated because it is not self-evident which acidic dissociation constant has to be chosen for the comparison. In looking for a relationship in the case of the derivatives of 8-hydroxyquinoline (Fig. 11.9) the dissociation constant of the phenolic hydroxy group was considered, but according to Williams *et al.*

[62] it is more reasonable to compare the product of the two acidic dissociation constants with the stability constants. Looking for a relationship in the case of complexes of derivatives of dipyridyl and terpyridyl, Brandt and Wright [63] compared the logarithms of the stability constants with the ratio of the pK of the most basic donor group to the number of nitrogen atoms in the ligand.

The validity of these relationships was questioned in some cases [62, 64] and even inverse correlations were found [65, 66]. Such behaviour is exemplified [66] by Fig. 11.10.

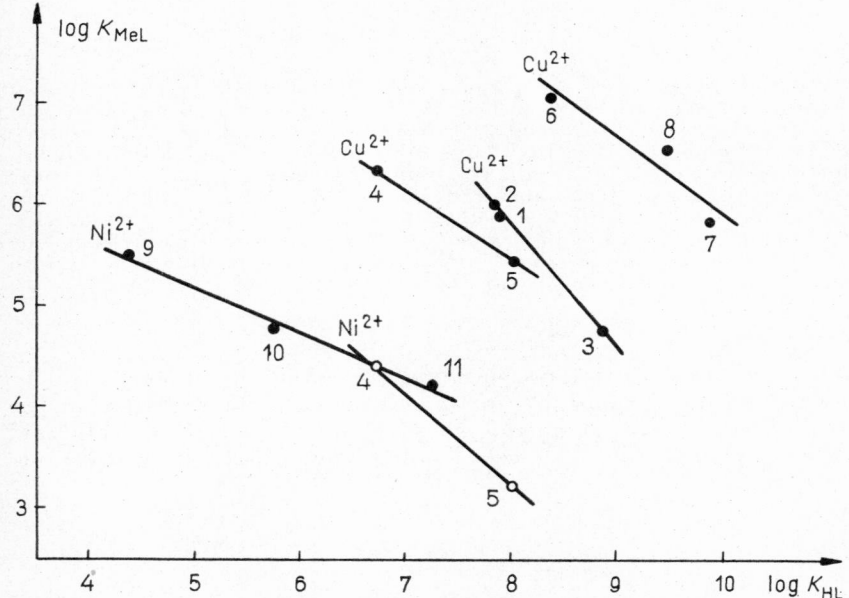

Fig. 11.10 Correlation between $\log K_{CuL}$, $\log K_{NiL}$ and $\log K_{HL}$ for the following ligands: *1* — 3-hydroxythiophen-2-carboxylic acid ethyl ester; *2* — 3-hydroxy-5-methylthiophen-2-carboxylic acid ethyl ester; *3* — 2-methyl-4-hydroxythiophen-3-carboxylic acid ethyl ester; *4* — 2-acetyl-3-hydroxythiophen; *5* — 3-acetyl-4-hydroxythiophen; *6* — 2-hydroxynaphthaldehyde-1; *7* — 2-hydroxy-naphthaldehyde-3; *8* — salicylaldehyde; *9* — 8-hydroxy-6-methyl-1,6-napthyridinium iodide; *10* — 5-hydroxyl-1-methylquinoxalinium iodide; *11* — 8-hydroxy-6-methylquinazolinium iodide. (Reproduced with permission from *Helv. Chim. Acta 49*, 1618 (1966))

Sigel and Kaden [66] attributed this behaviour for these ligands to π-bonding, which is influenced in the cases of ligands (9)–(11) by the positive charge on the quaternary nitrogen atom.

In certain cases a linear correlation was found between the Hammett σ constants of the ligands and the logarithms of the stability constants of the complexes [65]. Irving and Da Silva [67] introduced a stability factor, S_f, as the logarithm of the equilibrium constant for the reaction

$$n\mathrm{H}_m\mathrm{L}' + \mathrm{MeL}_n = n\mathrm{H}_m\mathrm{L} + \mathrm{MeL}'_n \qquad (11.8)$$

where L′ is more suitable for π-bonding than L. They think that S_f is a measure of the stabilization due to π-bonding. They found again a linear relationship between the Hammett σ constants and the stability constants of certain complexes. However, Yingst and McDaniel [68] obtained very convincing experimental data which invalidate these relationships, so the Hammett equation cannot be applied to judge the extent of π-bonding character in metal complexes.

11.3.2.3 Comparison of the stability constants of the complexes of two similar ligands formed with a number of metal ions

Irving and Rossotti pointed out that the logarithms of the stability constants of complexes of a certain ligand with a number of metals (log K_{MeP}), plotted as a function of the logarithms of the stability constants of the corresponding metal complexes formed with another but analogous ligand (log K_{MeQ}), give a straight line of unit slope and of intercept (log $K_{\mathrm{HP}} -$ $-$ log K_{HQ}), as is shown in Fig. 11.11.

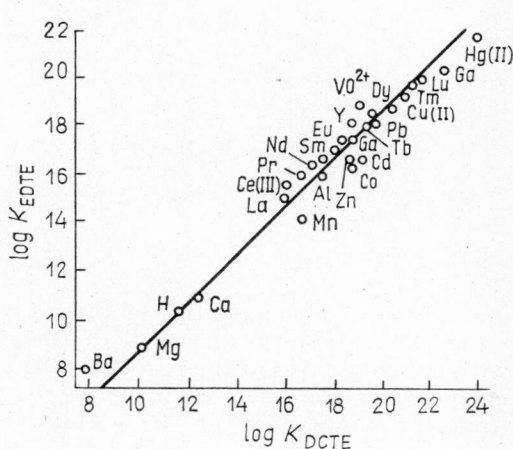

Fig. 11.11 Log K_{Me} of the stability constants of complexes of ethylenediaminetetra-acetate as a function of the logarithms of the stability constants of the corresponding complexes of 1,2-diamine-cyclohexane-tetra-acetate. The constants were obtained at 20 °C, at 0.1 M ionic strength. All ions shown are in their usual valency state

The same relationship was found for a number of other complexes. The theoretical prediction of the intercept is not quite unambiguous if multidentate ligands are considered. It seems to be more reasonable to expect that the intercept is log $\sum_1^N \log K_{\mathrm{H}_i\mathrm{P}} - \sum_1^N \log K_{\mathrm{H}_i\mathrm{Q}}$, that is, the difference of the sums of the successive protonation constants of the two ligands. For the EDTA and DCTA complexes of bismuth(III) the difference of the logarithms of the stability constants is just equal to the difference of the sum of the logarithms of the six protonation constants of DCTA and EDTA.

11.3.2.4 Effects of structural changes and steric factors

The effect of structural changes of ligands which do not involve the donor atom(s) is partially reflected in the altered proton affinity and can sometimes be characterized by the Hammett constants [69]. If bulky substituents are introduced into the ligand molecule, steric hindrance may result in additional decrease in the stability of complexes. This is clearly shown by the stability constants of complexes of the alkyl derivatives of ethylenediamine (Table 11.9) [70].

TABLE 11.9

Acidic dissociation constants of substituted ethylenediamines
$R-NH-CH_2-CH_2-NR'R''$ *and the stability constants*
of their copper(II) and nickel(II) complexes

R	R'	R''	pK_1	pK_2	Cu(II)		Ni(II)	
					$\log K_1$	$\log K_2$	$\log K_1$	$\log K_2$
H	H	H	7.47	10.18	10.73	9.30	7.60	6.48
Me	H	H	7.56	10.40	10.55	8.56	7.36	5.74
Et	H	H	7.63	10.56	10.19	8.38	6.78	5.30
i-Pr	H	H	7.70	10.62	9.07	7.45	5.17	3.47
Me	H	Me	7.40	10.16	9.69	6.65	6.65	3.85
Me	Me	H	6.63	9.53	9.23	6.73	—	—
Et	Et	H	7.07	10.02	8.17	5.55	—	—

Even when the basicities of the derivatives exceed that of ethylenediamine, the stability constants of the complexes decrease in comparison with the complexes of the parent ligand. The effect of steric hindrance overcompensates the effect of increased basicity. This is particularly evident from the comparison of the successive constants: the steric hindrance becomes more pronounced on co-ordination of the second ligand. The effect of steric hindrance is even more important in the case of rigid cyclic ligands such as 8-hydroxyquinoline or 1,10-phenanthroline.

It was claimed [70] that the steric effect of the introduction of a methyl group into the molecule of EDTA produces a dramatic effect on the relative stabilities of the calcium and strontium complexes. In all cases the stability of the calcium complexes is greater than that of the strontium complexes. Smith [70] stated that the logarithms of the stability constants of the calcium and strontium complexes of 1,2-diaminopropane–N,N,N',N'-tetra-acetate are 10.4 and 10.7 respectively. This difference might find important application in the decontamination of the body from radiostrontium. However, the measurements by Grimes, Huggard and Wilford [71] showed that the stability of the strontium complex is less than that of the calcium complex.

The effect of structural changes is illustrated by Fig. 11.12, referring to a number of alkali earth polyaminepolyacetate chelates [72]. It appears that the effect of the various structural changes is reflected to different ex-

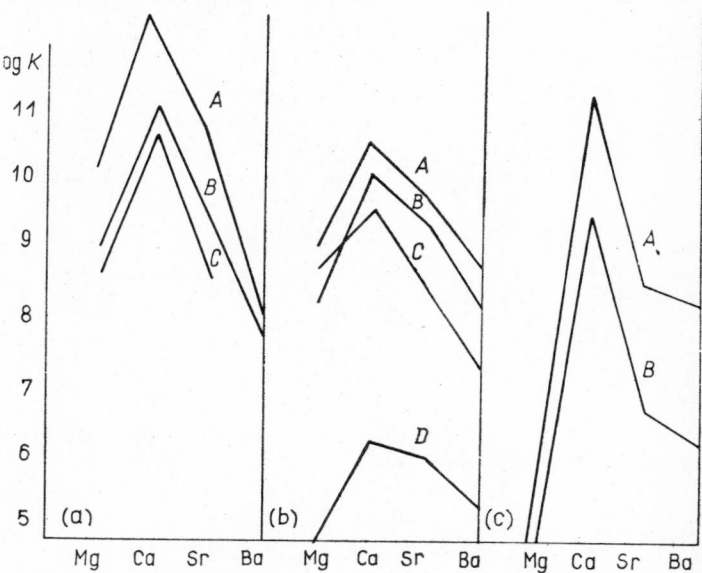

Fig. 11.12 Logarithms of the stability constants of several alkali earth metal complexes.
(a) A — cyclohexane-*trans*-1,2-diaminetetra-acetic acid; B — cyclopentane-*trans*-1,2-diaminetetra-acetic acid; C — ethylenediaminetetra-acetic acid.
(b) A — diethylenetriaminepenta-acetic acid; B — bis[(dicarboxymethyl) aminoethyl] ether; C — bis[(dicarboxymethyl) aminoethyl]methylamine; D — bis[(dicarboxymethyl)aminoethyl] sulphide.
(c) A — bis[(dicarboxymethyl)aminoethoxy] ethane; B — bis [(dicarboxy-methyl)-aminoethyl] methylamine (BDAM). (Reproduced with permission from *Ann. New York Acad. Sci.* **88**, 341 (1960))

tents in the stability of different metal complexes. Such comparative studies may lead to the discovery of selective complexing agents. Selective complex formation can evidently be achieved more easily in the case of transition or post-transition metal ions than with metal ions of noble gas configuration. Recently Bayer *et al.* [73, 74] synthesized some organic ligands which form complexes surprisingly selectively and — in principle — make it possible to concentrate the uranium and gold content of sea water.

11.3.2.5 *Effect of chelate formation on the stability of complexes*

The high kinetic and thermodynamic stability of chelates in comparison with the corresponding non-chelated complexes was indicated by early preparative experiments. The equilibrium data obtained in the last few decades have made it possible to characterize the difference of the stabilities of complexes of unidentate and multidentate ligands quantitatively. A quantitative measure of the increased thermodynamic stability due to the formation of chelate rings can be obtained by comparing the stability constants of the complexes of the unidentate and the corresponding multidentate ligands occupying the same number of co-ordination sites. In the simplest case:

$$\text{Me} + 2\text{L} = \text{MeL}_2; \qquad \beta_2 = \frac{[\text{MeL}_2]}{[\text{Me}][\text{L}]^2} \tag{11.9}$$

$$\text{Me} + \text{L} - \text{L} = \text{Me}\overset{\text{L}}{\underset{\text{L}}{\diagup}}\ ; \qquad \beta^* = \frac{\left[\text{Me}\overset{\text{L}}{\underset{\text{L}}{\diagup}}\right]}{[\text{Me}][\text{L}-\text{L}]} \tag{11.10}$$

The ratio of the two stability constants is the equilibrium constant of the following displacement reaction

$$\text{MeL}_2 + \text{L} - \text{L} = \text{Me}\overset{\text{L}}{\underset{\text{L}}{\diagup}} + 2\text{L} \tag{11.11}$$

The term "chelate effect" means the logarithms of the equilibrium constant of this displacement reaction.

Before dealing with the factors determining the chelate effect it must be mentioned that it depends on the standard state chosen. Adamson [75] pointed out that if the unit mole fraction standard state is considered, chelates are not favoured over the corresponding simple complexes. It follows then that the greater the dilution the greater is the chelate effect.

There are two basic possibilities for increasing the stabilities of chelates. Either the formation enthalpy or the formation entropy favours the stability of the chelate. The contribution of the formation entropy to the chelate effect is plausible and Schwarzenbach [76] offered a simple model to explain it. According to this model the formation of the chelate proceeds in distinct steps. First the ligand occupies one co-ordination site and then in the second step the ring is formed. However, the actual activity of the unattached donor group is very high and is determined by the volume of the ligand. Therefore it is more probable that in the second step a ring is formed than that a second ligand molecule is co-ordinated. It would follow that the stability of chelates decreases monotonically as the ring size increases. However, three- and four-membered rings are not stable because of the strain. It can therefore be expected that the stability shows a maximum with five- or six-membered rings. Martell [77] pointed out that the precise thermodynamic meaning of the effect described by Schwarzenbach is the change in the entropy of translation that accompanies metal chelate ring formation. The statistical calculations by Cotton and Harris [78] gave the same result as Schwarzenbach's approach, but there is only a qualitative agreement with the experimental data.

Spike and Parry [79, 80] have determined the stability of several zinc(II), cadmium(II) and copper(II) complexes at different temperatures and calculated the formation enthalpies and entropies.

TABLE 11.10

Thermodynamic functions of some cadmium(II) complexes at 25°C

Complex	$\log K$	ΔH kcal/mole	ΔG kcal/mole	$T \Delta S$ kcal/mole
$Cd(NH_3)_2^{++}$	4.950	7.12	6.75	0.37
$Cd(NH_2CH_3)_2^{++}$	4.808	7.02	6.56	0.46
$Cd(en)^{++}$	5.836	7.03	7.96	−0.93
$Cd(NH_3)_4^{++}$	7.44	12.7	10.16	2.53
$Cd(NH_2CH_3)_4^{++}$	6.55	13.7	8.94	4.77
$Cd(en)_2^{++}$	10.62	13.5	14.50	−0.98

As appears from Table 11.10, in the case of cadmium(II) amine complexes, entropy favours only chelate formation. The situation is similar with the analogous zinc(II) complexes, but in the corresponding copper(II) complexes the increase in stability resulting from chelate formation is distributed about equally between entropy and enthalpy terms.

According to Williams [22] the favourable heat of chelate formation is the consequence of the reduction of repulsion forces between neighbouring donor atoms in a chelate as compared with these forces in a complex formed with monodentate ligands. At present there are not enough data to permit a sound generalization.

As mentioned earlier the stability of a chelate depends on the size of the ring. The enthalpy effect also may be reflected in the change of the stability with the size of the ring. This appears from Table 11.11, where data obtained by Anderegg [81] are summarized.

TABLE 11.11

Thermodynamic functions of the metal chelates of the homologues of EDTA at 20°C and $\mu = 0.1$ (KNO$_3$)

Metal ion	ΔG, kcal/mole				ΔH, kcal/mole				ΔS, e.u.			
	EDTA	TMTA	TETA	HDTA	EDTA	TMTA	TETA	HDTA	EDTA	TMTA	TETA	HDTA
Mg^{2+}	−11.65	− 8.33	− 8.34	− 6.44	3.49	9.09	8.50	—	51.0	59	54.0	—
Ca^{2+}	−14.35	− 9.76	− 7.59	− 6.17	− 6.55	− 1.74	0.88	—	26.6	27.35	29.7	—
Mn^{2+}	−18.51	−13.40	−12.78	−12.07	− 4.56	− 0.72	3.41	0.87	46.6	52.9	55.2	44.2
Co^{2+}	−21.87	−20.80	−21.0	−17.5	− 4.2	− 2.6	− 1.56	− 4.56	60.3	62.2	66.3	44.1
Ni^{2+}	−24.97	−24.34	−23.28	−18.53	− 7.55	− 6.76	− 6.95	− 8.5	59.4	60.3	55.5	34.2
Cu^{2+}	−25.21	−25.37	−23.25	—	− 8.15	− 7.74	− 6.52	—	58.2	60.1	57.0	—
Zn^{2+}	−22.13	−20.38	−20.13	−17.0	− 4.85	− 2.27	− 3.48	− 4.00	59.0	61.77	56.8	44.4
Cd^{2+}	−22.07	−18.64	−16.12	−15.96	− 9.05	− 5.44	− 2.88	− 4.26	44.4	45.0	45.2	39.9
Pb^{2+}	−24.19	−18.37	−14.12	−13.93	−13.20	− 6.4	− 4.85	− 7.53	37.5	40.8	31.6	21.8
Hg^{2+}	−29.23	−26.71	−28.13	−28.94	−18.9	−18.9	−19.1	−20.97	35.5	26.6	30.8	27.25
La^{3+}	−20.79	−15.06	−12.24	—	− 2.8	3.76	0.1	—	61.4	64.2	42.0	—

It can be expected that the extra stability of chelates increases with the number of chelate rings. The stability data referring to some amine complexes of copper(II) show that this is true:

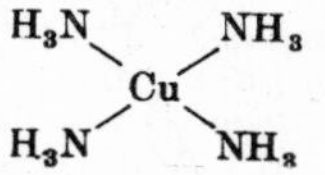

$$
\begin{array}{ccc}
H_3N & & NH_3 \\
 & Cu & \\
H_3N & & NH_2
\end{array}
$$

Ammonia: $\log K_1 = 4.13$; $\log K_2 = 3.48$; $\log K_3 = 2.87$; $\log K_4 = 2.11$

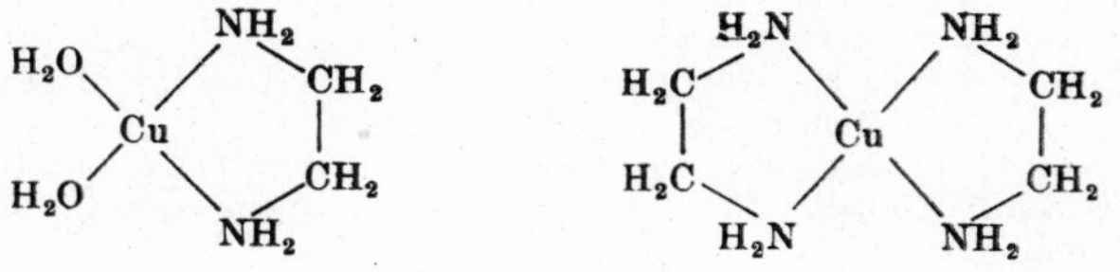

Ethylenediamine: $\log K_1 = 10.72$; $\log K_2 = 9.31$

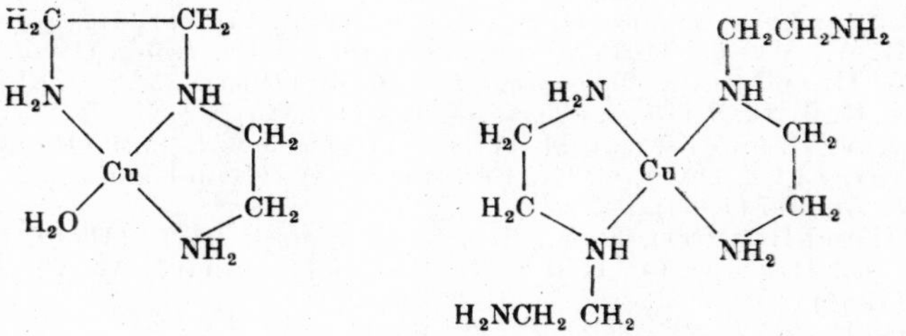

Diethylenetriamine: $\log K_1 = 15.9$; $\log K_2 = 5.4$

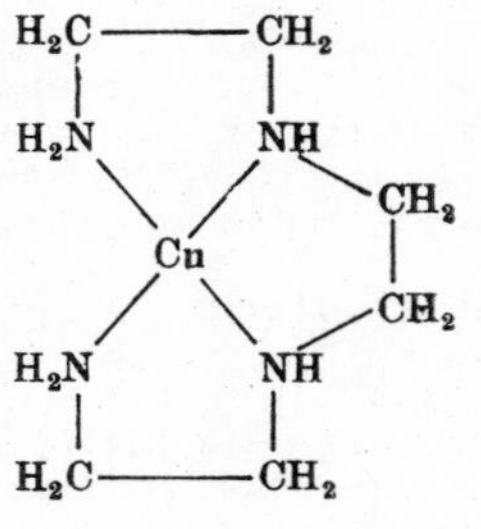

Triethylenetetramine: $\log K_1 = 20.5$

As a rule the preferred formation of chelates is the consequence of a number of factors. In certain cases one of these factors may be predominant, but usually their effects cannot well be distinguished.

REFERENCES

1. SILLÉN, L. G. and MARTELL, A. E., *Stability Constants of Metal-Ion Complexes*, 2nd Ed. Special Publication No. 17, Chemical Society, London, 1964.
2. MASAKI, K., *Bull. Chem. Soc. Japan 6*, 233 (1931).
3. SARTORI, G., *Gazz. Chim. Ital. 66*, 688 (1936).
4. LATIMER, W. M., *Oxidation Potentials*, Prentice-Hall, New York, 1952.
5. HUME, D. N. and KOLTHOFF, I. M., *J. Am. Chem. Soc. 72*, 4423 (1950).
6. DUODOROFF, P., *Sewage and Ind. Wastes 28*, 1020 (1956).
7. FREUND, H. and SCHNEIDER, C. R., *J. Am. Chem. Soc. 81*, 4780 (1959).
8. MARGERUM, D. W., BYDALEK, T. J. and BISHOP, J. J., *J. Am. Chem. Soc. 83*, 1791 (1961).
9. KOLTHOFF, I. M. and AUERBACH, C., *J. Am. Chem. Soc. 74*, 1452 (1952).
10. SCHWARZENBACH, G. and HELLER, J., *Helv. Chim. Acta 34*, 576 (1951).
11. BECK, M. T. and GÖRÖG, S., *Acta Chim. Acad. Sci. Hung. 22*, 159 (1960).
12. SELMER-OLSEN, A. R., *Acta Chem. Scand. 15*, 2052 (1961).
13. BECK, M. T. and GERGELY, A., *Acta Chim. Acad. Sci. Hung. 50*, 155 (1966).
14. BJERRUM, J., *Metal Ammine Formation in Aqueous Solutions*, p. 35. Haase, Copenhagen, 1941.
15. NAGYPÁL, I. and GERGELY, A., Lecture at 4th Colloquium on Complex Chemistry, Dobogókő, Hungary, 1 June 1968, *J. Inorg. Nucl. Chem.* in press.
16. CABANI, S., *J. Chem. Soc.* 5271 (1962).
17. BUCHANAN, J. and HAMANN, S. D., *Trans. Faraday Soc. 49*, 1425 (1953).
18. EWALD, A. H. and HAMANN, S. D., *Aust. J. Chem. 9*, 54 (1956).
19. ELLIS, A. J. and ANDERSON, D. W., *J. Chem. Soc.* 1765 (1961).
20. HORNE, R. A., MYERS, B. R. and FRYSINGER, G. R., *Inorg. Chem. 3*, 452 (1964).
21. EWALD, A. H. and SINN, E., *Inorg. Chem. 6*, 40 (1967).
22. WILLIAMS, R. J. P., *J. Phys. Chem. 58*, 121 (1954).
23. KENNEDY, M. B. and LISTER, M. W., *Can. J. Chem. 44*, 1709 (1966).
24. VAN UITERT, L. G., HAAS, C. G., FERNELIUS, W. C. and DOUGLAS, B. E., *J. Am. Chem. Soc. 75*, 455 (1953).
25. IRVING, H. and ROSSOTTI, H. S., *Acta Chem. Scand. 10*, 72 (1956).
26. IZATT, R. M., HAAS, C. G., BLOCK, B. P. and FERNELIUS, W. C., *J. Phys. Chem. 58*, 1133 (1954).
27. MATHESON, R. A., *J. Phys. Chem. 71*, 1302 (1967).
28. CHOPPIN, G. R., KELLY, D. A. and WAND, E. H., in *Solvent Extraction Chemistry*, (edited by Dyrssen, D., Liljenzin, J. O. and Rydberg, J.) p. 46. North-Holland, Amsterdam, 1967.
29. FIALKOV, YA. A. and SPIVAKOVSKII, V. B., *Zh. Neorgan. Khim. 4*, 1501 (1959).
30. TATE, J. F. and JONES, M. M., *J. Phys. Chem. 65*, 1661 (1961).
31. MELLOR, D. P. and MALEY, L., *Nature 159*, 370 (1947).
32. ACKERMANN, H., PRUE, J. E. and SCHWARZENBACH, G., *Nature 163*, 723 (1949).
33. FYFE, W. S., *J. Chem. Soc.* 2018 (1952).
34. VAN PANTHALEON VAN ECK, C. L., *Rec. Trav. Chim. 72*, 50 (1953).
35. VAN UITERT, L. G., FERNELIUS, W. C. and DOUGLAS, B. E., *J. Am. Chem. Soc. 75*, 2736 (1953).
36. IZATT, R. M., FERNELIUS, W. C., HAAS, C. G. and BLOCK, B. P., *J. Phys. Chem. 59*, 170 (1955).
37. CHAPMAN, D., *Nature 174*, 887 (1954).
38. UUSITALO, E., *Ann. Acad. Sci. Fenn. A87*, (1957).
39. SWINARSKI, A. and LODZINSKA, A., *Proc. 7th Intern. Conf. Coord. Chem. Stockholm*, 1962. p. 320.
40. IRVING, H. and WILLIAMS, R. J. P., *Nature 162*, 746 (1948).
41. IRVING, H. and WILLIAMS, R. J. P., *J. Chem. Soc.* 3192 (1953).
42. BJERRUM, J. and JØRGENSEN, C. K., *Rec. Trav. Chim. 75*, 658 (1956).
43. ORGEL, L. E., *Proc. 13th Solvay Conference in Chemistry, Brussels*, 1956, p. 289.
44. MOELLER, T., MARTIN, D. F., THOMPSON, L. C., FERRUS, R., FEISTEL, G. R. and RANDALL, W. J., *Chem. Rev. 65*, 1 (1965).
45. AHRLAND, S., CHATT, J. and DAVIES, N. R.. *Quart. Rev. 12*, 265 (1958).

46. AHRLAND, S. and LARSSON, R., *Acta Chem. Scand. 8*, 354 (1954).
47. CARLESON, B. G. F. and IRVING, H., *J. Chem. Soc.* 4390 (1954).
48. SIDGWICK, N. V., *J. Chem. Soc.* 433 (1941).
49. GRINBERG, A. A., *Introduction to the Chemistry of Complex Compounds* Pergamon, London, 1962. (Translation of 2nd Ed.; original Russian published in 1951).
50. PEARSON, R. G., *J. Am. Chem. Soc. 85*, 3533 (1963).
51. *Chem. Eng. News* 31st May, 1965, p. 90.
52. WILLIAMS, R. J. P. and HALE, J. D., *Structure and Bonding I*, 249 (1966).
53. FREISER, H., FERNANDO, Q. and CHENEY, G. E., *J. Phys. Chem. 63*, 250 (1959).
54. YATSIMIRSKII, K. B., *Zh. Neorgan. Khim. 5*, 264 (1960).
55. DILLON, K. B. and ROSSOTTI, F. J. C., *Chem. Commun.* 768 (1966).
56. LARSSON, E., *Z. Physik. Chem. Leipzig A169*, 207 (1934).
57. BJERRUM, J., *Chem. Rev. 46*, 381 (1950).
58. MARTELL, A. E. and CALVIN, M., *Chemistry of the Metal Chelate Compounds*, Prentice-Hall, New York, 1952.
59. LEBERMAN, R. and RABIN, B. R., *Proc. 7th Internat. Conf. Coord. Chem.* Stockholm, 1962, p. 121.
60. TUCCI, E. R., KE, C. H. and LI, N. C., *J. Inorg. Nucl. Chem. 29*, 1657 (1967).
61. BECK, M. T. and HALMOS, M., *Nature 191*, 1090 (1961).
62. JONES, J. G., POOLE, J. B., TOMKINSON, J. G. and WILLIAMS, R. J. P., *J. Chem. Soc.* 2001 (1958).
63. BRANDT, W. W. and WRIGHT, J. P., *J. Am. Chem. Soc. 76*, 3082 (1954).
64. TATE, J. F. and JONES, M. M., *J. Am. Chem. Soc. 83*, 3024 (1961).
65. MAY, W. R. and JONES, M. M., *J. Inorg. Nucl. Chem. 24*, 511 (1962).
66. SIGEL, H. and KADEN, T., *Helv. Chim. Acta 49*, 1617 (1966).
67. IRVING, H. and DA SILVA, J. J. R. F., *Proc. Chem. Soc.* 250 (1962)
68. YINGST, A. and McDANIEL, D. H., *J. Inorg. Nucl. Chem. 28*, 2919 (1966).
69. IRVING, H. and GRIFFITH, J. M. M., *J. Chem. Soc.* 213 (1954).
70. SMITH, R. L., *The Sequestration of Metals*, p. 80. Chapman and Hall, London, 1959.
71. GRIMES, J. H., HUGGARD, A. J. and WILFORD, S. P., *J. Inorg. Nucl. Chem. 25*, 1225 (1963).
72. KROLL, H. and GORDON, M., *Ann. New York Acad. Sci. 88*, 341 (1960).
73. BAYER, E. and SCHENK, G., *Ber. 93*, 1184 (1960).
74. BAYER, E., *Angew. Chem. Intern. Ed. 3*, 325 (1964).
75. ADAMSON, A. W., *J. Am. Chem. Soc. 76*, 1578 (1954).
76. SCHWARZENBACH, G., *Helv. Chim. Acta 35*, 2344 (1952).
77. MARTELL, A. E., *Werner Centennial Symposium* (ed. by G. Kaufmann) p. 272. American Chemical Society, Washington, 1967.
78. COTTON, F. A. and HARRIS, F. E., *J. Phys. Chem. 60*, 1451 (1956).
79. SPIKE, C. G. and PARRY, R. W., *J. Am. Chem. Soc. 75*, 2726 (1953).
80. SPIKE, C. G. and PARRY, R. W., *J. Am. Chem. Soc. 75*, 3770 (1953).
81. ANDEREGG, G., *Proc. 8th Intern. Conf. Coord. Chem.* (ed. by V. Gutmann), p. 34. Springer, Vienna, 1964.

Chapter 12

FUTURE DEVELOPMENTS

A few decades ago the formation of complexes was considered more as an exception than a rule. Owing to the very intensive research in the last two decades the ubiquitous nature of complex formation has become evident. As discussed in previous chapters, we now have reliable experimental methods and calculation procedures for the determination of equilibrium constants for all types of complex equilibrium. Consequently it cannot be regarded as forgivable now to determine and publish erroneous constants as it was ten or twenty years ago.

Three main directions of future equilibrium studies can be expected.

(*i*) Determination of formation enthalpy and entropy of different species.

(*ii*) Study of protonated, mixed ligand, polynuclear and outer-sphere complexes besides that of mononuclear binary species.

(*iii*) Study of complex equilibria in non-aqueous media.

In the determination of formation enthalpies and entropies the application of direct calorimetry is preferable to the evaluation of these quantities from equilibrium measurements at different temperatures. Nevertheless the two approaches can be regarded as complementary.

In most of the solutions which chemists use, it is generally not mononuclear binary complexes that exist but protonated, polynuclear, mixed ligand and outer-sphere complexes. These species used to be regarded as special types of complex but it is easy and necessary to realize that these very species are the normally occurring constituents of solutions, and that the simple parent complexes exist only under carefully selected experimental conditions. Therefore equilibrium studies of these types of complex require much more attention in the future. The various approaches to equilibrium studies of these species have been treated in the relevant Chapters of this book. It must be stressed, however, that reliable results in the investigation of these fairly complicated systems can only be expected if a given problem is attacked with as many different techniques as possible. It must always be carefully considered whether a calculated constant can be assigned to one well defined equilibrium, or is a "global" constant referring to isomeric species, or refers to a set of stepwise equilibria. A particularly important task is to distinguish co-ordination in the inner and in the outer sphere in the case of labile complexes.

Studies of complex equilibria in non-aqueous solutions are of manifold importance. The comparison of results obtained in different solvents may even give valuable information about the chemistry of aqueous solutions. The application of non-aqueous solvents makes possible the study of special reactions which cannot occur in aqueous solutions because of limited solubility and interfering side-reactions. From the practical point of view such studies are important because of the potential applications of non-aqueous media in industrial processes.

There is already an extremely wide variety of non-aqueous solvents applied in co-ordination chemistry and the possibilities are by no means exhausted. A readable book by Gutmann [1] gives an excellent survey of the field. Solvents with donor properties can be well characterized by the so-called donor number [2] which is the enthalpy of the reaction

$$SbCl_{5(dissolved)} + D_{(dissolved)} = D.SbCl_{5(dissolved)}$$

determined calorimetrically in an inert medium (D stands for the donor solvent molecule).

Much information can be obtained from studies of complex equilibria in solvent mixtures, including aqueous mixtures.

From both a theoretical and a practical point of view the investigation of co-ordination in molten salts is becoming of increasing importance. The same techniques can be applied as for aqueous or non-aqueous solutions (see e.g. [3–6]), however, considerable experimental difficulties sometimes arise. A most promising and versatile technique is the solvent extraction from molten salts reviewed recently by Marcus [7]. Low-temperature aqueous melts provide a medium intermediary between molten salts and aqueous solutions [8].

Complex equilibria are not restricted to the liquid phase. Recent studies indicate that different types of complex equilibrium occur in the gas phase. Besides charge-transfer complexes involving volatile organic donor and acceptor molecules [9] the following types were found: (i) dimerization of potassium bromide [10]; (ii) interaction between various dichlorides and trichlorides of aluminium and iron [11]; (iii) displacement of ligands in mixed cobalt carbonyls [12]. Undoubtedly mixed ligand complexes are formed in the vapours of covalent halides such as $AsCl_3 - AsBr_3$, $PCl_3 - PBr_3$. In the case of the volatile β-diketone metal complexes the formation of mixed ligand complexes also can be expected, but the thermal decomposition of the complexes may interfere.

Although the investigation of complex equilibria in the gas phase will lead to important discoveries and even to the development of practical applications, evidently the liquid phase will remain the centre of interest for co-ordination chemists.

REFERENCES

1. GUTMANN, V., *Coordination Chemistry of Non-aqueous Solutions*, Springer, Berlin, 1968.
2. GUTMANN, V. and WYCHERA, E., *Inorg. Nucl. Chem. Letters 2*, 257 (1967).
3. BRAUNSTEIN, H., BRAUNSTEIN, J. and INMAN, D., *J. Phys. Chem. 70*, 2726 (1966).
4. TIEN, H. T., *J. Phys. Chem. 69*, 3763 (1965).
5. JORDAN, J. and PENDERGAST, J., *Proc. 7th Int. Conf. Coord. Chem.* Stockholm, 1962, p. 102.
6. ANGELL, C. A. and GRUEN, D. M., *J. Phys. Chem. 70*, 1601 (1966).
7. MARCUS, Y., in *Solvent Extraction Chemistry* (D. Dyrssen, J.-O. Liljenzin and J. Rydberg eds.) p. 555. North-Holland, Amsterdam, 1967.
8. BRAUNSTEIN, J., ALVAREZ-FUNES, A. R. and BRAUNSTEIN, H., *J. Phys. Chem. 70*, 2734 (1966).
9. KROLL, M. and GINTER, M. L., *J. Phys. Chem. 69*, 3671 (1965).
10. HAGEMARK, K., BLANDER, M. and LUCHSINGER, E. B., *J. Phys. Chem. 70*, 276 (1966).
11. DEWING, E. W., *Nature 214*, 483 (1967).
12. MARKÓ, L., *D. Sc. Thesis*, Veszprém, 1968.

AUTHOR INDEX*

* Numbers in *italics* refer to names in REFERENCES

SUBJECT INDEX